In Situ Testing Methods in Geotechnical Engineering

In Situ Testing Methods in Geotechnical Engineering

Alan J. Lutenegger

CRC Press
Taylor & Francis Group
Boca Raton London New York

CRC Press is an imprint of the
Taylor & Francis Group, an **informa** business

First edition published 2021
by CRC Press
6000 Broken Sound Parkway NW, Suite 300, Boca Raton, FL 33487-2742

and by CRC Press
2 Park Square, Milton Park, Abingdon, Oxon, OX14 4RN

First issued in paperback 2022

**Visit the Taylor & Francis Web site at
http://www.taylorandfrancis.com**

**and the CRC Press Web site at
http://www.crcpress.com**

Library of Congress Cataloging-in-Publication Data
Names: Lutenegger, A. J., author.
Title: In situ testing methods in geotechnical engineering / Alan
Lutenegger.
Description: First edition. | Boca Raton, FL : CRC Press, 2021. |
Includes bibliographical references and index.
Identifiers: LCCN 2020050402 (print) | LCCN 2020050403 (ebook) |
ISBN 9780367432416 (hardback) | ISBN 9781003002017 (ebook)
Subjects: LCSH: Soils—Testing. | Engineering geology. | Building
sites—Evaluation. | Soil penetration test. | Geotechnical engineering.
Classification: LCC TA710.5 .L88 2021 (print) | LCC TA710.5 (ebook) |
DDC 624.1/51—dc23
LC record available at https://lccn.loc.gov/2020050402
LC ebook record available at https://lccn.loc.gov/2020050403

ISBN: 978-0-367-75874-5 (pbk)
ISBN: 978-0-367-43241-6 (hbk)
ISBN: 978-1-003-00201-7 (ebk)

DOI: 10.1201/9781003002017

Typeset in Sabon
by codeMantra

Contents

5 Field Vane Test (FVT)   167

Author

Alan J. Lutenegger has more than forty years of experience in geotechnical engineering and the use of *in situ* tests in soils. He is Emeritus Professor of Geotechnical Engineering at the University of Massachusetts-Amherst, where he has taught for over thirty years. His extensive research and publications in the use of *in situ* tests in soils include field investigations using nearly all the tests described in this book. He is a registered professional engineer and a Fellow of the American Society of Civil Engineers and has been involved in many consulting projects using *in situ* tests.

Chapter I

Introduction to *In Situ* Testing

1.1 INTRODUCTION

Over 40 years ago, Mitchell et al. (1978) gave a number of reasons for the growing interest in the use of *in situ* testing techniques:

1. Ability to determine properties of soils such as sands and offshore deposits that cannot easily be sampled in the undisturbed state;
2. Ability to test a larger volume of soil than can conveniently be tested in the laboratory;
3. Ability to avoid some of the difficulties of laboratory testing, such as sample disturbance, the proper simulation of *in situ* stresses, temperature, and chemical and biological environment; and
4. Increased cost effectiveness of an exploration and testing program using *in situ* methods.

Engineers should not expect a single *in situ* test to provide the answer to all geotechnical problems. Just as different laboratory tests are used to obtain specific soil properties, different *in situ* tests have been developed for the same purpose.

1.2 ROLE OF *IN SITU* TESTING IN SITE INVESTIGATIONS

Like all soil tests, *in situ* tests provide a way of obtaining additional information about subsurface conditions at a site. They are used to give a more complete picture of site conditions and soil behavior and reduce uncertainties inherent in most projects. Geotechnical engineering often requires the use of many tools, and Figure 1.1 shows the various tools available for geotechnical design. *In situ* tests are rarely used as a complete replacement for test borings and laboratory tests for a site investigation but are typically used to compliment a traditional subsurface exploration program in order to enhance the information regarding site conditions.

1.3 ADVANTAGES AND LIMITATIONS OF *IN SITU* TESTS

In situ tests can provide a number of advantages over the traditional drilling, sampling, and laboratory testing approach used in many geotechnical projects. However, like all tests, *in situ* tests also have a number of limitations. It is important that engineers understand both the advantages and the limitations of *in situ* tests.

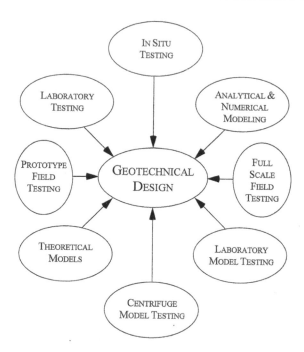

Figure 1.1 Tools used in the practice of geotechnical engineering.

1.3.1 Advantages of *In Situ* Tests

Advantages of *in situ* tests include the following:

1. Tests may be conducted in soil deposits that are difficult or impossible to sample or test;
2. Soil properties that cannot be easily determined by conventional laboratory tests can be determined from *in situ* tests;
3. A larger total volume of soil that may influence the design can be tested;
4. *In situ* tests avoid some of the difficulties inherent in a conventional lab testing program;
5. Some tests provide a near continuous record of vertical variations in soil conditions;
6. There is often a reduction in the time of the site investigation;
7. Some tests allow for real time or rapid data reduction;
8. *In situ* tests may be used to assess the influence of scale and macrofabric on soil behavior;
9. Tests are performed in a field environment; and
10. There is almost often substantial cost savings to a project.

1.3.1.1 Testing Soils that are Difficult to Sample

Often times, subsurface explorations encounter soils that are difficult or impossible to sample using conventional drilling and sampling methods. Typical examples include loose sands and silts below the water table, very soft or highly sensitive clays, and highly weathered or structured materials, such as surficial crusts or residual soils. In some cases, artesian or other unusual groundwater conditions may also create drilling and sampling difficulties. In these cases, the use of Cone Penetration Test (CPT)/Piezocone (CPTU) or Dilatometer Test (DMT) may provide results that are more reliable than laboratory tests conducted on samples of poor quality.

1.3.1.2 Determining Soil Properties that are Difficult to Measure by Laboratory Methods

Using common laboratory testing techniques, it is sometimes difficult to obtain accurate measurements of certain key soil properties that may be essential in a geotechnical engineering design. For example, the small-strain shear modulus or horizontal stress can be extremely difficult to measure in the laboratory and require special equipment and high-quality undisturbed samples. In these cases, an alternative may be to select an appropriate *in situ* test that can provide these measurements.

1.3.1.3 Testing a Larger Volume of Soil

For most typical geotechnical problems, a traditional subsurface exploration is usually set up to obtain samples at some preselected depth interval (often 5 ft (1.5 m)) or whenever a stratigraphic change occurs. For routine projects, the number of test borings and the number of samples are relatively small, which gives a limited view of the engineering behavior of the soil. A stratigraphic profiling test, such as the CPT, CPTU, or DMT, can increase the amount of soil investigated by two to three orders of magnitude. Important strata that might otherwise be missed during conventional drilling and sampling may be identified.

1.3.1.4 Avoiding Difficulties with Sampling and Laboratory Testing

One of the limitations of obtaining soil samples from the field, transporting them to the laboratory, and performing tests is that there are always some unavoidable changes in the environment relative to field conditions. Changes in pore water pressure, stress field, composition of the pore fluid, temperature, and sample disturbance may have differing degrees of influence on the behavior of different soils. *In situ* tests avoid many of these issues. In addition, laboratory tests are not without their own set of problems, related to differences in test equipment, test procedures, sample sizes, and rate of loading.

1.3.1.5 Obtaining Near Continuous Profiling

It is possible to obtain a near-continuous record of the vertical variations in soil conditions using penetration tests such as the CPT/CPTU and DMT. Data acquisition systems used with the CPT/CPTU can provide detailed information about changes in stratigraphy that occur with depth. The use of stratigraphic profiling tools is one of the main areas where *in situ* tests can enhance site investigations for geotechnical engineers.

1.3.1.6 Reduced Testing Time

The use of *in situ* tests often provides a substantial reduction in the time necessary to complete the site investigation in comparison to conventional site investigation practice. Rapid determination of specific soil properties during field exploration also allows the engineer to make a preliminary assessment of the subsurface conditions. There is also the opportunity to evaluate any problematic areas, e.g., unstable or soft ground conditions, excavation problems, etc.

1.3.1.7 Rapid Data Reduction

Some *in situ* tests provide rapid and automatic data acquisition and reduction in the field, often as the test proceeds. This means that the technician or engineer can track the progress of the test and troubleshoot any difficulties that may arise during the test. Judgement may then be used to decide whether more tests need to be conducted at the site.

1.3.1.8 Assessing the Influence of Scale or Macro-Fabric on Soil Behavior

In some soils, such as highly fissured or fractured clays or highly weathered residual deposits, the macrofabric can exert a significant influence on soil behavior. These soils present a problem to the engineer in obtaining high-quality undisturbed samples for laboratory testing and in evaluating how the laboratory response reflects the behavior of the soil in the field. In these materials, it is important to test a sufficiently large volume of soil in order to obtain a reasonably accurate indication of soil behavior.

1.3.1.9 Conducting Tests in a Field Environment

In situ tests are performed in the field under conditions that are closer to "undisturbed" than laboratory tests. That is, the vertical and horizontal stress conditions, pore water pressure, temperature, and pore fluid chemistry are more representative than conditions that are typically used in laboratory tests.

1.3.1.10 Cost Savings

The reduction in testing time and increase in information provided by many *in situ* tests usually result in substantial economic benefit to many projects. The conventional approach to geotechnical design involves a high degree of uncertainty that evolves as a result of the highly variable nature of geologic deposits and the small level of investigation generally possible. This in turn leads to a design approach that is often more conservative than if more detailed information could be obtained. A reduction in the level of uncertainty that can be achieved by incorporating *in situ* tests into the subsurface exploration program may result in cost savings.

1.3.2 Limitations of *In Situ* Tests

There are limitations inherent in most *in situ* tests. In order to correctly use the results obtained from *in situ* tests, the understanding of these limitations is important. Limitations of *in situ* tests include the following:

1. Boundary conditions of the test in terms of stresses and/or strains are often poorly defined;
2. Drainage conditions are generally unknown and cannot be controlled;
3. The level of soil disturbance is generally unknown;
4. Stress paths, modes of deformation, and/or modes of failure imposed on surrounding soil by the test may be different than full-scale structures;
5. Strain rates are usually higher than either laboratory tests or anticipated in full-scale structures;
6. The specific nature of soil being tested is often unknown;
7. Effects of environmental changes on soil behavior are difficult to assess; and
8. Typical difficulties associated with performing field work.

The overall effect of these limitations is that the interpretation of results from *in situ* tests often requires an empirical approach. While it is important to be cautious of empirical procedures, it should also be remembered that the quality of an empirical approach is most often a function of the quality of the reference parameters, i.e., how good are the laboratory or field measurements being used for reference values and for the development of the empirical correlations?

1.3.2.1 Unknown Boundary Conditions

The installation of most *in situ* tests usually requires either (1) full displacement of the soil, i.e., the soil is forced apart to make room for the probe or (2) a prebored hole, i.e., a borehole is prepared, and a probe is lowered into the cavity. As a result of either method of installation, the soil stresses and pore water pressures acting on the surface of the instruments are often unknown. The level of strain imposed on the surrounding soil is usually unknown, as often times only deformations are measured.

1.3.2.2 Unknown Drainage Conditions

In some cases, pore water pressures may dominate some measurements in a test. Unlike most laboratory tests in which the exact location of drainage boundaries are known and can be controlled, the drainage boundaries are usually unknown during the performance of most *in situ* tests. Tests in freely draining materials such as sands and silty sands are usually interpreted as drained tests even if the test is performed rapidly. On the other hand, tests performed rapidly in saturated clays are usually interpreted as undrained tests.

1.3.2.3 Unknown Disturbance

Soil disturbance should not be confused with soil displacement. Most penetration tests produce a repeatable level of displacement when inserted into the soil, but because the stress-strain-relaxation behavior of soils is different, different levels of soil disturbance will result in different soils using the same instrument. Full-displacement techniques produce a zone of disturbed and possibly remolded soil immediately around the probe.

1.3.2.4 Modes of Deformation and Failure May be Unique

In most geotechnical testing, it is often difficult to perform a test that represents the actual mode of deformation or failure that is anticipated in the field. Estimating soil properties from tests that impose deformations or failure that are unrealistic in terms of full-scale problems usually requires assumptions in order to interpret the test. An example is determining undrained shear strength in clays from the field vane test (FVT). Few real problems involve a cylindrical failure surface as is obtained with the FVT.

1.3.2.5 Strain Rates or Loading Rates are Higher than Laboratory and Full-Scale

Most soils are strain-rate sensitive; i.e., the stress-strain-strength behavior depends to some degree on the rate of strain. For practical reasons, the strain rates or loading rates used in most *in situ* tests are higher than laboratory testing or full-scale loading. In view of the potential problems that might arise because of differences in test procedures, attempts have been made to standardize the procedures used in most *in situ* tests so that constant rates are used to conduct the test.

1.3.2.6 Nature of the Soil Being Tested is Unknown

With the exception of the SPT, no other *in situ* test provides a sample for visual identification or laboratory index or classification testing. This is often considered by some engineers as a drawback to the use of *in situ* tests, since many engineers still want a sample. Procedures for soil identification or classification have been developed for the CPT/CPTU and DMT by using combinations of measured values from the test.

1.3.2.7 Effects of Environment Change on Soil Behavior are Difficult to Assess

Occasionally, it is of interest to determine how environmental changes affect soil properties. These variables can be systematically controlled in the laboratory. Field tests are not well suited to allowing environmental changes to be introduced into the testing sequence.

1.3.2.8 Typical Difficulties with Field Work

The nature of field work often involves the need for robust equipment, not prone to breakage or other damage. Field work involves a certain amount of finesse and creativity to complete the work. Usually the problems encountered with performing field tests are not present in the lab, and therefore, many engineers see this as a drawback to using *in situ* tests.

1.4 APPLICATIONS OF *IN SITU* TESTS

Wroth (1984) suggested the following main applications for using *in situ* tests:

1. Site investigations;
2. Measurement of a specific property of the ground;
3. Control of construction; and
4. Monitoring of performance and back analysis.

Given the possible range of geologic conditions and the specific requirements of a particular project, the application of particular tests may be different for different projects and different ground conditions. *In situ* tests in geotechnical practice generally fall into one of the following categories:

1. Tests that provide information about changes in subsurface materials (stratigraphic profiling tests);
2. Tests that provide soil behavior at a specific point in the subsurface (specific property tests);
3. Tests that provide direct information relative to full-scale foundation behavior (prototype tests).

1.4.1 Stratigraphic Profiling

Stratigraphic profiling tests are used primarily to determine the stratigraphic profile. They are most often penetration tests that are simple and fast. With only minimal interpretation, the results can be used qualitatively to provide an indication of material changes, e.g., decrease or increase in penetration resistance with depth. Table 1.1 summarizes the most common stratigraphic profiling tests. Since these are penetration tests, they are generally economical to perform and are best integrated into the early stage of a site investigation.

Table 1.1 Common *in situ* stratigraphic profiling tests

Test	Comments
Cone penetrometer (CPT)	Tip resistance and sleeve friction used with various classification charts
Piezocone (CPTU)	Tip resistance, sleeve friction, and pore pressure used with various soil classification charts. Usually more detailed than CPT
Drive cone penetrometer (DCP)	Relative changes in driving resistance usually obtained over an interval of 100–150 mm. Less continuous than CPT/CPTU but very simple
Dilatometer (DMT)	Testing interval typically 250–300 mm. Relative changes in pressure readings or soil classification chart. Thrust measurement enhances profiling
Resistivity cone (RCPT)	Changes in electrical resistivity need to be sufficient to indicate stratigraphic changes. Pore water chemistry can significantly influence results

1.4.2 Specific Property Measurement

Specific property tests are used to provide a measurement of a desired soil property identified as being important for design. These are usually tests performed at a specific point and are therefore more specialized tests. They can also be part of the stratigraphic profiling test in which the results obtained during profiling are interpreted for a specific property. Specific property tests tend to be slower and more expensive to perform than stratigraphic profiling tests. They usually provide a limited amount of data as a result of time and budgetary constraints and the reduced frequency of testing. Specific property tests are often used in critical zones defined by stratigraphic profiling methods to provide estimates of important soil properties such as shear strength, stress history, compressibility, stress-strain behavior, state of stress, or flow characteristics. There are usually several tests that may be used to provide an estimate of the same soil property. Tables 1.2–1.6 provide summaries of the most common *in situ* tests used to estimate specific soil properties and their applicability to different soils.

Table 1.2 Common *in situ* tests for estimating undrained shear strength in fine-grained soils

Soil type	FVT	CPT	CPTU	DMT	PMT	SBPMT	SPT	PLT
V. soft	X	X	X	X	–	X	–	–
Soft	X	X	X	X	–	X	X	–
Medium	X	X	X	X	X	X	X	X
Stiff	–	X	X	X	X	X	X	X
V. stiff	–	X	X	X	X	X	X	X

FVT – generally considered the most direct and reliable estimate of undrained strength in medium to very soft clays; difficult to advance in stiffer clays; drainage may be a question in stiff clays.

CPT/CPTU – in very soft clays, the reliability of tip resistance is questionable; in very stiff clays, the penetration resistance may be too high for conventional pushing rigs; CPTU best suited in medium to very soft clays; most reliable estimates of undrained strength are obtained from pore water pressure measurements; requires special attention of deairing piezoelement.

DMT – applicable to a wide range of soil; requires about twice the pushing thrust as a CPT/CPTU.

PMT – borehole needs to remain open; best suited to medium to very stiff clays.

SBPMT – specialized equipment; may be difficult to deploy.

SPT – not suited to very soft clay; empirical approach in other clays; hammer energy measurements needed.

PLT – plate tests at shallow depth or excavation required.

Table 1.3 Common *in situ* tests for estimating stress history in fine-grained soils

Soil type	FVT	CPT	CPTU	DMT	PMT	SPT
V. soft	X	–	X	X	–	–
Soft	X	X	X	X	–	–
Medium	X	X	X	X	X	X
Stiff	–	X	X	X	X	X
V. stiff	–	X	X	X	X	X

FVT – use of normalized strength provides an estimate of stress history.

CPT – tip resistance correlation to stress history widely applicable.

CPTU – redundancy in estimates of stress history from both pore water pressure and tip resistance.

DMT – wide range of application; in very stiff clays, penetration may be difficult.

PMT – relationship to PMT creep pressure or limit pressure.

SPT – empirical approach.

Table 1.4 Common *in situ* tests for estimating lateral stress in soils

Soil type	PMT	SBPMT	DMT
Stiff clay	X	–	X
Medium stiff clay	X	X	X
Soft clay	–	X	X
Silt	X	X	X
Dense sand	X	X	X
Loose sand	X	X	X

PMT – graphical interpretation of pressure/volume or pressure/strain curve.

SBPMT – graphical interpretation of pressure/strain curve.

DMT – empirical correlation between OCR and K_o may be site-specific.

Table 1.5 Common *in situ* tests for estimating shear strength of coarse-grained soils

Soil	CPT	DMT	PMT	SPT	BST
Loose sand	X	X	X	X	X
Medium dense sand	X	X	X	X	X
Dense sand	X	X	X	X	X
Loose gravel	X	–	–	X	–
Medium dense gravel	–	–	–	X	–
Dense gravel	–	–	–	–	–

CPT – penetration may be difficult in gravelly materials.

DMT – gravel may damage the membrane.

PMT – need good borehole for testing.

SPT – spoon may become plugged in gravelly material.

BST – borehole stability may be difficult.

Table 1.6 Common *in situ* tests for estimating soil modulus in soils

Soil type	CPT/CPTU	DMT	PMT	SBPMT	SPT	PLT
Soft clay	–	X	–	X	–	–
Medium stiff clay	X	X	X	X	X	X
Stiff clay	X	X	X	X	X	X
silt	X	X	X	X	X	X
Loose sand	X	X	X	X	X	X
Dense sand	X	X	X	X	X	X

CPT/CPTU – empirical correlations to tip resistance; seismic cone (SCPT/SCPTU) for shear wave velocity.

DMT – empirical correlation to DMT modulus; seismic DMT (SDMT) for shear wave velocity.

PMT – direct stress-strain curve; empirical correlations between PMT modulus and settlement; unload-reload curve provides an estimate of shear modulus.

SBPMT – direct stress-strain curve; unload-reload curve provides an estimate of shear modulus.

SPT – empirical relationships for fine-grained soils; good database for modulus for use in shallow foundation settlements in granular soils. Seismic SPT (SSPT) newly developed for shear wave velocity.

PLT – direct load-deformation curve for determining subgrade reaction modulus.

Table 1.7 Summary of prototype *in situ* tests

Test	Design application
CPT	End bearing and side resistance of driven piles
PMT	Lateral load behavior of drilled shafts
BST	Side resistance of drilled shafts/grouted anchors in axial compression or tension
PLT	Load/settlement of shallow foundations
DMT	Lateral load behavior of driven piles
SPT-T	Side resistance of driven piles

1.4.3 Prototype Modeling

Tests that fall into this category are used for a direct design approach. This approach gives the opportunity to move directly from the *in situ* measurements to the performance of foundations without the need to evaluate intermediate soil parameters. An example is the use of the CPT/CPTU to predict axial capacity of driven or jacked piles. A summary of different prototype tests used for direct design applications is given in Table 1.7.

1.5 INTERPRETATION OF *IN SITU* TEST RESULTS

The correct application of *in situ* tests to solve geotechnical problems requires correct interpretations of the test results. For the most part, the interpretation of *in situ* tests can generally be divided into three classes (Jamiolkowski et al. 1988):

1. Soil elements follow very similar effective stress paths. Therefore, with appropriate assumptions on drainage conditions and stress-strain relationships, the solution of a more or less complex boundary value problem can lead to the determination of stress-strain and strength characteristics. This category of tests includes the PMT, especially the SBPT, and seismic tests.

2. Soil elements follow different effective stress paths depending on the geometry of the problem and on the magnitude of the applied load. In this case, a rational interpretation is often very difficult. Even with appropriate assumptions concerning the drainage conditions and soil model, the solution of a complex boundary value problem leads to more or less "average" soil characteristics. Comparisons between these average values and the behavior of a typical soil element tested in the laboratory or their use in the specific design calculation are far from straightforward. A typical example is using the CPT for estimating soil shear strength.
3. Soil elements follow different effective stress paths, and the *in situ* test results are empirically correlated to selected soil properties. Typical examples are the widely used correlations between penetration resistance measured in the SPT and deformation modulus. Because of the purely empirical nature of these correlations, they are subject to many limitations, which are not always fully recognized by potential users. In addition, it is important to recognize that the empirical correlations are usually formulated for either fully drained or fully undrained conditions.

The major sources of uncertainty in the interpretation of many *in situ* tests are related to the following issues:

1. Complex boundary value problems;
2. Complex and often unknown drainage conditions;
3. Complex variations in stress and strain levels;
4. Complex influence of stress path-dependent soil behavior, i.e., anisotropy and plasticity.

As a result of these uncertainties, interpretation of most penetration tests is based on empirical correlations to selected soil properties. Because of the purely empirical nature of these correlations, it is important to be aware of their many limitations. Often, the correlations are only partly able to account for soil nonlinearity and plasticity, as well as other complexities in natural soils, such as mineralogy, *in situ* stress state, stress-strain history, cementation, sensitivity, ageing, anisotropy, and structure (fabric). At the present time, only a few of these factors are accounted for in even the simplest methods of interpretation.

Wroth (1984) suggested that any empirical correlation developed for *in situ* tests that can be used with confidence outside the immediate context in which it was established should ideally be as follows:

1. Based on a physical appreciation of why the properties can be expected to be related;
2. Set against a background of theory, however idealized this may be;
3. Expressed in terms of dimensionless variables so that advantages can be taken of the scaling laws of continuum mechanics.

    Jamiolkowski et al. (1988) suggested that the following additional criteria could be included:
4. Validation of the correlation using large calibration chambers or field performance observations.

Wroth (1988) suggested that empirical correlations should be thought of in terms of a hierarchy such that "primary" and "secondary" correlations are possible. For example, the tip resistance measured in the CPT is some measure of the strength – but not the stiffness – of the soil being tested. Therefore, the primary correlation in the test must be between q_c and s_u. Additionally, the single observed quantity q_c can only lead to *one* independent interpretation of soil properties; any additional interpretation, if truly independent, must depend on some other information obtained from the test or based on a relationship with the primary soil property. For example, any correlation that is suggested between q_c and the undrained Young's modulus of the soil, E, should be considered a secondary relationship that is dependent on some other relationship between s_u and E. This implicit dependence on another correlation means that the secondary relationship is a weaker one, with more scatter in the data and on which less reliance can be placed.

1.6 USING *IN SITU* TESTS IN DESIGN

There are two approaches for using results from *in situ* tests in geotechnical design: (1) indirect design and (2) direct design.

1.6.1 Indirect Design

The indirect design approach relies on the interpretation of *in situ* tests to obtain conventional design parameters of soils and subsequent application of these parameters in more traditional design methodology. An example would be to use the field vane to estimate the undrained shear strength of a clay from the torque measurement, which would then be used in a bearing capacity equation to predict the undrained side resistance of a driven pile.

A drawback to this approach is that a transformation must be made between the measurement obtained in the test and the property needed for the design. In the example of the field vane, the torque measured in the test is used to obtain the undrained shear strength by making a series of assumptions relative to the behavior of the soil, drainage conditions, failure surface, shear stress distribution, strain rate, etc., all of which can influence the resulting estimate of undrained strength. Additionally, experience has shown that often times, the results of the test do not accurately predict performance, and an "adjustment" factor is needed to match test results and field performance, e.g., the correction factor introduced by Bjerrum (1972) for the application of field vane results for stability analysis of embankments on soft clay.

1.6.2 Direct Design

The direct design approach gives the engineer the opportunity to go directly from the *in situ* measurement to the performance of foundation without the need to evaluate intermediate soil parameters if the test procedure closely approximates the construction and load/deformation sequence of the full-scale member. An example would be to use the CPT/CPTU to design driven piles. The direct design approach eliminates most of the assumptions involved in the indirect approach since the results of the test are being used directly in the design. Usually, the direct design approach needs to be verified substantially with full-scale performance. The difference between these two design approaches is illustrated in Figure 1.2.

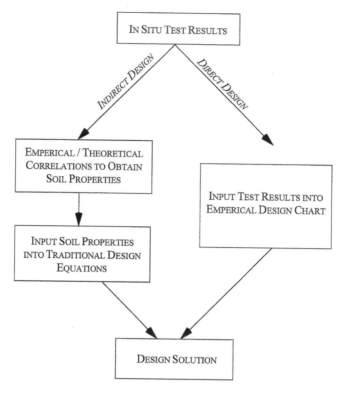

Figure 1.2 Comparison between indirect and direct design approaches using in situ tests.

REFERENCES

Bjerrum, L. 1972. Embankments of Soft Clay. *Performance of Earth and Earth-Supported Structures,* *ASCE*, Vol. 2, pp. 1–54.

Jamiolkowski, M., Ghionna, V.N., Lancellotta, R. and Pasqualini, E., 1988. New Correlations of Penetration Tests for Design Practice. *Proceedings of the 1st International Symposium on Penetration Testing*, Vol. 1, pp. 263–296.

Mitchell, J.K., Guzilkowski, F. and Villet, W., 1978. The Measurement of Soil Properties In-Situ. 1978. Lawrence Berkeley Laboratories Report LBL-6363, University of California, Berkeley, 67 pp.

Wroth, C.P., 1984. Interpretation of In Situ Tests. *Geotechnique*, Vol. 34, No. 4, pp. 449–489.

Chapter 2

Standard Penetration Test (SPT)

2.1 INTRODUCTION

Penetration tests are the oldest and most common form of *in situ* tests and involve pushing or driving a rod, point, or sampler into the ground to delineate soft zones from firm zones. In Chapter 3, dynamic cone penetration tests (DCPs) are covered, and in Chapter 4, the Cone Penetrometer (CPT) and Piezocone (CPTU) are described. In this chapter, the Standard Penetration Test (SPT) will be examined. Other tests that are similar to the SPT are also briefly described at the end of this chapter and include the Large Penetration Test (LPT) and the Becker Hammer Test (BHT). Despite its perceived shortcomings, the SPT is still the most commonly used *in situ* test in North America and probably throughout the world. Engineers need the SPT to do two things: (1) provide a soil sample for field identification or laboratory index classification tests and (2) provide an indication of soil behavior.

2.2 BACKGROUND

The SPT is a Dynamic Penetration Test in which the resistance to the driving and installation of a thick-walled tube is measured. At the same time, the test is a sampling technique in which a sample of the soil through which the penetration takes place is obtained. It is the only *in situ* test that provides a sample for soil classification and other index testing, an attribute that many engineers feel is a distinct advantage of the test and one that sets the test apart from all others.

Fletcher (1965), Broms & Flodin (1988), and Rogers (2006) present histories of the SPT beginning with its introduction in about 1902 by Col. Charles R. Gow who used a 25 mm (1 in.) open pipe driven into the ground to collect a soil sample. Up to this time, soil investigation in the U.S. had primarily been conducted using wash boring techniques based on portable tripod equipment erected at the site. The open pipe was driven into the ground with a 50 kg (110 lbs) weight to recover soil samples. In 1922, the Gow Co. became a subsidiary of the Raymond Concrete Pile Co. and further development and modifications to the sampling procedure apparently took place over the next several years by Gow and H.A. Mohr.

The 50 mm (2 in.) split spoon sampler was designed around 1927, and about the same time, the 63.5 kg (140 lbs) weight and 0.76 m (30 in.) drop were more or less standardized by the company and others. The number of blows required to drive the sampler a distance of 0.3 m (12 in.) constituted the record of the test. The sampler was only driven a total of 0.3 m (12 in.) using a 25 mm (1 in.) drive pipe until about 1945 when standard "A" size drill rods were introduced in the industry. Around this time period, a ball check valve was added to the top of the sampler in an attempt to help prevent sample loss.

One of the most significant modifications to the test in these early days was made by J.D. Parsons in 1954 when the blows required for each of three consecutive 0.15 m (6 in.) increments were noted and the sum of the two increments giving the *lowest* total for a penetration distance of 0.3 m (12 in.) was taken. This technique is no longer used, and the first 0.15 m (6 in.) increment is still taken; it is considered by many as a "seating" increment. The blows from the second and third 0.15 m (6 in.) drive increments are added to give the measurement from the test, known as the "N-value".

According to Fletcher (1965), the original purpose of the SPT was to measure the density of soil formations by a standard procedure in order to give a correlation with experience in the design and installation of caisson foundations. The equipment described by Fletcher (1965) showed some clear differences from modern SPT equipment; (1) the inside diameter of the sample barrel was the same as the inside diameter of the shoe with both having a constant diameter of 34.9 mm (1.375 in.); i.e., there is no internal relief inside the barrel; (2) a 24 in. long spoon was used; (3) a pin weight hammer was shown as standard equipment; and (4) a hardwood cushion block was used between the hammer and drive rods.

Modifications to the equipment and procedure to bring the test to the present day (2020) configuration will be discussed subsequently and should be obvious to the reader in comparison to the 1950s and 1960s arrangement of the test. The equipment and procedure used to conduct the SPT was standardized by ASTM in 1958 in test procedure D1586 and is also an international reference test.

Some geotechnical engineers feel that the SPT has outlived its usefulness for site investigations and geotechnical design and perhaps should be retired given that there are other options available. Some reasons given for this are that the test is outdated, the test results are too variable, and there are more advanced *in situ* techniques available such as the CPT/CPTU or DMT. For many years, a number of issues plagued the SPT:

1. The test was considered highly variable (i.e., equipment and procedures varied too much);
2. Test results were historically too dependent on the operator; and
3. Control of the test has generally been taken away from engineers and given to drillers.

However, the SPT has some advantageous attributes that make it useful for many routine site investigations:

1. The test concept, arrangement, and equipment are relatively simple, robust, and inexpensive;
2. The equipment is readily available from most drillers around the world and is easily adaptable to most drill rigs;
3. The procedure is relatively easy to carry out, and testing may be performed at reasonably frequent intervals, often being performed continuously in the upper layers of soil or the primary zone of influence for foundations;
4. A soil sample is usually obtained for visual/manual identification and index property evaluation;
5. The test has a wide range of applicability, from weathered rock and gravelly sands to soft insensitive clay;
6. The test data are simple to collect and the test results are reduced rapidly in the field.

There is also an argument by some engineers that the SPT is a "one-number test", that is, the SPT only gives a single number to use in assessing soil behavior.

2.3 MECHANICS OF THE TEST

The test procedure calls for the input of specified impact energy to drive a split tube soil sampling barrel of specified dimensions a required distance. A schematic of the current test is shown in Figure 2.1. The test consists of a falling weight or hammer of mass 63.5 kg (140 lbs) that is allowed to drop (free-fall) a distance of 0.76 m (30 in.) to strike an anvil. The anvil is connected to a set of drill rods that extend to the depth of testing/sampling. A sampling barrel, usually called a split spoon, is attached to the end of the drill rods and is advanced into the soil with each impact of the hammer.

In typical practice, chalk marks are made on the drill string to mark off three 0.15 m (6 in.) increments; the hammer is raised and dropped; the number of hammer blows to advance the spoon each 0.15 m (6 in.) increment are recorded as N_{0-6}; N_{6-12}; and N_{12-18}. The initial 0.15 m (6 in.) penetration is considered by ASTM as a "seating" penetration and the sum of the hammer blows for the second two 0.15 m (6 in.) increments is called the SPT N value (with units of blows per 0.3 m or blows per ft) and is the reference measurement obtained from the test. It is important that the incremental blow count values for each 0.15 m (6 in.) be recorded and reported, which is actually required by ASTM D1586.

In the event the full 0.46 m (18 in.) of penetration cannot be achieved, ASTM D1586 allows the test to be terminated if

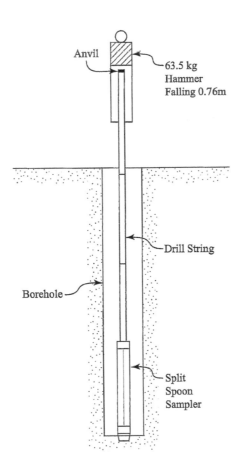

Figure 2.1 Schematic of SPT.

1. A total of 50 hammer blows have been applied during any one of the three 0.15 m (6 in.) increments;
2. A total of 100 hammer blows have been applied; or
3. There is no observed advance of the sampler during the application of ten successive hammer blows.

If only partial penetration occurs, the penetration resistance is recorded as the number of hammer blows for the penetration increment achieved, for example, "50 for 2 in.". In very soft clays, the spoon and drill rods may advance on their own without driving. This is often referred to as "Weight of Rod" (WOR). If the spoon and rods advance after the hammer has been attached, this is referred to as "Weight of Hammer" (WOH). When this occurs, it is important that the water level onside the boring be noted on the boring logs and the size of drill rods being used are recorded. The total mass of the spoon and rods may include partially buoyant weight for that length below the water level.

After driving, the spoon is brought to the surface and opened, and the amount of soil retrieved for that drive is recorded. This is known as recovery. The recovery ratio, R, is defined as

$$R = L_s/L_D \times 100\% \tag{2.1}$$

where:

L_s = length of recovered soil in the spoon
L_D = length of spoon drive

Recording the recovery is considered a part of the ASTM procedure and should be recorded on the boring logs. The recovery ratio may be used qualitatively to help interpret the SPT results. For example, if the recovery ratio is consistently low in coarse-grained deposits, this may suggest that abundant gravel or cobbles may be present with particles too large to enter the spoon. After recording the Recovery, the soil is usually placed into water-tight containers such as glass jars or bags or is wrapped in plastic and aluminum foil for preservation and then transported back to the office or laboratory.

2.4 EQUIPMENT

The mechanics of the SPT described in the previous section represent a relatively simple concept; however, because the test is perceived as being so simple, the execution of the test may vary widely. This is a result of the variations in test equipment that have been available and are currently being used in the field to conduct the test. As indicated in Figure 2.1, the test equipment consists of four basic components: the drop weight or hammer with an anvil, a string of drill rods connecting the sampler to the hammer, and a barrel sampler.

2.4.1 Hammer

The purpose of the hammer system is to provide the specified amount of energy to the rods in order to advance the sampler. Because this energy may be created in a number of ways, SPT hammer systems typically represent the largest equipment variation observed in the test. Historically, different styles of drop hammers have been used to perform the SPT. In fact, this has caused considerable consternation among users and represents the largest source of variability to test results. Up to about 1970 pin weight hammers were used, but in the past 40 years, three primary hammer types have been used in North America. Figure 2.2 shows the schematics of (1) donut hammer; (2) safety hammer; and (3) automatic hammer. At present, most hand-operated drop hammer systems have largely been retired and are no longer used in routine practice. The automatic hammer is the single most important improvement to the SPT in the past 50 years and now considered the preferred and in many cases, the required drop hammer system for use in performing the SPT.

In other parts of the world, similar "free-fall" hammers are used. In European countries, these include the Pilcon trip monkey and the Borros AB drop-hammer. In Japan and other parts of Southeast Asia, the Tombi method (e.g., Shi-Ming 1982) is used.

A typical automatic hammer is a self-contained, totally enclosed device that operates using a hydraulically powered chain lift mechanism. The drive weight-lifting and dropping sequence is actuated by the operator using a hydraulic valve switch. This makes the hammer fully automatic, and the test is essentially operator-independent. This type of hammer appears to give the most reproducible results and avoids problems associated with the friction losses of the rope, cathead, pulley, etc. and the variations of hammer drop height. The author recommends that, when possible, only automatic hammers be used to conduct the SPT.

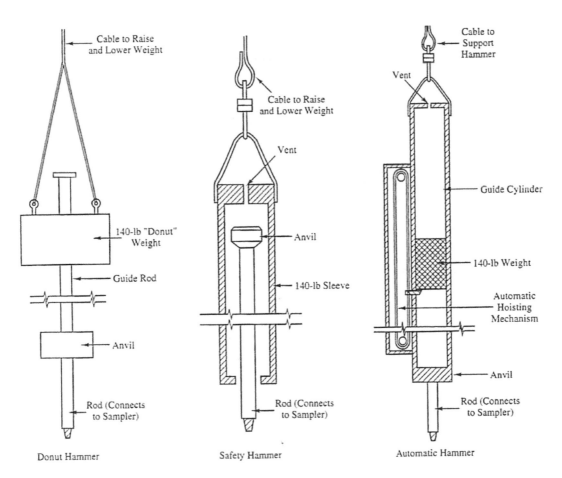

Figure 2.2 SPT drop hammers.

2.4.2 Drill Rods

Rods used to conduct the SPT are usually standard drill rods used for routine test borings. ASTM D1586 requires that flush joint drill rods having a stiffness (moment of inertia) greater than that of A rods be used. Drill rods are available in a wide variety of lengths, diameters, and thread designs. Drillers have traditionally used square thread drill rods that have been in use for many years. However, in more recent years, tapered thread drill rods have become increasingly popular since they can be assembled and disassembled more quickly.

2.4.3 Split Barrel Sampler

The split barrel or spoon used for the SPT is an open-ended thick-walled drive sampler. The barrel is designed in two halves and split along its longitudinal axis, hence the name split spoon. The two halves are held together by a hardened drive shoe threaded onto the lower end and a drill rod adapter head threaded onto the upper end. A ball check valve and ports are contained in the head of the spoon so that if the spoon is lowered into a borehole filled with water or drilling fluid, the fluid will exit out the top end. This allows the soil to enter the spoon during driving and prevents fluid from becoming trapped in the spoon.

Most split spoons are available from suppliers in lengths varying from 0.46 to 0.61 m (18 to 24 in.). A longer spoon allows some room for loose soil cuttings that may be at the bottom of the borehole to enter the spoon and not affect the penetration resistance. The configuration of the split-spoon is shown in Figure 2.3. Typically, the sample is held in the spoon with a small metal or plastic catcher that is equipped with spring action "fingers" and positioned just inside the shoe. To remove the sample from the spoon, the drive head and shoe are removed (usually with a pair of pipe wrenches), and the two halves of the spoon are separated.

The standard size split spoon sampler recommended for use with the SPT in the U.S. has an outside barrel diameter of 50.8 mm (2 in.) and an inside barrel diameter of 38.1 mm (1.5 in.). In the U.S., the drive shoe has an inside diameter of 34.9 mm (1.375 in.). This I.D. is smaller than the I.D. of the barrel for a standard 50.8 mm (2 in.) spoon and creates a slight relief inside the barrel. This relief of the barrel is thought to have originated sometime

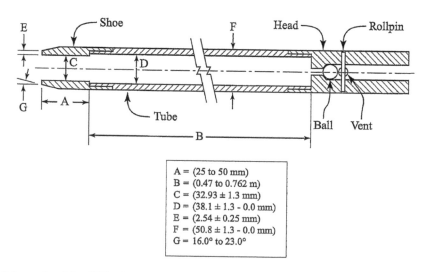

A = (25 to 50 mm)
B = (0.47 to 0.762 m)
C = (32.93 ± 1.3 mm)
D = (38.1 ± 1.3 - 0.0 mm)
E = (2.54 ± 0.25 mm)
F = (50.8 ± 1.3 - 0.0 mm)
G = 16.0° to 23.0°

Figure 2.3 Schematic of the SPT split spoon sampler. (After ASTM D1586.)

around the latter 1960s to allow the use of a series of brass liners to be added to the spoon. (Note that Fletcher (1965) indicated that the inside diameter of the barrel was the same as the inside diameter of the shoe as shown in Figure 2.1.) Ireland et al. (1970) also described the "standard" barrel sampler as having a constant internal diameter of 34.9 mm (1.375 in.), i.e., the same I.D. as the shoe. This appears to be a routine practice in the U.K.

It is important to know whether the spoon is manufactured with this relief or if liners are being used during the SPT since this can affect the N-value by changing the frictional forces as the soil enters the spoon. The international reference test procedure for the SPT shows the I.D. of the sampler barrel to be the same as that of the shoe, i.e., 34.9 mm (1.375 in.). In the UK, Japan, and other parts of the world, it appears that samplers with a constant I.D. equal to that of the drive shoe are more common.

2.5 TEST PROCEDURES

As previously noted, the test procedures and equipment for conducting the SPT are described in detail in ASTM D1586 *Standard Test Method for Penetration Test and Split-Barrel Sampling of Soils* and in *ASTM D6066 Standard Practice for Determining the Normalized Penetration Resistance of Sands for Evaluation of Liquefaction Potential*. The SPT is also an international reference test and described in ISO 22476-3:2005 *Geotechnical Investigation and Testing–Field Testing–Part 3: Standard Penetration Test.*

2.6 FACTORS AFFECTING TEST RESULTS

The goal of the SPT is to have the response from the test reflect differences in soil behavior, not differences in the test procedure. There are a number of factors that can affect the results of the SPT. Because of the historical variability in drilling equipment, techniques, personnel, etc., and the more or less crude fashion in which the test was performed, the results tended to show a high degree of variability. Drop hammer systems using a rope and cathead tend to give erratic results simply because the energy is largely uncontrolled and varies widely from drop to drop. However, many of these issues have been eliminated by using a calibrated automatic hammer. There are three reasons that only calibrated hammers should be used to conduct the SPT:

1. Automatic hammers provide a repeatable operator-independent known energy level;
2. A fully enclosed automatic hammer is much safer for the drill crew;
3. Automatic hammers provide increased productivity in the drilling operation.

2.6.1 Energy Delivered to the Sampler

A number of early investigations (e.g., Schmertmann 1979; Schmertmann & Palacios 1979; Kovacs & Salomone 1982) showed that the most significant factor affecting the N-value from the SPT as the amount of energy delivered to the sampler. The energy delivered by the hammer to the drill rods and reaching the sampler as an initial compression wave was defined by Schmertmann & Palacios (1979) using the term ENTHRU, E_i. The theoretical energy of the SPT hammer (474 J) is termed E^* and thus the ratio E_i/E^* equals the ENTHRU efficiency delivered by the SPT system to the sampler. Therefore, ENTHRU efficiency may be dependent on the combined effect of the hammer/drop system, the drill rod size and length, the mass of the anvil, etc.

Table 2.1 Reported historic SPT hammer system energy efficiencies

Hammer	Release	Efficiency (%)	References
Donut	Rope and cathead	31	Schmertmann & Palacios (1979)
		45	Kovacs et al. (1981)
		43	Robertson et al. (1983)
	Tombi	80	Kovacs & Salomone (1982)
		80–90	Tokimatsu & Yoshimi (1983)
Safety	Rope and cathead	52	Schmertmann & Palacios (1979)
		55	Schmertmann & Palacios (1979)
		61	Kovacs et al. (1981)
		52	Kovacs & Salomone (1982)
		62	Robertson et al. (1983)
		55–115	Riggs et al. (1983)
		71–91	Riggs et al. (1984)
Automatic	–	56–115	Riggs et al. (1983)
		90	Riggs et al. (1983)
		86–91	Schmertmann (1984)
		84–106	Frost (1992)
		68.5–81.4	Batchelor et al. (1994)
		76–94	Lamb (1997)
		83.2	Davidson et al. (1999)

Because of the variation in drill rigs hammer designs, drop mechanisms, and operator variables, it should be recognized that a wide range in operating efficiencies might be observed in SPT practice. Additionally, the type of soil may also have a significant influence on the SPT energy with hard soils producing more energy than soft soils (Bosscher & Showers 1987). A summary of some reported measurements of hammer efficiencies is given in Table 2.1.

2.6.2 SPT Hammer Energy Calibration

ASTM D4633 provides a recommended procedure for performing the energy calibration of the SPT. An example of an energy calibration is described by Schmertmann (1978). Hall (1982) and Robertson et al. (1992) also described the use of equipment for performing the energy calibration using a load cell in the drill string and different recording equipment. A downhole calibration device has also been described by Van der Graaf & Van den Heuvel (1992). Matsumoto et al. (1992) also describe a SPT energy measurement system using strain gages attached to the drill rods at two points. Figure 2.4 shows a hammer calibration in progress using strain gauges and accelerometers attached to the top of the drill string.

A side-by-side comparison of SPT N-values obtained between a safety hammer with a rope and cathead and an automatic hammer using the same drill rig, driller, drilling method, and drill rods is shown in Figure 2.5. The measured energy for each hammer drop is shown. As expected, the mean N-value from the automatic hammer is lower than that from the safety hammer because of the higher energy level provided by the automatic hammer ($E_{Auto} = 79.1\%$; $E_{Safety} = 60.8\%$). More obvious is that the energy delivered by the safety hammer is more variable than from the Automatic Hammer ($COV_{Safety} = 8.7\%$; $COV_{Auto} = 4.0\%$) for these measurements.

Since about 2005, it has become increasingly easy to perform energy calibrations on SPT hammers. The equipment has become more readily available and easier to operate.

Figure 2.4 Energy calibration of SPT automatic hammer in progress.

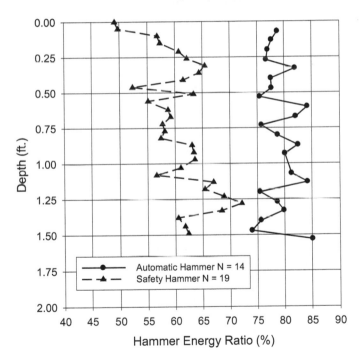

Figure 2.5 Comparison of measured energy from each hammer drop between automatic hammer and safety hammer with rope and cathead.

Most state highway agencies require or strongly recommend the use of automatic hammers for performing SPT work. A large number of measurements using various drill rigs and automatic hammers in a range of soils have been summarized by Biringen & Davie (2008) and Honeycutt et al. (2014). Table 2.2 gives a summary of some reported energy calibrations on automatic hammers since 2005.

Table 2.2 Some reported automatic hammer energy since 2005

Location	Hammer/rig	Efficiency (%)	References
California	CME	84.5	Liebich (2005)
S. Carolina, Tennessee, Georgia, Texas, Maryland	Various CME, Failing, Mobil, Diedrich	68.5–80.0	Biringen & Davie (2008)
New York	CME	73.0–88.0	Akbas & Kulhawy (2008)
Vermont	CME 75	85.6	Kelley & Lens (2010)
	CME 55	87.5–90.5	
	CME 45c	79.6–84.2	
North Carolina	N/A	78.6	Valiquette et al. (2010)
Illinois	CME-75	91.3–96.4	Stark et al. (2017)
	CME-55	97.5	
	CME-45c	85.8	
	CME-550x	80.4	

2.6.3 Other Factors Affecting SPT Results

A number of investigations have summarized factors other than energy that can affect the SPT N-value (Fletcher 1965; Schmertmann 1978; Decourt 1989; Kulhawy & Trautmann 1996). Nearly all of these factors can be placed in two categories: (1) equipment variations and (2) operator or procedural variations. If a calibrated automatic hammer is used, the remaining factors are summarized in Table 2.3.

2.6.3.1 Diameter of Drill Rods

ASTM D1586 specifies that the rods must have a stiffness equal to or greater than that of parallel wall "A" size rods. In the U.S. and elsewhere in the world, the SPT is most often performed using either "A" size rods or "N" size rods. Different styles of rod threads (square and tapered) may be used, but typically, a single rod size will be used in the string. Most reported comparisons between SPTs conducted with both A and N rods at the same site do not show significant difference in N values (Gibbs & Holtz 1957; Palmer & Stuart 1957; Brown 1977; DeGodoy 1971; Yokel 1982; Matsumoto & Matsubara 1982). It appears that up to depths of about 30 m (100 ft), the use of drill rod sizes in the range most often used in the field and recommended by ASTM D1586 produces a negligible effect on N-values. ASTM D6066 suggests that for depths greater than 15 m (50 ft), BW and NW rods are preferred to avoid whipping or buckling. Figure 2.6 show a side-by-side comparison of N-values obtained using AWJ and NWJ rods in sand.

2.6.3.2 Drill Rod Length

The total length of drill rods may influence the energy reaching the spoon from the hammer drop (e.g., Schmertmann & Palacois 1979; Yokel 1982; Decourt 1989; Morgano & Liang 1992; Odebrecht et al. 2004). A secondary effect of longer rods is the tenancy for increased rod whip and loss of energy from multiple rod connections, both of which tend to reduce the energy reaching the sampler. A correction factor is often used to adjust N-values for the total length of rods used, i.e., depth (e.g., Skempton 1986).

Table 2.3 Factors other than hammer energy that may influence SPT results

Equipment variations	Sampler dimensions	Variations in exact sampler dimensions vary around the world. Sampler should conform to the latest ASTM standard and should be measured before use
	Liners/no liners	Use of liners vs. no liner but spoon with internal relief increases blow counts
	Use of damaged or deformed tip on sample spoon	Damaged shoe may change blow counts
	Using damaged drill rods	Drill rods that are slightly bent or otherwise damaged tend to reduce energy transfer giving artificially high N-values
Procedural variations	Inadequate cleaning of the borehole	SPT is only partially made in original soil. Sludge may be trapped in the sampler and compressed as the sampler is driven, increasing the blow count
	Failure to maintain sufficient hydrostatic head in the boring	The water table in the borehole must be at least equal to the piezometric level in the sand, otherwise the sand at the bottom of the borehole may be transformed into a loose state
	Using a too large pump	Too high a pump capacity will loosen the soil at the base of the hole causing a decrease in blow count
	Over-washing ahead of casing	Low blow count may result in dense sand since sand may be loosened by over-washing
	Drilling method	Drilling technique (e.g., cased holes vs. mud stabilized holes) may result in different N-values for the same soil. The SPT was originally developed from wash boring techniques. Drilling procedures which seriously disturbs the soil will affect the N-value, e.g., drilling with cable tool equipment
	Rate of testing	In saturated soils, a fast rate of testing may increase pore water pressures.
	Plugged casing	High N-values may be recorded for loose sand when sampling below groundwater table. Hydrostatic pressure causes sand to rise and plug casing
	Loose drill rod connections	Energy losses can occur from loose rod connection giving artificially high N-values
	Marking drive increments	Drive marks should be made after the spoon and drill string have been just set on the bottom of the borehole but before the hammer is attached or rods are released
	"Seating" the spoon before marking the rods	There is no such thing as "seating" of the spoon before marking the three 0.15 m (6 in.) incremental drive lengths
	Sampler plugged by gravel	Artificially high blow counts result when gravel plugs sampler; resistance of loose sand could be highly overestimated
	Carelessness in counting the blows and measuring penetration	Poor observations of incremental blow counts may produce errors in N-values.
	Using drill holes that are too large	Holes greater than 100 mm (4 in.) in diameter are not recommended. Use of larger diameters may result in decreases in the blow count from stress relief at bottom of hole
	Attitude of operators	Blow counts for the same soil using the same rig can vary, depending on who is operating the rig, and perhaps the mood of operator and time of day

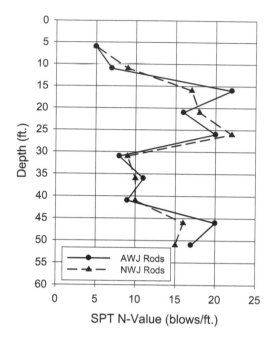

Figure 2.6 Comparison of N-values from AWJ and NWJ drill rods.

2.6.3.3 Sampler Dimensions

The dimensions of the standard split spoon sampler are defined in ASTM D1586 as previously shown in Figure 2.5. In the U.S., the inside diameter of the barrel is slightly larger than the inside diameter of the shoe i.e., 38.1 mm (1.5 in.) vs. 34.9 mm (1.375 in.) to accept liners. Schmertmann (1979) showed that the use of liners or split spoons with an inside diameter equal to that of the shoe can have a significant influence on the recorded N value as a result of the additional friction that can build up as the sample moves along the inside of the barrel. For internal relief samplers used without liners, the only internal resistance on the sample comes from a short segment of the shoe. The result is a higher N value (from increased resistance to driving), and the overall effect is a false impression that the soil is denser or stiffer than it actually is.

Seed et al. (1985) summarized data comparing N-values using a spoon with liners and a spoon with no liner. As shown in Figure 2.7, there appears to be no difference in loose sands up to about N = 10; however, for N = 10–50, the results suggest approximately $N_{unlined} = 0.8\ N_{lined}$. Data collected at a stiff clay site in Missouri by the author gave a mean ratio of $N_{unlined} = 0.72\ N_{lined}$ from six side-by side comparison tests using a CME automatic hammer.

Early SPT practice routinely used a split spoon sampler with a constant inside diameter of 34.9 mm (1.375 in.), i.e., no internal relief in the barrel and is shown in most period textbooks (e.g. Terzaghi & Peck 1948; Peck et al. 1953, 1974; Spangler 1960; Sowers & Sowers 1961) and manuals (e.g., Bureau of Reclamation Earth Manual 1974; AASHTO Standard Specifications 1981). In some cases, a note is provided that indicates that the barrel may have an inside diameter of 38.1 mm (1.5 in.) provided it contains a liner. Even as late as 1996, some texts (e.g. (Terzaghi et al. 1996) still showed no internal relief. The current edition of ASTM D1586 shows a split spoon with an internal relief between the shoe and the barrel. It appears that a barrel with a constant diameter of 34.9 mm (1.375 in.) is currently routinely used in Japan (Matsumoto et al. 2015).

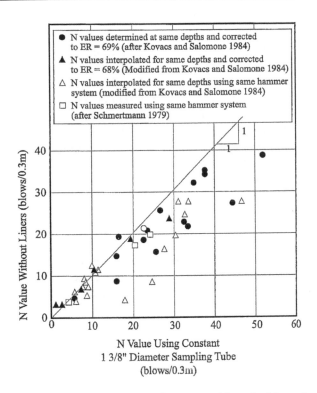

Figure 2.7 Comparison between N-values from a split spoon with and without liners. (After Seed at al. 1985.)

The consequence of using modern practice which uses a split spoon with an internal relief as compared to using a split spoon with no relief (circa 1948–1973) is that current SPT blow counts would be lower since there is no internal side resistance to overcome. Correlations between SPT N-values and different sol properties developed before about 1980 were most likely based on using a spoon with no internal relief and therefore need to be adjusted.

2.6.3.4 Diameter of Borehole

The size of the borehole may influence the test results since larger diameter holes tend to produce stress relief at the base of the hole. The influence of borehole diameter is likely to be less significant in stiff fine-grained soils than in granular soils and probably most important in loose fine sands below the water table. Most studies (e.g., Lake 1974; Jain & Handa 1979; Sanglerat & Sanglerat 1982) have indicated that there appears to be no significant effect on N values provided the borehole diameter is less than about 150 mm (6 in.).

2.6.3.5 Method of Drilling/Drilling Fluid

Parsons (1966) found that in coarse sand, N-values obtained in mudded holes were about 2.5 times higher than those obtained in holes only partially filled with water. Seed et al. (1988) found no difference in N-values at several sites for water-filled vs. drilling mud-filled boreholes with N-values ranging from about 15 to 40. Whited & Edil (1986) found no effect on SPT results in fine-grained soils using 57.1 mm (2.25 in.) I.D. hollow stem augers, drilling mud with an open borehole, or 63.5 mm (2.5 in.) casing. However, simply using hollow-stem augers without drilling mud in sands below the water table can produce lower blow counts in sands.

In coarse-grained soils below the water table, the SPT is typically performed in conjunction with either mud rotary wash boring techniques or with hollow stem augers. It is important to maintain an equivalent head of fluid in the borehole to counteract the tendency for the upward flow of water at the base of the borehole. This is especially important when drilling with hollow stem augers or casing. In loose sands, the upward gradient of flow can exceed the critical gradient of the soil so that a "quick" condition develops, and the soil tends to "boil". This condition can be accentuated by a close-fitting drill bit and drill rods that are pulled too quickly, producing a slight suction at the base of the hole. In order to reduce the potential for sands entering the bottom of the borehole, it is important to maintain a head of drilling fluid inside casing or hollow-stem augers. Even while the drill rods are being removed, the fluid level should be adjusted to prevent blow-in at the bottom of the hole.

In addition to maintaining a sufficient equilibrium head of drilling fluid inside the borehole, the method of drilling can influence SPT results especially in loose granular deposits if a bottom discharge bit is used and high fluid pressures are used. This may result in loosening of soil at the bottom of the borehole. In fine-grained soils, there is not sufficient evidence to indicate that the drilling method significantly influences the test results.

2.6.3.6 Cleanout of the Borehole

Drill cuttings left at the bottom of a borehole because of inadequate cleaning may affect the test results in three ways: (1) the cuttings may cause the spoon to get hung up at the base of the hole and not rest on virgin material for the first 0.15 m (6 in.) penetration increment; (2) excess cuttings may enter the spoon and then restrict the free movement of the actual soil sample into the spoon; and (3) the cuttings may create additional internal and external side frictional resistance on the spoon. The first of these conditions would tend to produce lower N values, while the latter two conditions would tend to produce higher N values. The borehole should always be cleaned out as much as possible before inserting the spoon which will reduce these effects.

2.6.3.7 Rate of Testing

The rate at which hammer blows are applied to the drill rods may influence the test results. A standard rate of 30–40 blows per minute is recommended by ASTM D1586. If the test is performed very rapidly, there may be a progressive buildup of excess pore water pressures with each hammer blow resulting in weaker conditions. Whereas if the test is performed very slowly, or if a waiting period is allowed midway in the test, pore water pressures may dissipate, resulting in an increase in effective stress. In order to reduce these effects, the rate of testing should be held as close as possible to the recommended rate.

2.6.3.8 Seating of the Spoon

The SPT was initially conducted by driving the spoon sampler a distance of 0.3 m (12 in.). After about 1954, the practice of first driving a "seating" distance of 0.15 m (6 in.) at the bottom of the borehole before driving 0.3 m (12 in.) became common. This practice probably developed because of problems with disturbed soil or soil cuttings at the base of the borehole and the general feeling that blow counts obtained over the first 0.15 m (6 in.) increment in a 0.3 m (12 in.) drive were often not representative of the in-place density of the soil.

Technically, there is no "seating" of the spoon. The spoon and rods are simply set on the bottom of the borehole while still being held, and three 0.15 m (6 in.) increments are marked on the rods. The test then begins after the spoon and rods have been released but before the hammer is attached to the top of the rods. Any "seating" practice used by drillers as local practice should not be allowed.

2.6.3.9 Condition of the Drive Shoe

ASTM D1586 gives specific dimensions of the drive shoe used on the end of the split spoon. Shoes often become worn or damaged through excessive use or driving on hard materials such as gravel or rocks, for example, as shown in Figure 2.8. A damaged shoe may introduce unknown errors in the SPT N-values (e.g. Das 2014) and therefore should not be used. New shoes are relatively expensive, and so, it is always best to use a shoe that is not damaged.

2.6.3.10 Summary

The practice for performing the SPT should be as consistent as possible from one project to the next. Engineers need to make careful observations and records of the procedures and equipment used for all tests. This helps eliminate most perceived problems with the test.

2.7 CORRECTIONS TO SPT BLOW COUNTS

Because the SPT blow count is directly related to the hammer system energy, it is necessary to adjust the results of the test to a standard energy level. This will allow a proper comparison of test results so that different hammer systems can be compared, and proper

Figure 2.8 Photo of damaged split spoon drive shoe.

interpretation of the test data can be made. Corrections to the field SPT N-value are made to account for hammer energy, rod length, borehole diameter, and sampler geometry.

2.7.1 Corrections for Hammer Energy, Equipment, and Drilling: N to N_{60}

Based on several suggestions, a reference value of 60% of the theoretical free-fall energy is used to adjust N-values to a common reference point. The use of a 60% energy level also represents a likely average level of energy that has been in use since the 1960s using traditional SPT equipment and therefore was the basis for a number of correlations for different soil properties. A number of correction factors have been introduced to adjust the field-measured N-values to a reference energy level of 60% and taking into account the sampler geometry, drill rod length, and borehole diameter. In this way, we can define the energy-corrected blow count as follows:

$$N_{60} = N \times ER \times C_B \times C_S \times C_R \tag{2.2}$$

where
   N_{60} = energy and procedure-corrected blow count
   N = field-measured blow count
   ER = energy ratio = ES/E_{60}
   ES = energy of the system
   C_B = correction factor for borehole diameter
   C_S = correction factor for sampler geometry
   C_R = correction factor for rod length

Table 2.4 presents recommended values for these correction factors. Field measured SPT N-values should always be corrected according to Equation 2.2 and reported as corrected blow counts, N_{60}.
   Figure 2.9 shows a side-by-side comparison between uncorrected and corrected N-values in sand.

2.7.2 Correction for Overburden Stress in Sands: N_{60} to $(N_1)_{60}$

In a uniform sand deposit with a constant void ratio or relative density, the SPT N-value will increase with depth as the mean effective stress increases with depth. Therefore, the N-value must be corrected for the influence of this changing stress level to give a single characteristic value that describes a single relative density. To account for this, a correction factor is usually incorporated to provide a consistent effective stress reference. A number of overburden correction factors have been suggested as summarized in Table 2.5. The application of a correction factor has the general form:

$$\left(N_1\right)_{60} = C_N \times N_{60} \tag{2.3}$$

where
   N_{60} = energy-corrected blow count
   $(N_1)_{60}$ = corrected blow count to a standard vertical effective stress level
   C_N = vertical effective stress correction factor

Table 2.4 Recommended average SPT correction factors: N to N_{60}

Borehole diameter			C_B
65–115 mm (2.5–4.5 in.)			1.00
150 mm (6 in.)			1.05
200 mm (8 in.)			1.15
Sampler			C_S
Sampler without liner			1.00
Sampler with liner or barrel diameter same as shoe diameter			0.83
Drill rod length			C_R
< 3 m (10 ft)			0.75
3–4 m (10–13 ft)			0.80
4–6 m (13–20 ft)			0.85
6–10 m (20–30 ft)			0.95
>10 m (>30 ft)			1.0
Hammer and drop mechanism			ER
North America	Automatic	–	1.40
	Safety	Rope and cathead	1.00
	Donut	Rope and cathead	0.75
Japan	Donut	Trip	1.3
	Donut	Rope and cathead	1.1
China	Donut	Trip	1.0
	Donut	Rope and cathead	0.9
United Kingdom	Safety	Trip	1.0
	Safety	Rope and cathead	0.8

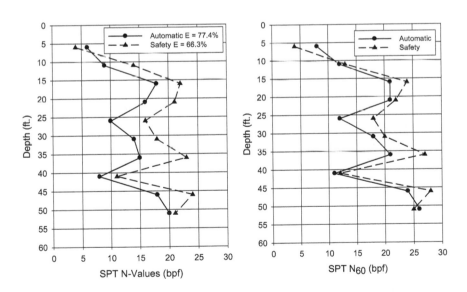

Figure 2.9 Comparison of uncorrected and corrected N-values.

As noted in Table 2.5, Skempton's suggested correction factors are the only ones to take into account gradation. The correction factor suggested by Liao & Whitman (1986) appears to be the most popular in use. It is simple and generally falls in the middle of the rest of the suggested correction factors. The correction factor is equal to 1 for a vertical effective stress of 1 kg/cm^2 (1 tsf), hence the term $(N_1)_{60}$.

Table 2.5 Suggested SPT overburden correction factors for sands: N_{60} to $(N_1)_{60}$

C_N	Units of σ'_{vo}	References
$C_N = 50/(10 + \sigma'_{vo})$	psi	Gibbs & Holtz (1959)
$C_N = 4/(1 + 2\sigma'_{vo})$	ksf	Bazaraa (1967)
$(\sigma'_{vo} \leq 1.5)$	ksf	
$C_N = 4/(3.25 + 0.5\sigma'_{vo})$		
$(\sigma'_{vo} > 1.5)$		
$C_N = 0.77 \log_{10}(20/\sigma'_{vo})$	kg/cm²; tsf	Peck et al. (1974)
$C_N = 1 - 1.25 \log_{10} \sigma'_{vo}$	kg/cm²; tsf	Seed (1976)
$C_N = 1.7/(0.7 + \sigma'_{vo})$	kg/cm²; tsf	Tokimatsu & Yoshimi (1983)
$C_N = (1/\sigma'_{vo})^{0.5}$	kg/cm²; tsf	Liao & Whitman (1986)
$C_N = 2/(1 + \sigma'_{vo})$	kg/cm²; tsf	Skempton (1986)
(NC medium loose fine sands)	kg/cm²; tsf	
$C_N = 3/(2 + \sigma'_{vo})$	kg/cm²; tsf	
(NC dense coarse sand)		
$C_N = 1.7/(0.7 + \sigma'_{vo})$		
(OC fine sands)		

There is no substantial evidence to suggest the use of overburden correction factors when using the SPT in fine-grained soils. However, Oskorouchi & Mehdibeigi (1988) suggested that an overburden correction factor be applied to SPT N-values in medium to stiff clays (N>10) for estimating undrained shear strength.

2.8 INTERPRETATION OF SOIL PROPERTIES

A number of suggestions have been made for using SPT results to predict individual soil properties. The SPT has applications in a wide range of soils; however, correlations between the test results and soil properties can be divided into either coarse-grained or fine-grained soils. In the following sections, methods to predict soil properties are presented. Because of unknown SPT practices and differences in SPT energies used up to the 1980s when many original SPT correlations were developed and the present day (2020) energies, there is the potential for considerable error in many of the empirical correlations. In general, because current energy levels are typically higher than those of the 1950s, the results generally lead to conservative estimates of soil properties using older correlations. The best correlations are generally those that use N_{60} in fine-grained soils and $(N_1)_{60}$ in coarse-grained soils.

2.8.1 SPT in Coarse-Grained Soils

2.8.1.1 Relative Density

A number of early correlations of SPT N-values and relative density, D_r, of coarse-grained soils were presented at a time when either donut or safety hammer were used to perform the SPT. Differences in these correlations are related to variations in the composition, geologic origin, stress history, and moisture conditions of coarse-grained soils and the procedures and equipment used. For example, the correlation given by Gibbs & Holtz (1957) used a spoon with a constant internal diameter and no relief. Therefore, the blow counts may be higher than those obtained if a spoon with an internal relief had been used. A number of correlations developed after 1975 are summarized in Table 2.6.

It is clear that no single expression can be used to describe the relationship between SPT blow counts and relative density for all sands considering differences in grain-size distribution, age, stress history, geologic origin, etc., and differences in field practices used to obtain N-values. Cubrinovski & Ishihara (2001) suggested an approach for estimating relative density from N-values accounting for gradation as characterized by the void ratio difference ($e_{max} - e_{min}$). A compilation of several correlations using energy and stress-corrected blow counts, $(N_1)_{60}$, is presented in Figure 2.10.

Table 2.6 Some reported correlations between N and relative density, D_r

Correlation	Notes	References
$D_r = 8.6 + 0.83 \left((N + 10.4 - 3.2 \, OCR - 0.24 \, \sigma'_{vo})/0.0045\right)^{0.5}$ OCR = over-consolidation ratio	Fine sand σ'_{vo} in psi	Marcuson & Beiganousky (1977a)
$D_r = 12.2 + 0.75 \left(222N + 2311 - 711 \, OCR - 53 \, \sigma'_{vo} - 50 \, C_u^2\right)^{0.5}$ OCR = over-consolidation ratio C_u = uniformity coefficient	Coarse sand σ'_{vo} in psi	Marcuson & Beiganousky (1977b)
$D_r = 11.7 + 0.76 \left(222N + 1600 - 53 \, \sigma'_{vo} - 50 \, C_u^2\right)^{0.5}$	NC σ'_{vo} in psi	Marcuson (1978)
$D_r = 0.118 + 0.441 \log N$		Borowczyk & Frankowski (1981)
$D_r/100 = (N)^{0.5} / \left(4.188 + 0.639 \, \sigma'^{0.606}_{vo}\right)$	σ'_{vo} in metric tons/m^2	
$D_r = 16(N_1)^{0.5}$ $C_n = 1.7/(\sigma'_{vo} + 0.7)$ $D_r = 16(N_1 + \Delta N_f)^{0.5}$ Fines content (FC)　　　ΔN_f 　0 – 5%　　　　　0 　5 – 10%　　　Interpolate 　>10%　　　0.1FC + 4	Clean sands σ'_{vo} in kg/cm^2 or tsf Sands with fines	Tokimatsu & Yoshimi (1983)
$(N_1)_{60}/D_r^2 = 60$	$D_r > 0.35$	Skempton (1986)
$N/D_r^2 = 17 + 22 \, \sigma'_{vo}$	$\sigma'_{vo} = 0.5$ to 1.5 kg/cm^2 $D_r = 0.4$ to 0.9	Skempton (1986)
$D_r = 22(N)^{0.57} \, \sigma'^{-0.14}_{vo}$ $D_r = 18(N)^{0.57} \, \sigma'^{-0.14}_{vo}$ $D_r = 25(N)^{0.44} \, \sigma'^{-0.13}_{vo}$ $D_r = 25(N)^{0.46} \, \sigma'^{-0.12}_{vo}$	Fine sand Gravel content 25% Gravel content 50% All soils σ'_{vo} in kPa	Yoshida et al. (1988)
$D_r = 100 \left[(N_1)_{60}/60\right]^{0.5}$	Normally consolidated, unaged sands	Kulhawy & Mayne (1990)
$D_r = 1.55N_1 + 40.0$ $D_r = 0.84N_1 + 57.8$ $N_1 = N(98/\sigma'_v)^{0.5}$ $\left(\sigma'_v \text{ in kPa}\right)$	$0 \leq N_1 \leq 25$ $25 \leq N_1 < 50$	Hatanaka & Feng (2006)

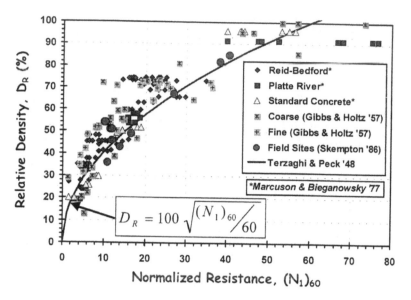

Figure 2.10 Correlation between SPT $(N_1)_{60}$ and relative density. (From NHI 2002.)

2.8.1.2 Friction Angle

The results of the SPT may be useful in estimating the shear strength of granular soils by providing an estimate of the drained friction angle, φ'. It should be recognized that the value of φ' is not unique but depends on stress level, stress path, loading conditions, etc., and therefore, any estimate does not take these factors into account. A number of suggestions have been proposed by various researchers for evaluating φ' from SPT N-values. Direct correlations between N and φ' have been presented and are summarized in Table 2.7. In most cases, the reference values for φ' have not been given.

Table 2.7 Reported correlations between N and φ' for coarse-grained soils

Correlation	References
$\varphi' = (10N)/35 + 27^0$	Meyerhof (1956)
$\varphi' = (20N)^{0.5} + 15^0$	Kishida (1967)
$\varphi' = 3.5(N)^{0.5} + 20^0$	Muromachi et al. (1974)
$\varphi' = \left(N/\sigma'_{vo}\right)^{0.5} + 26.9^0$ $\left(\sigma'_{vo} \text{ in MN/m}^2\right)$	Parry (1977)
$\varphi' = (15N)^{0.5} + 15^0$	Shioi & Fukui (1982)
$\varphi' = \left[15.4(N_1)_{60}\right]^{0.5} + 20^0$	Kulhawy & Mayne (1990)
$\varphi' = (12N)^{0.5} + 23.7^0$	Bergado et al. (1993)
$\varphi' = (20N_1)^{0.5} + 20^0$ $N_1 = N/\left(\sigma'_{vo}/98\right)^{0.5}$ $\left(\sigma'_{vo} \text{ in kPa}\right)$	Hatanaka & Uchida (1996)

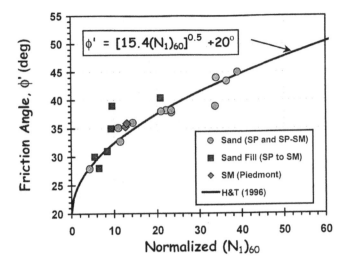

Figure 2.11 Correlation between $(N_1)_{60}$ and φ'. (From NHI 2002.)

Figure 2.11 shows a comparison between SPT $(N_1)_{60}$ values and φ' values obtained from triaxial compression tests. This correlation appears to be in popular use and is given in a number of FHWA manuals.

Engineers should use caution when estimating the friction angle in sands from N values. Local experience with foundation performance should be used to corroborate any correlations.

2.8.1.3 Soil Elastic Modulus

The stiffness of granular soils, represented by the drained elastic Young's Modulus, E, or constrained modulus, M, is also a function of the mean effective stress level for a constant void ratio such that the variation in modulus with depth will be similar to the increase in N; i.e., in a uniform deposit at constant void ratio, soil modulus increases with depth. Therefore, with stiffness, we should look for a relationship with uncorrected N_{60} values. In this way, the increase in N with depth in the uniform relative density deposit will reflect an increase in stiffness with depth.

Numerous suggestions have been made to use the SPT for estimating the elastic modulus, E_S, of granular soils. Although a number of different equations have been used, most of these correlations have the general form of

$$E_S = a(N+b) \tag{2.4}$$

where
   E_S = soil modulus
   N = SPT blow counts
   a and b = constants (empirical factors)

Suggested correlations between N and E_S are presented in Table 2.8.

Table 2.8 Some reported expressions for soil modulus from N.

Correlation	Soil type	References
E_S (tsf) = 5 (N + 15)	Sand	Webb (1969)
E_S (tsf) = 10/3 (N + 5)	Clayey sand	
E_S (MPa) = 7 $(N)^{0.5}$	Sand	Denver (1982)
E_S (MPa) = 3.5N to 40N	Sand	Clayton et al. (1985)
E_S (MPa) = 2.22$N^{0.888}$	Partially saturated gravels	Wrench & Nowatzki (1986)
E_S (kPa) = 1200 (N + 6)	Gravelly sand and gravel	Bowles (1988)
E_S (kPa) = 600 (N + 6) N<15		
E_S (kPa) = 600 (N + 6) + 2000 N>15		
E_S (kPa) = 320 (N + 15)	Clayey sand	Bowles (1988)
E_S (kPa)= 300 (N + 6)	Silty sand	Bowles (1988)
E_S (MPa) = 1.6N	Residual	Jones & Rust (1989)
E_S (MPa) = 7.5 + 0.8N	Sand	Papadopoulos (1992)
E_S (MPA) = 2.5N_{60}	Saprolite	Decourt (1994)

2.8.1.4 Constrained Modulus

Other attempts have been made to correlate the results of the SPT to the constrained modulus of the soil (M) as a function of overburden stress (e.g., Schultze & Melzer 1965; D'Appolonia et al. 1970). Since the constrained modulus, M, is related to the elastic Young's modulus, E_S, as

$$M = \left[E_S(1-\mu)\right]/\left[(1+\mu)(1-2\mu)\right], \tag{2.5}$$

an estimate of Poisson's ratio is required to estimate E_S from M. For most granular soils in drained loading conditions, the constrained modulus generally varies in the range of $1.2E_S$–$1.5E_S$.

There is considerable scatter in suggested correlations between soil stiffness or modulus and SPT blow counts. Wroth (1984) considered correlations between N and soil compressibility as "secondary" and dependent on the relationship between strength and stiffness of granular soils. Since soil modulus is strain level-dependent, the correlations include comparisons at a range of strain levels.

2.8.1.5 Small-Strain Shear Modulus

The shear modulus of soils is also strain level-dependent and reaches a maximum value at strain levels in the order of 10^{-4}%. The value at small strains is often referred to as G_o or G_{max} and in recent years, it has become customary to use this value as a reference point. The value of G_{max} is often obtained from dynamic measurements, which produce small dynamic strains, by measuring the dynamic shear wave velocity, V_S. The shear modulus, G_{max}, may then be obtained as

$$G_{max} = V_S^2 \rho \tag{2.6}$$

where

G_{max} = shear modulus (MPa)
V_S = shear wave velocity (m/s)

ρ = total soil unit weight (kg/m^3)
g = acceleration of gravity (9.8 m/s^2)

Correlations between SPT blow counts and both shear wave velocity, V_S, and G_{max} have been suggested by a number of investigators (e.g., Sykora & Koester 1988; Lee 1992; Fabbrocino et al. 2015) and are summarized in Tables 2.9 and 2.10. Most empirical correlations between V_S and N are in the form of

$$V_S = aN^b \qquad (2.7)$$

2.8.1.7 Liquefaction Potential

The SPT has a long history of use for evaluating the liquefaction potential of sites composed predominantly of sands, sands and gravel, and silty sand. A detailed test procedure for performing the test is given in ASTM D6066 *Standard Practice for Determining the Normalized Penetration Resistance of Sands for Evaluation of Liquefaction Potential*. Seed et al. (1985) had recommended the following SPT procedure for use in liquefaction correlations of coarse-grained soils:

1. Borehole 10–15 cm (4–5 in.) diameter; rotary hole with bentonite drilling mud for stability;
2. Drill bit – upward deflection of drilling mud (tricone or baffled drag bit);
3. Sampler O.D. = 50.8 mm (2.0 in.) I.D. = 34.9 mm (1.375 in.) constant, i.e., no room for liners;
4. Drill rod A or AW rods for depths less than 15.2 m (50 ft); N or NW rods for depths greater than 15.2 m (50 ft);
5. Energy delivered to sampler 60% theoretical;
6. Blow count rate = 30–40 blows/min;
7. Penetration resistance Count 0.15–0.46 m (6–18 in.).

While there has been a significant shift in this area of work from the use of the SPT to the use of the CPT, there is still considerable routine work being performed with the SPT.

Based on field performance, Seed et al. (1985), Seed and De Alba (1986) had presented charts representing the relationship between the cyclic stress ratio (CCR) causing liquefaction and $(N_1)_{60}$ values for clean sands and sand with varying amounts of fines for M = 7.5 earthquakes. Based on field performance, three approximate ranges of liquefaction damage potential could be established as follows:

$(N_1)_{60}$	Potential damage
0–20	High
20–30	Intermediate
>30	No significant damage

Figures 2.12 and 2.13 show typical charts for both clean sands and silty sands for earthquake magnitude M = 7.5. Note that the presence of fines increases the CSR needed to cause liquefaction for the same SPT $(N_1)_{60}$ value. Adjustments to CSR have been suggested to account for factors such as high confining stress and nonlevel ground conditions and are available elsewhere (e.g., Kraemer 1996).

Table 2.9 Reported correlations between N and V_S for coarse-grained soils

Correlation[a]	Soil	References
$V_S = 80.6 \ (N)^{0.331}$	Alluvial sand	Imai (1977)
$V_S = 97.2 \ (N)^{0.323}$	Dilluvial sand	
$V_S = 91.0 \ (N)^{0.337}$	All	
$V_S = 15 \ (N)$	Fine sands above water table	Schmertmann (1978)
$V_S = 85 \ (N)^{0.341}$	All	Ohta & Goto (1978)
$V_S = 88 \ (N)^{0.340}$	Sands	
$V_S = 75.3 \ (N)^{0.351}$	Gravels	
$V_S = 87 \ (N)^{0.333}$	Clays	
$V_S = 97 \ (N)^{0.314}$	All	Imai & Tonouchi (1982)
$V_S = 56 \ (N)^{0.5}$	Sands	Seed et al. (1983)
$V_S = 107 \ (N)^{0.29}$	Granular	Sykora & Stokoe (1983)
$V_S = 61 \ (N_1)^{0.5}$	Sands & silty sands	Seed et al. (1985)
$V_S = 80 \ (N)^{0.33}$	Alluvial sand	Towhata & Ronteix (1988)
$V_S = 53.5 (N_{60})^{0.17} \ z^{0.193} \ f_a \ f_G$	Sands	Jamiolkowski et al. (1988)
(z = depth (m))		
(f_a = age factor)		
Holocene = 1.0		
Pleistocene =1.3		
(f_G = grading factor)		
Fine sand =1.09		
Medium sand =1.07		
Coarse sand = 1.14		
Sand and gravel =1.45		
Gravel = 1.45		
$V_S = 49(N_1)^{0.25} \ \sigma_{vo}'^{-0.14}$	Fine sand	Yoshida et al. (1988)
$V_S = 56(N_1)^{0.25} \ \sigma_{vo}'^{-0.14}$	Gravel content 25%	
$V_S = 60(N_1)^{0.25} \ \sigma_{vo}'^{-0.14}$	Gravel content 25%	
$V_S = 55(N_1)^{0.25} \ \sigma_{vo}'^{-0.14}$	All	
$\left(\sigma_{vo}' \text{ in kPa}\right)$		
$V_S = 104.7 \ (N)^{0.296}$	Sandy soils	Lee (1992)
$V_S = 93.1 \ (N + 1)^{0.329}$	Silts	
$V_S = 76.2 \ (N)^{0.076} \ (D)^{0.313}$		
$V_S = 86.1 \ (N)^{0.075} \ (D + 1)^{0.340}$		
(D = depth in meters)		
$V_S = 49.1 \ (N)^{0.502}$	Noncohesive Greece	Kalteziotis et al. (1992)
$V_S = 90 \ (N)^{0.34}$	Misc. soils from Singapore	Veijayaratnam et al. (1993)
$V_S = 123 \ (N_{60})^{0.286}$	Loose sands and silts	Raptakis et al. (1994)
$V_S = 100 \ (N_{60})^{0.237}$	Medium and dense Sands	
$V_S = 192 \ (N_{60})^{0.131}$	Gravelly soil mixtures	
$V_S = 85.3 \ (N)^{0.42}$	Gravelly soils	Athanasopoulos (1994)
$V_S = 55.6 \ (N)^{0.5}$	Sand and rock	Akino & Sahara (1994)
$V_S = 145 \ (N_{60})^{0.178}$	Silts and sands	Pitilakis et al. (1998)
$V_S = 63 \ (N_{60})^{0.43}$	Holocene gravels	Rollins et al. (1998)
$V_S = 132 \ (N_{60})^{0.32}$	Pleistocene gravels	
$V_S = 53(N_{60})^{0.19} \ (\sigma_{vo}')^{0.18}$	Holocene gravels	
$V_S = 115(N_{60})^{0.17} \ (\sigma_{vo}')^{0.12}$	Pleistocene gravels	
$\left(\sigma_{vo}' \text{ in kN/m}^2\right)$		

(Continued)

Table 2.9 (Continued) Reported correlations between N and V_S for coarse-grained soils

Correlation[a]	Soil	References
$V_S = 145 \ (N)^{0.178}$	Silts	Pitilakis et al. (1999)
$V_S = 132 \ (N)^{0.271}$	Clays	
$V_S = 22 \ (N)^{0.770}$	Silts	Jafari et al. (2002)
$V_S = 27 \ (N)^{0.730}$	Clays	
$V_S = 131 \ (N_{60})^{0.205}$	Sands	Hasancebi & Ulusay (2007)
$V_S = 107.6 \ (N_{60})^{0.237}$	Clays	
$V_S = 73 \ (N)^{0.330}$	Sands	Dikmen (2009)
$V_S = 44 \ (N)^{0.480}$	Clay	
$V_S = 60 \ (N)^{0.360}$	Silt	
$V_S = 100.5 \ (N)^{0.265}$	Sands	Uma Maheswari et al. (2010)
$V_S = 79.7 \ (N_{60})^{0.365}$	Cands	Tsiambaos & Sabatakakis (2011)
$V_S = 112.2 \ (N_{60})^{0.324}$	Clay	
$V_S = 88.8 \ (N_{60})^{0.370}$	Silt	
$V_S = 107.2 \ (N)^{0.34}$	Sands	Esfehanizadeh et al. (2015)
$V_S = 77.1 \ (N)^{0.355}$	Sands	Fatehnia et al. (2015)
$V_S = 100.3 \ (N)^{0.338}$	Sandy	Kirar et al. (2016)
$V_S = 78.7 \ (N)^{0.352}$	Sand	Gautam (2017)

[a] V_S in m/s.

Table 2.10 Reported correlations between N and G_{max} for coarse-grained soils

Correlation[a]	Soil	References
$G_{max} = 11.5 \ (N)^{0.78}$	All	Ohsaki & Iwasaki (1973)
$G_{max} = 6.1 \ (N)^{0.94}$	Cohesionless	
$G_{max} = 5000 \ (N)^{0.3}$ (G_{max} in psf)	Data from Stokoe & Woods (1972)	Kovacs (1975)
$G_{max} = 94 \ (N)^{0.715}$	Alluvial sand	Imai (1977)
$G_{max} = 170 \ (N)^{0.650}$	Diluvial sand	
$G_{max} = 120 \ (N)^{0.737}$	All	
$G_{max} = 14 \ (N)^{0.68}$	All	Imai & Tonouchi (1982)
$G_{max} = 6.2 \ (N)$	Sands	Seed et al. (1983)
$G_{max} = 20{,}000 \ (N_1) 0.33 \left(\sigma'_{vo}\right)^{0.5} \left(G_{max} \text{ and } \sigma'_{vo} \text{ in psf}\right)$	Gravels	Seed et al. (1985)
$G_{max} = 7 \ (N)$	Data from Imai & Tonouchi (1982)	Stroud (1989)
$G_{max} = 47.5 \ (N)^{0.72}$	Lateritic soils	Decourt (1994)
$G_{max} = 5 \ (N)$	Misc. soils	Hirayama (1994)
$G_{max} = 62.8 \ (N)^{0.30}$	Gneissic residual soil	Pinto & Abramento (1997)
$G_{max} = 55.2 \ (N)^{0.665}$	Lateritic soils	Barros & Pinto (1997)
$G_{max} = 56 + 20.3 \ (N)$	Saprolitic soils	
$G_{max} = 43.8 \ (N)^{0.419}$		
$G_{max} = 94 + 2.3 \ (N)$		
$G_{max} = 98 + 0.42 \ N_{60}$	Granitic saprolite	Viana da Fonseca et al. (1998)
$G_{max} = 57 \ (N)^{0.2}$		
$G_{max} \ (MN/m^2) = 24.3 \ (N)^{055}$	Mixed soils	Anbazhagan & Sitharam (2010)
$G_{max} \left(MN/m^2\right) = 29.2 \left(N_1\right)_{60}^{0.57}$		
$G_{max} = 15.1 \left(N_1\right)_{60}^{0.74}$	All soils	Anbazhagan et al. (2012)

[a] G_{max} in MPa except as noted.

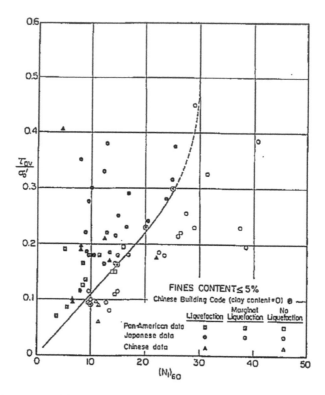

Figure 2.12 Liquefaction potential based on SPT results for clean sands. (From Seed et al. 1985.)

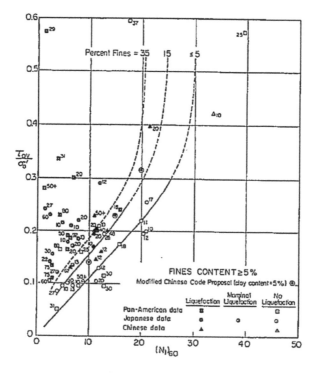

Figure 2.13 Liquefaction potential based on SPT in silty sands. (From Seed et al. 1985.)

2.8.2 SPT in Fine-Grained Soils

The SPT is used in fine-grained soils ranging from very soft clays and silty clays to very dense over-consolidated clays or glacial tills. The use of the SPT in these materials has often been to give an estimate of the undrained strength or the undrained elastic properties of the soil. More recently, it has been shown that the SPT may also be used to indicate the stress history of fine-grained soils. Clearly, the SPT is not the most preferred technique for evaluating engineering properties of fine-grained soils, especially soft to very soft clays. Other tests, such as the CPT, CPTU, or DMT are preferred, however, the SPT may be helpful in certain situations for providing preliminary estimates of a number of pertinent parameters.

As previously noted, in soft clays, the spoon may advance after the drill string is lowered into the borehole, or after the hammer is placed on top of the drill string, referred to as "weight of hammer" (WOH). However, not all hammer assemblies have the same mass, which means that the term "weight of hammer" is not equal in all cases. Engineers should be careful to record the style, model, serial number, and manufacturer of the hammer in use.

2.8.2.1 Undrained Shear Strength

One of the practical uses of the SPT in fine-grained soils is to use N to make an estimate of the undrained shear strength. In soft to very soft normally consolidated clays, it might be expected that the results of the SPT would show a more or less linear increase in N values with increasing depth. This would be consistent with increasing effective stress and undrained strength. Calculation of the normalized N value, by dividing by the vertical effective stress, σ'_{vo}, would produce a constant value of N/σ'_{vo}, indicative of a constant normalized undrained shear strength.

Such an observed trend in SPT results would be fine, except for the fact that in most soft clays, SPT values are typically very low and often result in Weight of Rod (WOR) or Weight of Hammer (WOH) penetration. Most correlations established between undrained shear strength and blow counts are for stiff over consolidated clays. Reference values of undrained strength are typically obtained from unconfined compression tests or unconsolidated-undrained triaxial compression tests. In general, local correlations are required to establish any reasonable degree of confidence. As DeMello (1971) and others have suggested, it appears that correlations between N and s_u should take into account the sensitivity of the clay to have any significant meaning.

Most soil mechanics or foundation engineering texts provide a simple chart correlating SPT N-values to the consistency (unconfined compressive strength) of fine-grained soils. Figure 2.14 shows a compilation of a number of correlations. Several correlations between SPT N and undrained shear strength are given in Table 2.11. The problem with many of these correlations is that they give no indication as to the type of SPT Hammer that was used to perform the test. Since many of them are from the 1970s and 1980s, it is reasonably safe to assume that they were not conducted using an automatic hammer which was not used until about 1995. Therefore, one may expect considerable scatter in the correlations.

An alternative is to use the SPT N-value to estimate the unconfined compressive strength (UCS) as noted in Table 2.12. Note that most of these correlations use the "uncorrected" SPT N-values and therefore are subject to some variability, depending on the system used in developing the correlation. Only limited work has been performed to correlate strength of clays from the energy-corrected SPT N-value; N_{60}.

The undrained strength of stiff intact clays can be related to N-values as proposed by Stroud (1974) using the relationship:

$$s_u = f_1 N \tag{2.8}$$

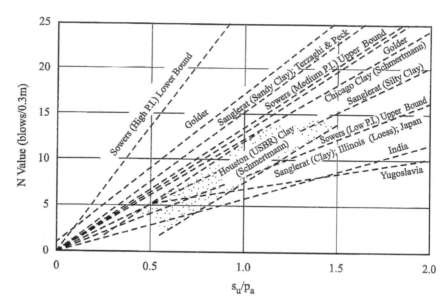

Figure 2.14 Comparison of historic reported correlations between SPT N-value and undrained shear strength. (After Kulhawy & Mayne 1990.)

Table 2.11 Reported correlations between SPT N and undrained shear strength

Correlation	Units of s_u	Soil type	References
$s_u = 29N^{0.72}$	kPa	Japanese cohesive soils	Hara et al. (1974)
$s_u/N_{60} = f(P.I.)$	kPa	Stiff UK soils	Stroud (1974)
$s_u = N/15$	Tsf	Stiff clays in Houston	Reese et al. (1976)
$s_u = 8N$ $N<10$ $s_u = 7N$ $10<N<20$ $s_u = 6N$ $20<N<30$ $s_u = 5N$ $30<N<40$	kPa	Guabirotuba Clay	Tavares (1988)
$s_u =1.39N + 74.2$	kN/m²	Tropical soil	Ajayi & Balogun (1988)
$s_u = 12.5N$ $s_u = 15N_{60}$	kPa	Sao Paulo Over-consolidated clay	Decourt (1989)
$s_u = 0.059N + 0.2$	tsf	Clay and soft shale	Nevels & Laguros (1993)
$s_u = 7.8N_{60}$ (CH) $s_u = 5.35N_{60}$ (CL)	kPa	Claysi – Turkey	Sivrikaya & Togrol (2006)
$s_u/p_a = 0.17N^{0.58}$	–	Ankara (Turkey) clay	Akbas & Kulhawy (2010)
$s_u = 2.1N_{60} + 17.6$	kPa	Iran clay	Nassaji & Kalantari (2011)
$s_u/N = 14.7 - 0.35w -2.4 \log (PI)$ (w = water content in %) (PI = Plasticity Index in %)	kPa	2 clays in Greece	Plytas et al. (2011)
$s_u = 6.932N_{70}$	kN/m²	Silty clays – Turkey	Cangir & Dipova (2017)
$s_u = 8.32N_{60}$	kPa	Stiff glacial till in Canada	Balachandran et al. (2017)
$S_u = 5.7N$	kN/m²	London Clay	White et al. (2019)

Table 2.12 Reported correlations between SPT N and unconfined compressive strength

Correlation	Units of q_u	Soil type	References
$q_u = 12.5N$	kPa	Fine-grained	Terzaghi & Peck (1967)
$q_u = N/8$	tsf	Clay	Golder (1961)
$q_u = 25N$	kPa	Clay	Sanglerat (1972)
$q_u = 20N$	kPa	Silty clay	
$q_u = 25N$	kPa	Highly plastic clay	Sowers (1979)
$q_u = 15N$		Medium plastic clay	
$q_u = 7.5N$		Low plasticity clay	
$q_u = 24N$	kPa	Clay	Nixon (1982)
$q_u = 62.5\,(N - 3.4)$	kPa		Sarac & Popovic (1982)
$q_u = 1.37N$	t/m^2	CH Bangkok clay	Sambhandharaksa & Pitupakorn (1985)
$q_u = 1.04N$		CL Bangkok clay	
$q_u = 15N$	kPa	CL and CL-ML	Behpoor & Ghahramani (1989)
$q_u = 58N^{0.72}$	kPa	Fine-grained	Kulhawy & Mayne (1990)
$q_u = 14.3N\ (LL \le 35)$	kPa	Bangladesh clays	Serajuddin & Chowdhury (1996)
$q_u = 16.9N\ (LL\ 36-50)$			
$q_u = 17.8N\ (LL > 50)$			
$q_u = 13.6\,N_{60}$	kPa	CH	Sivrikaya & Togrol (2002)
$q_u = 9.8N_{60}$		CL	
$q_u = 8.6N_{60}$		Fine-grained	
$q_u = (0.19 P.I. + 6.2)N_{60}$		Fine-grained	

where

f_1 = an empirical factor

The SPT results from which the N-values were derived were based on the modern UK practice so that the parameter f_1 is more properly defined as:

$$f_1 = s_u / N_{60} \tag{2.9}$$

In the U.K., the SPT split spoon sampler is not recessed for liners and therefore has a constant diameter of 34.9 mm (1.375 in.). In order to make use of Equation 2.8, an appropriate correction factor must be applied (to decrease N) if a standard U.S. split spoon sampler is used without liners. The value of f_1 with soil plasticity (P.I. 15–60) only varies from about 4 to 6. An average value of about $f_1 = 5$ (for s_u in KN/m^2) appears reasonable for most insensitive materials.

An interesting approach to estimating the unconfined compressive strength of soft and very soft clays was presented by Saiki (1983) for situations in which the split spoon may penetrate by either the weight of the hammer, or the blow counts may be very low (i.e., N<5). Two situations are considered: self-penetration and blow-penetration, and a corrected N value, N', is defined.

Spoon self-penetration

$$N' = 1 - \alpha\,(S/W) + \beta \tag{2.10}$$

where

    N' = corrected blow count

    α = coefficient = ratio of a static weight to a standard amount of self-penetration (kN/m)

    S = total amount of self-penetration measured in the field (m)

    W = total static weight applied to the test (kN)

    β = modification factor

The value of the modification factor β was taken as 0 for $\sigma'_{vo} < 50\,\text{kPa}$ and $\beta = (\sigma'_{vo}/50) - 1$ for $\sigma'_{vo} > 50\,\text{kPa}$. The value of α was taken for cases in which the weight was taken from the known mass of a standard hammer, standard spoon, and 40.5 mm (1.6 in.) drill rods for various test depths giving a value of $\alpha = 3$.

Spoon blow penetration

$$N' = N(\sigma'_{vo}/\sigma'_{vs}) + \beta \tag{2.11}$$

where

    N' = corrected blow count

    N = measured blow count

    σ'_{vo} = effective vertical stress (kPa)

    σ'_{vs} = reference value of effective stress (kPa)

    β = modification factor

In this case, the value of $\beta = 0.406$ was obtained and $\sigma'_{vs} = 91.4$ kPa was determined.

Using both Equations 2.10 and 2.11, Saiki (1983) obtained excellent results compared with laboratory unconfined compression tests using the expression

$$q_u = 52 + 23N' \tag{2.12}$$

where

    q_u = unconfined compressive strength (kPa)

    N' = corrected blow count (Equations 2.10 or 2.11)

More recently, there have been some attempts to develop theoretical models for estimating undrained shear strength in clays based on energy balance (Hettiarachchi & Brown 2009) and energy transfer (Schnaid et al. 2009). These methods as based on separating the end bearing and side resistance acting on the spoon and then using the energy to predict the undrained shear strength. These approaches are fundamentally more sound than simple empirical correlations based on site-specific observations.

2.8.2.2 Stress History

Mayne & Kemper (1988) suggested that the results of the SPT could be used to provide a rough estimate of stress history in fine-grained soils by correlating the normalized blow count to over-consolidation ratio (OCR). The results of tests collected from a number of sources are shown in Figure 2.15 and indicated that OCR could be related to normalized N-value as follows:

$$OCR = K_s (N/\sigma'_{vo}) p_a \tag{2.13}$$

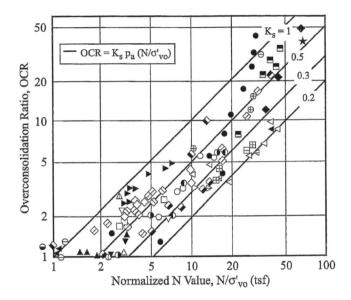

Figure 2.15 Correlation between OCR and N. (After Mayne & Kemper 1988.)

where
σ'_{vo} = *in situ* vertical effective stress (kPa)
K_s = an empirical factor (varies from 0.2 to 1.0)
p_a = atmospheric pressure (100 kPa)

Since these data show a wide range in values of OCR for a single normalized N-value, caution should be exercised in the use of Equation 2.13. The SPT is intended to only provide an indication of the range in OCR and probably can do no better to obtain design values, except where SPT practice is carefully controlled and local correlations for specific soil deposits have been developed. For example, Decourt (1989) found a correlation between σ'_p and N_{60} for over-consolidated Sao Paulo clays as follows:

$$\sigma'_p = 27.8 \, N_{60} \ (kPa) \tag{2.14}$$

Mayne (1995) suggested a statistical relationship between N_{60} and the preconsolidation stress for a number of intact clays as follows:

$$\sigma'_p = 0.47 \, N_{60} \, p_a \tag{2.15}$$

These data are shown in Figure 2.16. Note that fissured clays do not fit well in this correlation.

2.8.2.3 In Situ Lateral Stress

Estimates of the *in situ* state of stress in soils from the results of *in situ* tests are usually indirect; i.e., some estimate of the at-rest lateral stress ratio, K_o, is first made, and then the horizontal effective stress, σ'_{ho}, is calculated using local estimates of the unit weight of the soil and water table or pore pressure measurements. Another approach may be to first make an estimate of the stress history (OCR) and then rely on correlations established between K_o and OCR for the appropriate type (e.g., Brooker & Ireland 1965; Mayne & Kulhawy 1982).

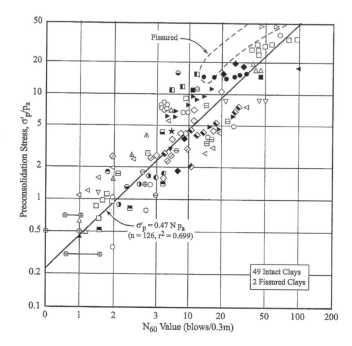

Figure 2.16 Correlation between N_{60} and preconsolidation stress for clays. (After Mayne 1995)

This approach may suffer from too many uncertainties; first associated with the accuracy of the estimate of OCR from the SPT and second associated with the assumed relationship between OCR and K_o. This approach may also be limited to those soils in which K_o has developed from simple mechanical unloading and is not related to other more complex factors.

Kulhawy et al. (1989) compiled a set of data for clays and suggested a simple correlation between K_o and SPT N-values in clays:

$$K_o = 0.423 + (Np_a)/(16.95\sigma'_{vo})$$ (2.16)

Kulhawy & Mayne (1990) presented the data as shown in Figure 2.17 where K_o values have been obtained by a variety of methods. This correlation is given as follows:

$$K_o = 0.073(Np_a)/\sigma'_{vo}$$ (2.17)

2.8.2.4 Soil Elastic Modulus

Stroud & Butler (1975) suggested a relationship between N and the drained deformation modulus determined from one-dimensional odometer tests, m_v, in fine-grained glacial materials. Using values of m_v for a pressure increment of 100 KN/m² in excess of the vertical effective overburden stress, they proposed the simple expression

$$f_2 = (1/m_v)N$$ (2.18)

where
   f_2 = an empirical factor

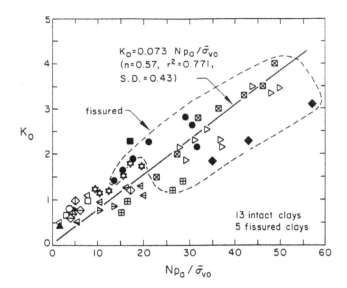

Figure 2.17 Correlation between N-values and K_o in clays. (From Kulhawy & Mayne 1990.)

Values of f_2 range from about 440 to 600 KN/m² with a slight decrease in f_2 with increasing P.I. and approximately

$$f_2 = 100\,f_1 \tag{2.19}$$

Behpoor & Ghahramani (1989) suggested that for clayey and silty clays soils with N<25, the modulus of elasticity could be obtained as follows:

$$E_S(MPa) = (0.17)\,N \tag{2.20}$$

A number of studies have been made correlating the results of the SPT to the elastic modulus, E_p, obtained from the initial loading curve of the pressuremeter test. Table 2.13 presents a summary of some reported correlations in different materials. Most comparisons between N and E_p show a linear correlation when plotted as log N vs. log E_p (e.g., Martin 1977; Toh et al. 1987). However, in some cases, e.g., Ohya et al. (1982), there is considerable scatter in the data.

2.8.2.5 Small-Strain Shear Modulus

As with granular soils, it may be desirable to make some initial estimate of the small-strain shear modulus of fine-grained soils from SPT results. Several correlations between either G_{max} or V_S and N have been suggested and are summarized in Table 2.14. In fine-grained soils, a relationship between N and G_{max} was suggested by Wroth et al. (1979) and was given in a dimensionless form by Kulhawy & Mayne (1990) as shown in Figure 2.18.

2.8.3 SPT in Soft/Weak Rock

In competent rock, the SPT usually will not develop full penetration, or in some cases, the spoon and rods will simply bounce during driving. In these cases, the N-value may have little or no significance beyond identifying refusal. In the U.S., most drillers stop the test if

Table 2.13 Correlations between N and pressuremeter modulus[a]

Correlation	Soil	References
$E_p = \log^{-1} [0.65180 \log N + 1.33355]$ (E_p in tsf)	Piedmont residual soil	Martin (1977)
$E_p = 7.7N$	Clay	Nayak (1979)
$E_p = 15N$ $E_p = 4N$	Clayey soil Sandy soil	Ohya et al. (1982)
$E_p = 6.84N^{0.986}$ (E_p in bar)	Misc. soil types[b]	Tsuchiya & Toyooka (1982)
$E_p = 22N + 160$ $E_p = 26N + 120$	Gneissic saprolite ($20 < N < 30$) Gneissic saprolite ($30 < N < 60$)	Rocha Filho et al. (1985)
$\ln E_p = 3.509 + 0.712 \ln N$ (E_p in ksf)	Residual soil	Barksdale et al. (1986)
$E_p = 15N + 240$	Lateritic or mature Gneissic residual soil ($7 < N < 15$)	Toledo (1986)
$E_p = \log^{-1} [0.70437 \log N + 1.17627]$ (E_p in tsf)	Piedmont residual soil	Martin (1987)
$E_p = 1.6N$ (E_p in MPa)	Residual soil	Jones & Rust (1989)
$\log E_p = 1.0156 \log N + 1.1129$ (E_p in tsf)	Clay and clay shale	Nevels & Laguros (1993)
E_p (kPa) $= 388.7N_{60} + 4554$	Sandy silty clay	Yagiz et al. (2008)
E_p (MPa) $= 1.33 (N_{60})^{0.77}$ E_p (MPa) $= 1.61 (N_{60})^{0.71}$	Sandy soils – Istanbul Clayey soils – Istanbul	Bozbey & Togrol (2010)
$E_p = 0.285 (N_{60})^{1.4}$ E_p in MPa)	Clayey soils – Turkey	Kayabasi (2012)
E_p (MPa) $= 2.22 + 0.0029(N_{60})^{2.5}$	Clayey soils – Turkey	Agan & Algin (2014)

[a] E_p in Kg/cm^2 unless noted.
[b] Individual equations given by authors for eight different soils types ranging from very soft organic soil to mudstone.

the penetration has not progressed beyond the initial 0.15 m (6 in.) after 100 blows. In other cases, where the bedrock may be soft or highly weathered, such as in the case of weathered shale or other fine-grained sedimentary deposits, the test may be carried out to completion, and the results may provide meaningful results for estimating the strength or deformation characteristics. For example, in the U.K., the SPT is used extensively in chalk and marl deposits.

Stroud (1989) summarized the use of the SPT to estimate the compressive strength of insensitive weak rock, as indicated in Figure 2.19. Stroud (1989) also demonstrated that the vertical compression elastic modulus, E_S, could also be correlated to N_{60} for similar materials as shown in Figure 2.20. Again, this approach, taken principally from settlement observations of foundations, is similar to that previously shown for granular materials. It can be seen that the variation is strongly related to the relative degree of loading.

In situations where SPT refusal is encountered, such as weathered or soft rock, it may be more useful to express the results of the SPT in terms of the penetrability, as suggested by Stamatopoulos & Kotzias (1974, 1993). Extrapolation of the blow counts when the penetration is less than 0.3 m (1 ft) is more or less arbitrary and essentially meaningless. Penetrability is defined as the penetration of the SPT spoon, in millimeters, produced by

Table 2.14 Reported correlations between V_S and G_{max} and N-values for Fine-grained soils

Correlation	Soil	References
$V_S = 121 \, (N +.027)^{0.22}$	Shanghai	Jinan (1985)
$V_S = 84.5N^{0.118} \, (D + 1)^{0.246}$ (D = depth in meters)	Taipei basin	Lee (1992)
$V_S = 76.5 \, (N)^{0.445}$	Cohesive Greece	Kalteziotis et al. (1992)
$V_S = 145 \, (N)^{0.178}$	Silts	Pitilakis et al. (1999)
$V_S = 132 \, (N)^{0.271}$	Clays	
$V_S = 22 \, (N)^{0.770}$	Silts	Jafari et al. (2002)
$V_S = 27 \, (N)^{0.730}$	Clays	
$V_S = 107.6 \, (N_{60})^{0.237}$	Clays	Hasancebi & Ulusay (2007)
$V_S = 44 \, (N)^{0.480}$	Clay	Dikmen (2009)
$V_S = 60 \, (N)^{0.360}$	Silt	
$V_S = 89.3 \, (N)^{0.358}$	Clay	Uma Maheswari et al. (2010)
$V_S = 112.2 \, (N_{60})^{0.324}$	Clay	Tsiambaos & Sabatakakis (2011)
$V_S = 88.8 \, (N_{60})^{0.370}$	Silt	
$V_S = 77.1 \, (N)^{0.355}$	Clay	Fatehnia et al. (2015)
$V_S = 94.4 \, (N)^{0.379}$	Clayey	Kirar et al. (2016)
$G_{max} = 14.0 \, N^{0.722}$	Clay	Ohsaki & Iwasaki (1973)
G_{max} (MPa) $= 15.8 N^{0.668}$	Clay	Hara et al. (1974)

V_S in m/s; G_{max} in MPa.

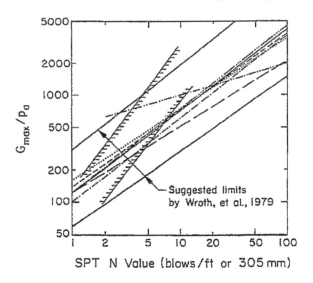

Figure 2.18 Correlation between G_{max} and SPT N-values. (From Kulhawy & Mayne 1990.)

60 standard blows. The general trend in uniaxial compressive strength with both N values and 1/N values suggested by Stamatopoulos and Kotzias (1993) is shown in Figure 2.21.

A similar approach has been suggested by Bosio (1992) for interpreting the results of the SPT in soft rock where full penetration of the spoon does not occur. After cleaning the borehole, a series of 50 hammer blows is applied, with the penetration obtained for each ten blows measured. The data are then plotted on a semi-log plot as penetration versus number

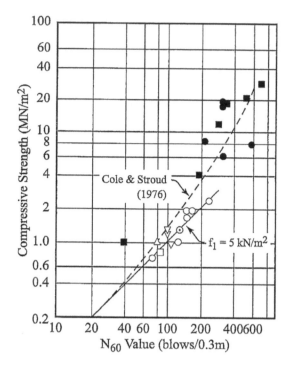

Figure 2.19 Correlation between compressive strength of weak rock and SPT N-values. (After Stroud 1989.)

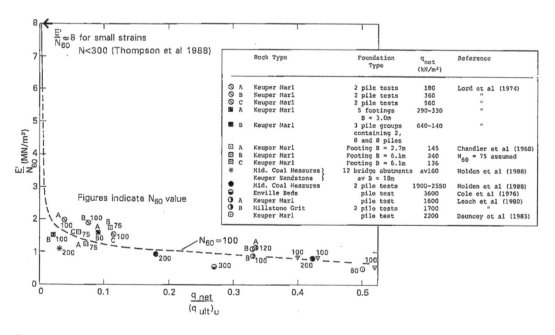

Figure 2.20 Correlation between modulus of weak rock and SPT N-values. (From Stroud 1989.)

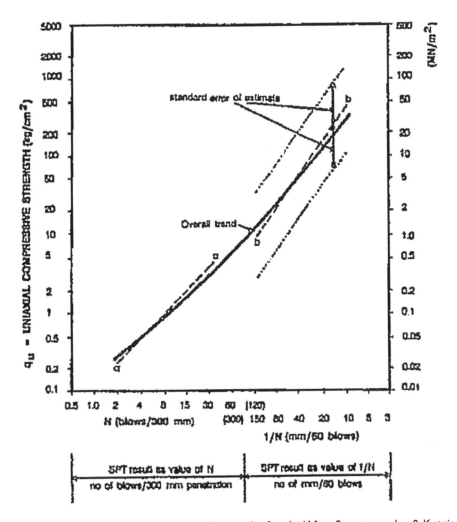

Figure 2.21 Correlation between SPT results and strength of rock. (After Stamatopoulos & Kotzias 1993.)

of blows as shown in Figure 2.22. The Penetration Index (N_p) is the slope of the straight part of the curve after the initial penetration as follows:

$$N_p = (P_{50} - P_{30})/(\log 50 - \log 30) \qquad (2.21)$$

or approximately

$$N_p = 4.51(P_{50} - P_{30}) \qquad (2.22)$$

This method has been found to be useful for determining the compressive strength of rock, as indicated in Figure 2.23.

Stark et al. (2013, 2017) described the use of a modified SPT (MSPT) procedure for use in the design of drilled shafts in weak rock. Like others, they recognized that the spoon will likely not penetrate the full 477 cm (18 in.) into rock. They suggested measuring the spoon

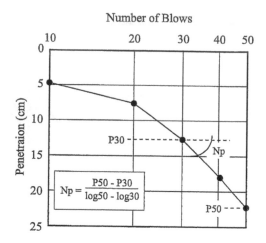

Figure 2.22 Determination of Penetration Index, N_p, in soft rock. (After Bosio 1992.)

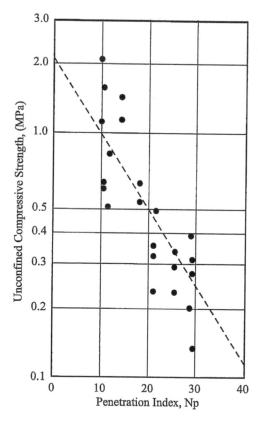

Figure 2.23 Correlation between Penetration Index and compressive strength of rock. (After Bosio 1992.)

penetration for each 10 drops of the hammer until 100 drops have been applied. A plot of penetration (in.) vs. number of hammer drops usually shows a straight line after some driving distance, as shown in Figure 2.24. The slope of the straight-line portion of the curve is defined as penetration rate, N^* and is obtained as follows:

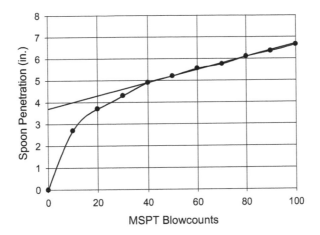

Figure 2.24 Determining penetration rate in rock using the modified SPT procedure. (Data from Stark et al. 2013.)

$$N^* = 1/\left[(\Delta\,\text{Penetration Distance})/(\Delta\,\text{MSPT Blowcount})\right] \tag{2.23}$$

Stark et al. (2017) showed a correlation between the penetration rate adjusted for 90% hammer efficiency, N_{90}^*, and unconfined compressive strength (UCS) of rock. In the range of UCS between 5 and 120 ksf

$$\text{UCS (ksf)} = 0.092\,N_{90}^* \quad \left(r^2 = 0.94\right) \tag{2.24}$$

2.9 IMPROVEMENTS TO SPT PRACTICE

It may be possible to gain additional information from the SPT without substantially altering the tests and still complying with the ASTM procedure.

2.9.1 SPT-T Test

According to Decourt (1989), Ranzine (1988) was the first to suggest a simple modification to the SPT in which a traditional SPT is complemented by torque measurements. That is, after driving the spoon, torque is applied to the drill rod of the SPT sampler to measure the friction between the sampler and the soil as shown in Figure 2.25. Decourt & Quaresma (1994) have shown that the ratio of torque to N value (T/N with T measured in kgf-m) has proved useful in practice. It can be argued that an advantage of the torque measurement is to add a static testing component to a test which results initially from a dynamic phase. While most of the soil structure may be destroyed during installation of the spoon, the torque measurement may act in a region where the soil retains much of its original fabric and is only partially remolded.

The ratio T/N appears to be useful in identifying highly structured soils and may also prove useful for design of driven piles (Decourt & Quaresma 1994; Peixoto et al. 2004; Winter et al. 2005). The torque measurement is a simple addition to the SPT, which does not detract from the standard test procedure and requires only minimal additional effort. In fact, the test only takes about another minute to perform. A small rod adapter is fabricated

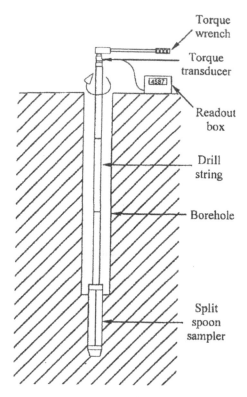

Figure 2.25 Schematic of SPT-T test.

to thread onto the ends of the drill string and a direct read mechanical or electronic torque wrench can be used.

It seems logical that while the actual N value obtained from the SPT may be subject to wide variations for all of the reasons previously discussed, the torque measurement should be less variable. In the sense, the soil may not really care how the spoon is advanced into the ground; the torque measurement depends on lateral stress acting on the outside of the spoon. This means that the T/N ratio would be dependent on the method used to obtain N and would be different in the same soil if different hammer systems are used. The torque would be affected if the spoon or rods wobble and contact between the spoon and soil is lost. Figure 2.26 shows typical results of SPT-T tests conducted at both a sand and clay site (Kelley & Lutenegger 2004).

The torque measurement may also have direct application for estimating skin friction on driven piles as described by Lutenegger & Kelley (1998) and pile setup. Using the moment arm as the distance from the center of the spoon (where torque is applied) to the outside diameter and neglecting any contribution from the soil at the end of the spoon, the unit side resistance, f_s, may be obtained as follows:

$$f_s = (2T)/(\pi d^2 L) \qquad (2.25)$$

where

   T = measured torque
   d = diameter of spoon
   L = length of the spoon driven

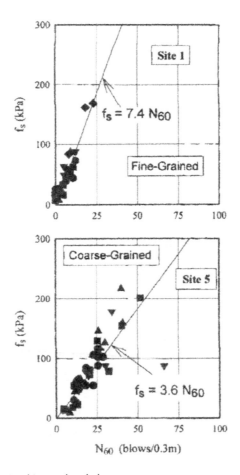

Figure 2.26 SPT-T results obtained in sand and clay.

A comparison of side resistance measurements using Equation 2.25 and tension tests on pipe piles in sand showed similar results (Lutenegger & Kelley 1998). Recent results also suggest that this side resistance may be similar in magnitude to the quasi-static penetration local friction required to advance the spoon and local skin friction from a CPT (e.g., Takesue et al. 1996). The test is gaining popularity around the world (e.g., Peixoto et al. 2004; Winter et al. 2005; Heydarzadeh et al. 2013). It has recently been suggested that the test may also be useful at estimating the undrained shear strength in clays (Ruge et al. 2018). The author now routinely performs torque measurement as a part of every SPT conducted and recommends that engineers include this in all routine site investigations.

2.9.2 Seismic SPT

An addition to the SPT to allow for the measurement of shear wave velocity was presented by Kim et al. (2004) and Bang & Kim (2007). The seismic SPT (S-SPT) is an "uphole" test that uses receivers on the surface and a source at depth; essentially the reverse configuration of the seismic CPT and seismic DMT. Vertical driving of the SPT split spoon generates a shear wave, which can be picked up by the surface sensors as shown in Figure 2.27.

Bang & Kim (2007) used the SPT drop hammer (140 lbs), while Pedrini & Giacheti (2013) suggest using a 2 kg sledgehammer to generate the waves after the traditional SPT

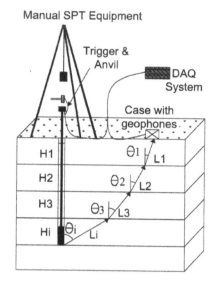

Figure 2.27 Schematic of the seismic-SPT. (From Bang & Kim 2007.)

is completed. Results obtained between S-SPT and SCPT at a number of sites have shown excellent agreement of shear wave velocity (e.g., Pedrini & Giacheti 2011; Giacheti et al. 2013; Rocha et al. 2015; Pedrini et al. 2018).

2.9.3 Measurement of Penetration Record

Measurement of the spoon penetration record was previously described in Section 2.8.3 related to soft rock and appears to have been first recommended by Granger (1963). The measurement of spoon advance is performed over the full 0.46 m (18 in.) of penetration and with a little practice can be accomplished without interrupting the test. A tape measure is held alongside the drill rod string, and the operator calls out the displacement to a recorder after each hammer blow. This gives the cumulative penetration of the spoon after each blow of the hammer. The conventional N value is still obtained by adding the number of blows to advance the last 0.30 m (12 in.), after the initial 0.15 m (6 in.) as is required by ASTM. The author has used this practice for a number of projects when recording SPT data in new area.

Vallee & Skryness (1979) suggested that the penetration record could be used to estimate the N-value in cases where the spoon becomes plugged after driving some distance. The artificially high N-values can be adjusted by using the initial straight-line portion of the penetration record as shown in Figure 2.28. A more recent analysis performed by Dung & Chung (2013) showed that this approach is valid in sands.

Obtained in this manner, the test results take on the form similar to that of a pile driving record and may be displayed as either incremental penetration or cumulative penetration resistance. Typical penetration records obtained in a soft clay and sand are shown in Figure 2.29.

2.9.4 Incremental Penetration Ratio

Incremental penetration ratios were suggested by Schmertmann (1979) as a way of extracting more information about the soil from the SPT. As required by the test, the resistance over three consecutive 0.15 m (6 in.) distances is recorded, the sum of the last two being N.

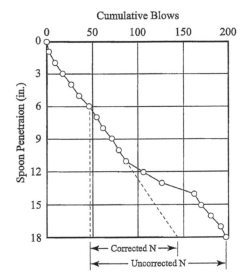

Figure 2.28 Suggested method of estimating SPT N-value from penetration record. (After Vallee & Skryness 1979.)

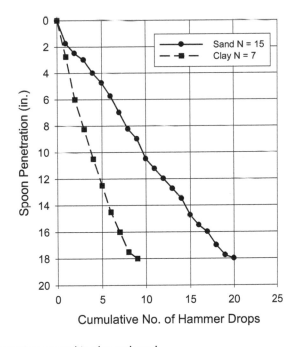

Figure 2.29 Spoon penetration record in clay and sand.

These data are rarely used for direct design applications but may offer a possible improvement to the test interpretation.

The ratio of individual 0.15 m (6 in.) resistance to the next consecutive value may help provide some insight into liquefaction resistance. Taking into account the changing geometry as the spoon advances and increases contact with the soil, it may be possible to quantify liquefaction or apparent strength loss by simply calculating the change in resistance as the spoon moves from 0 to 0.15 m (6 in.), 0.15m (6 in.) to 0.31 (12 in.), and 0.31 (12 in.) to 0.46 m (18 in.) of penetration.

2.9.5 Differential Penetration Record

One possibility that has not previously been explored is to evaluate the *difference* in incremental blow count values for each of the three 0.15 (6 in.) advances of the spoon. In this case, $\Delta N_1 = N_{6-12} - N_{0-6}$ and $\Delta N_2 = N_{12-18} - N_{6-12}$. The penetration difference may be useful in evaluating the influence of skin friction on the spoon (Lutenegger 2008).

2.10 LARGE PENETRATION TEST

Larger diameter spoons are allowed by ASTM D1586 but are not considered acceptable for determining N-values. Several drill rig manufacturers provide spoons with an outside diameter of 76 mm (3 in.) and an inside barrel diameter of 63.5 mm (2.5 in.). The inside diameter of the shoe is 44.5 mm (1.75 in.). A larger spoon is useful if a larger volume of soil is desired or if gravelly material is present at the site. Because of the difference in spoon sizes, the blow count values obtained with a larger spoon will be different than those obtained with a standard 50.8 mm (2 in.) spoon.

In soil deposits that contain significant amounts of gravel, the penetration resistance of the SPT may be artificially high since gravel particles may not enter the spoon or may get jammed in the shoe. Gravel is usually indicated by poor sample recovery for each SPT. In order to overcome this problem, the use of LPT was introduced. The LPT should not be confused with simply using a larger size spoon and the same SPT hammer. Various configurations of the LPT have been used in Japan (Kaito et al. 1971), Italy (Crova et al. 1993), and Canada (Koester et al. 2000; Daniel et al. 2003). Harrison et al. (2017) also described an LPT used in Canada with a 109 kg (240 lbs) hammer with a drop height of 450 mm (18 in.) and a spoon with an inside diameter of 102 mm (4 in.).

The LPT was apparently introduced by Kaito et al. (1971) for evaluating the liquefaction potential of granular soils with some gravel size particles. The test is performed much in the same way as the SPT but adopts a larger hammer and sampler and a larger fall height. In the Japanese LPT, a 100 kg (220 lbs) hammer is used with a fall height of 1.5 m (5 ft). The sampler has an inside diameter of 50 m (2 in.) and an outside diameter of 73 mm (2.9 in.) and is driven a distance of 30 cm (1 ft) to obtain the LPT blow count value, N_{LPT}. A variation of this test has been described by Crova et al. (1993). Table 2.15 presents a comparison between various components of the SPT and different LPTs.

Table 2.15 Comparison between SPT and large penetration tests

Drive method	SPT	LPT (Japan)	LPT (Italy)	LPT (Canada)
	Drop hammer	Drop hammer	Drop hammer	Drop hammer
Hammer mass (kg)	63.5	100	560	134
Drop height (m)	0.76	1.50	0.50	0.76
Sampler OD (mm)	51	73	140	76
Sampler ID (mm)	35	50	100	61

Most common casing size.

Since there is significant similarity between the SPT and LPT, it should be possible to establish a correlation between the two tests. Tokimatsu (1988) has suggested that provided the energies delivered to the unit surface of the two samplers are the same, the correlation between the two tests should be as follows:

$$N_{SPT}/N_{LPT} = 1.5 \tag{2.26}$$

For soils without gravel, this correlation appears to be satisfactory since soil can be ingested into both samplers. However, the ratio tends to become greater than 1.5 for gravelly soils.

Tokimatsu (1988) presented a comparison between N values obtained from the SPT and LPT for different soils. The results shown in Figure 2.30 indicate that the actual ratio N_{SPT}/N_{LPT} varies from about 1.5 to 2.5. Similar results were presented by Yoshida et al. (1988). The use of the LPT was also described by Suzuki et al. (1993) for the investigation of gravelly soil at several sites in Japan. Measured ratios of N_{SPT}/N_{LPT} ranged from 1.0 to 3.5.

The ratio N_{SPT}/N_{LPT} appears to be related to the mean grain size, at least for well-graded soils (e.g., Daniel et al. 2004). The tendency for the ratio to increase with increasing D_{50} probably reflects the higher SPT blow counts that occur as a result of gravel particles that are too large to enter the split spoon. This might also be reflected in low recovery ratios for the SPT. On the other hand, Crova et al. (1993) showed that on average, the ratio $(N_{1(60)})_{SPT}/(N_{1(60)})_{LPT}$ obtained using an Italian-style LPT was about 1.0 for several sites and also indicated that there appeared to be little or no relationship with D_{50}.

Engineers should consider using a 76 mm (3 in.) O.D. split spoon sampler with conventional SPT equipment when gravelly sands are encountered. It will be necessary to develop local correlations to convert the blow counts obtained to equivalent SPT N-values. Comparisons made by the author using a standard 50.8 mm (2 in.) barrel sampler without liners and a 76 mm (3 in.) sampler (I.D. = 63.5 mm (2.5 in.)) without liners at several sites are presented in Figure 2.31. At each of the sites, the same driller and SPT equipment were used to perform both tests. As can be seen for most of the sites, the ratio of $N_{(3)}/N_{(2)}$ varies from about 1 to 3. Based on the differences in end and side areas, the ratio $N_{(3)}/N_{(2)}$ should be on the order of 1.5. The difference is probably related to the relative contribution of end bearing and side resistance to N for each sampler and each soil.

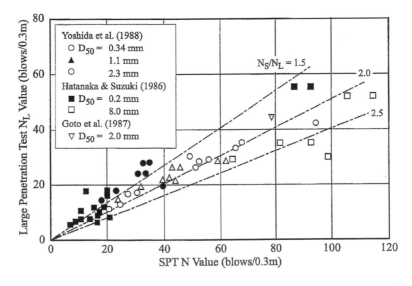

Figure 2.30 Comparison between N-values obtained from SPT and LPT. (After Tokimatsu 1988.)

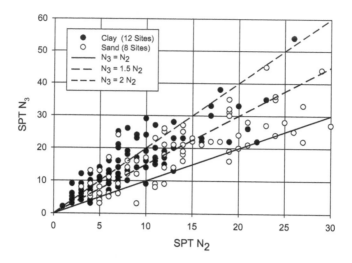

Figure 2.31 Comparison of N-values between 2 and 3 in. spoons.

2.11 BECKER PENETRATION TEST

In coarse-grained gravelly sands and gravels and cobbly soils, a large-scale Dynamic Penetration Test known as the Becker Hammer Test or Becker Penetration Test (BPT) has been used. The BPT resembles a closed end pipe pile and simulates the driving of a displacement pile. A double-acting diesel pile hammer is used to drive a closed-end casing. The number of blow counts for a penetration distance of 0.3 m (1 ft) is recorded. Casing lengths of 2.4 or 3.0 m (8 or 10 ft) and diameters of 140, 170, and 230 mm O.D. (5.5, 6.7 and 9.1 in.) are available depending on the coarseness of the materials or the hole to be drilled. Details of the BPT have been presented by Harder & Seed (1986), Sy & Campanella (1993, 1994), and Wightman et al. (1993). The test is highly specialized and requires specific equipment. The application of the BPT is restricted to large projects involving difficult soil materials.

An early correlation between the BPT and SPT in a variety of materials suggested an approximate 1:1 relationship although there was considerable scatter. Using a correction scheme to account for differences in diesel hammer energy, Harder & Seed (1986) developed a correlation between the BPT and SPT in which N_{60} has been related to the corrected BPT N values, N_{bc}.

2.12 SPT IN GEOTECHNICAL DESIGN

The SPT has a wide variety of applications in geotechnical design above the typical application for site investigation work and the evaluation of individual soil properties. Extensive use of SPT results have been reported for estimating the bearing capacity and settlement of shallow and deep foundations based on empirical correlations. The number of proposed approaches for using SPT results to estimate the behavior of foundations is too extensive for this volume; however, it may be appropriate to consider the typical approaches suggested and then consider more recent analyses.

2.12.1 Shallow Foundations

There are at least 15 different methods for estimating the bearing capacity of shallow foundations from SPT results and over 20 methods for estimating settlements. Most of these are applicable to sands. Early charts were developed giving the allowable foundation stress to produce a fixed settlement (e.g., Terzaghi & Peck 1967; Peck et al. 1974; Parry 1977; etc.). Settlement estimates of shallow foundations on sands typically rely on an indirect elastic approach by converting N-values to an elastic modulus and then calculating settlement using an elastic equation (e.g., Berardi et al. 1991; Stroud 1989; Anagnostopoulos et al. 1991).

Burland & Burbidge (1985) presented a method for using SPT results to estimate settlement of footings taking into account load intensity, shape of the footing, and depth of influence below the footing. The method also considers whether the sand is normally consolidated or over-consolidated and is based on nearly 200 published case histories of observed settlement of various size footings. Settlements are calculated as follows:

$$s = 0.14 C_S C_I I_C \left(B/B_R \right)^{0.7} \left(q'/\sigma'_R \right) B_R \quad \text{for NC soils} \tag{2.27a}$$

$$s = 0.047 C_S C_I I_C \left(B/B_R \right)^{0.7} \left(q'/\sigma'_R \right) B_R \quad \text{for OC soils \& } q' \leq \sigma'_C \tag{2.27b}$$

$$s = 0.14 C_S C_I I_C \left(B/B_R \right)^{0.7} \left(q' - 0.67/\sigma'_C \right)/\left(\sigma_R \right) B_R \quad \text{for OC soils \& } q' > \sigma'_C \tag{2.27c}$$

where
    s = settlement (mm)
    C_S = footing shape factor = $\left[\left(1.25 L/B \right)/\left(\left(L/B \right) + 0.25 \right) \right]^2$
    L = footing length (m)
    B = footing width (m)
    C_I = depth factor = $\left(H/Z_I \right) \left[2 - \left(H/Z_I \right) \right] \leq 1$
    H = depth from bottom of footing to bottom of compressible layer
    Z_I = depth of influence below footing = $1.4 \left(B/B_R \right)^{0.75} B_R$
    B_R = reference width = 0.3 m
    q' = net footing stress (kPa)
    σ'_R = reference stress (100 kPa)
    σ'_C = preconsolidation stress (kPa)
    I_C = soil compressibility index = $1.71(N_{60})^{1.4}$ (NC soils) or $0.57/(N_{60})^{1.4}$ (OC soils)

If the N-values decrease with depth, use $Z_I = 2B$ or the depth to the bottom of the loose layer, whichever is less. The average N-values between the base of the footing and Z_I are used. For very fine and silty sand below the water table, use adjusted N-values as $N' = 15 + 0.5(N - 15)$. For gravel or gravelly sand, use adjusted N-values as $N' = 1.25N$.

Viswanath & Mayne (2013) showed that the relative settlement of shallow footings on sands could be expressed in terms of SPT-normalized applied stress. For a number of case histories of loading tests of shallow footings on sands, they found

$$q/N_{\text{representative}} = 0.3 \left(s/B \right)^{0.5} \tag{2.28}$$

where
    q = applied stress (MPa)
    s = settlement

B = footing width

$N_{representative}$ = average N-value between the bottom of the footing and a depth of 1.5B

This means that the settlement for any applied stress may be estimated directly from SPT results. If the ultimate capacity is defined as the stress producing a settlement of 10% of the footing width (Lutenegger & Adams 1998, 2003), then by substituting into Equation 2.28, the ratio of bearing capacity to N-value (q_{ult}/N) is approximately 0.10.

2.12.2 Deep Foundations

Estimates of the bearing capacity of deep foundations using SPT results have typically been based on local empirical correlations between N-values and unit side resistance and unit end bearing. Correlations for both driven piles and bored piles in both sands and clays have been reported. Poulos (1989) summarized a number of reported correlations between side resistance and N-values and suggested that many of these correlations could be expressed using the general equation

$$f_s = \beta + \alpha N \qquad (2.29)$$

where
   f_s = unit pile side resistance

Lutenegger (2009) summarized many of the reported values for side resistance and found that the majority of the reported observations could be simplified to

$$f_s = \alpha N \qquad (2.30)$$

Values of α ranged from about 2.0 to 10 for both driven and bored piles with f_s in units of kPa, as shown in Table 2.16.

2.13 SUMMARY OF SPT

The SPT can be a useful *in situ* test, provided that engineers carefully control the test. The test results can be reliable if the equipment and procedures are controlled. The following procedure for performing the SPT is suggested:

1. Use a calibrated automatic hammer set to a test rate of 30–40 blows/min;
2. Use NW or AW (square or tapered thread) drill rods in lengths of 3.0 m (10 ft);
3. Use a 0.61 (24 in.)-long split spoon without liners with an internal shoe diameter of 34.9 mm (1.375 in.) and an internal barrel diameter of 38.1 mm (1.5 in.);
4. With the rods touching the bottom of the borehole but still being held, carefully mark three 0.15 m (6 in.) drive increments;
5. Drive the spoon a total of 0.46 m (18 in.) and record incremental blow counts for each 0.15 m (6 in.) increment;
6. Use only solid-stem or hollow-stem augers with an internal diameter not greater than 107.9 mm (4.25 in.) or 100 mm (4 in.) rotary wash boring to advance the borehole;
7. In coarse-grained soils below the water table, use drilling mud full depth inside the hollow stem augers or casing;
8. Always record the sample recovery;

Table 2.16 Summary of some reported correlations between SPT N-value and deep foundation side resistance

Pile type	Soil	β	α	References
Driven	Granular	0	2.0	Meyerhof (1976)
	Miscellaneous soils ($f_s < 170\,kPa$)	10	3.3	Decourt (1982)
	Cohesive	0	10	Shioi & Fukui (1982)
	Cohesive	0	3	Bazaraa & Kurkur (1986)
	Cohesionless	0	1.8	
	Sandy	29	2.0	Kanai & Yubuuchi (1989)
	Clayey	34	4.0	
	Misc	0	1.9	Robert (1997)
Bored	Granular	0	1.0	Meyerhof (1976)
	Cohesive	0	5.0	Shioi & Fukui (1982)
	Cohesive	0	1.8	Bazaraa & Kurkur (1986)
	Cohesionless	0	0.6	
	Residual soil and weathered rock	0	2.0	Broms & Flodin (1988)
	Clay	0	1.3	Koike et al. (1988)
	Sand	0	0.3	
	Sandy soil cohesive	35	3.9	Kanai & Yubuuchi (1989)
		24	4.9	
	Residual soil	0	4.5	Winter et al. (1989)
	Residual soils	0	2.0	Chang & Broms (1991)
	Clayey soil	0	10.0	Matsui (1993)
	Sandy soil	0	3.0	
	Misc.	0	1.9	Robert (1997)
	Sand	0	5.05	Kuwabara & Tanaka (1998)
	Weathered rock	0	4	Wada (2003)
	Cohesionless	0	5.0	Shioi & Fukui (1982)
	Cohesive	0	10.0	

Note: $f_s = \beta + \alpha N$ (f_s in units of kPa).

9. Field SPT blow counts should always be corrected for test procedures and the energy level of the hammer system. Blow count values should be reported as N_{60}.

10. Perform a torque test after completing the drive.

REFERENCES

AASHTO, 1981. Standard Specifications for Transportation Materials and Methods of Sampling and Testing.

Agan, C. and Algin, H., 2014. Determination of Relationships between Menard Pressuremeter Test and Standard Penetration Test Data by Using ANN Model: A Case Study on the Clayey Soil in Sivas, Turkey. *Geotechnical Testing Journal, ASTM*, Vol. 37, No. 3, pp. 1–12.

Ajayi, L.A. and Balogun, L.A., 1988. Penetration Testing in Tropical Lateritic and Residual Soils – Nigerian Experience. *Proceedings of the 1st International Symposium on Penetration Testing*, Vol. 1, pp. 315–328.

Akbas, S. and Kulhawy, F., 2008. Case History of SPT Energy Ratio for an Automatic Hammer in Northeastern U.S. Practice. *Proceedings of the 3rd International Symposium on Geotechnical and Geophysical Site Characterization*, Vol. 2, pp. 1241–1245.

Akbas, S.O. and Kulhawy, F.H., 2010. Characterization and Estimation of Geotechnical Variability in Ankara Clay: A Case History. *Geotechnical and Geological Engineering*, Vol. 28, pp. 619–631.

Akino, N. and Sahara, M., 1994. Strain-Dependency of Ground Stiffness Based on Measured Ground Settlement. *Proceedings of the International Symposium on Pre-Failure Deformation of Geomaterials*, Vol. 1, pp. 181–187.

Anagnostopoulos, A.G., Papadopoulos, B.P., and Kavvadas, M.J., 1991. Direct Estimation of Settlements on Sand, Based on SPT Results. *Proceedings of the 10th European Conference on Soil Mechanics and Foundation Engineering*, Vol. 2, pp. 293–296.

Anbazhagan, P., and Sitharam, T., 2010. Relationship between Low Strain Shear Modulus and Standard Penetration N Values. *Geotechnical Testing Journal, ASTM*, Vol. 33, No. 2, pp. 1–15.

Anbazhagan, P., Parihar, A, and Rashmi, H., 2012. Review of Correlations between SPT N and Shear Modulus: A New Correlation Applicable to any Region. *Soil Dynamics and Earthquake Engineering*, Vol. 36, pp. 52–69.

Athanasopoulos, G., 1994. An Empirical Correlation VS-NSPT and an Evaluation of Its Reliability. *Proceedings of the 2nd International Conference on Earthquake Resistant Construction and Design*, pp. 219–226.

Balachandran, K., Liu, K., Cao, L. and Peaker. Statistical Correlations between undtained shear strength (C$_u$) and both SPT N-value and net limit pressure (P$_L$) for cohesive glacial tlls. Proceedings GoeOttawa, Canadian Geotechncial Society, 8 pp.

Bang, E.-S. and Kim, D.-S., 2007. Evaluation of Shear Wave Velocity Using SPT Based Uphole Method. *Soil Dynamics and Earthquake Engineering*, Vol. 27, pp. 741–758.

Barksdale, R.D., Ferry, C.T., and Lawrence, J.D., 1986. Residual Soil Settlement from Pressuremeter Moduli. *Use of In Situ Tests in Geotechnical Engineering, ASCE*, pp. 447–461.

Barros, J. and Pinto, C., 1997. Estimation of Maximum Shear Moduli of Brazilian Tropical Soils from Standard Penetration Test. *Proceedings of the 14th International Conference on Soil Mechanics and Foundation Engineering*, Vol. 1, pp. 29–30.

Batchelor, C., Goble, G., Berger, J., and Miner, R., 1995. *Standard Penetration Test Energy Measurements on the Seattle ASCE Field Testing Program*. Seattle Section of the American Society of Civil Engineers.

Bazaraa, A.R., 1967. Use of the Standard Penetration Test for Estimating Settlements of Shallow Foundations on Sand. Thesis presented to University of Illinois at Urbana, Illinois in partial fulfillment of the requirements for the degree of Doctor of Philosophy.

Bazaraa, A.R. and Kurkur, M.M., 1986. N-Values Used to Predict Settlement of Piles in Egypt. *Use of In Situ Tests in Geotechnical Engineering, ASCE*, pp. 462–474.

Behpoor, L. and Ghahramani, A., 1989. Correlation of SPT to Strength and Modulus of Elasticity of Cohesive Soils. *Proceedings of the 12th International Conference on Soil Mechanics and Foundation Engineering*, Vol. 1, pp. 175–177.

Berardi, R., Jamiolkowski, M., and Lancellotta, R., 1991. Settlement of Shallow Foundations in Sands-Selection of Stiffness on the Basis of Penetration Resistance. *Geotechnical Engineering Congress, ASCE*, Vol. 1, pp. 185–200.

Bergado, D.T., Alfaro, M., Bersabe, N., and Leong, K., 1993. Model Test Results of Sand Compaction Pile (SCP) Using Locally-Available and Low-Quality Backfill Soils. *Proceedings of the 11th Southeast Asian Geotechnical Conference*, pp. 313–318.

Biringen, E. and Davie, J., 2008. SPT Automatic Hammer Efficiency Revisited. *Proceedings of the 6th International Conference on Case Histories in Geotechnical Engineering*, Paper No. 4.06, 8 pp.

Borowczyk, M. and Frankowski, Z.B., 1981. Dynamic and Static Sounding Results Interpretation. *Proceedings of the 10th International Conference on Soil Mechanics and Foundation Engineering*, Vol. 2, pp. 451–454.

Bosio, J.J., 1992. A New Way of Measuring the Soundness of Very Soft Rocks Using the "SPT" Test Equipment. *Proceedings of the US/Brazil Geotechnical Workshop on Applicability of Classical Soil Mechanics Principles to Structured Soil*, pp. 91–96.

Bosscher, P.J. and Showers, D.R., 1987. Effects of Soil Type on Standard Penetration Test Input Energy. *Journal of Geotechnical Engineering, ASCE*, Vol. 113, No. 4, pp. 385–389.

Bowles, J.E., 1988. *Foundation Analysis and Design*. McGraw-Hill, Inc., 4th Edition, New York.

Bozbey, I. and Togrol, E., 2010. Correlation of Standard Penetration Test and Pressuremeter Data: A Case Study from Turkey. *Bulletin of Engineering Geology and Environment*, Vol. 69, pp. 505–515.

Broms, B.B. and Flodin, N., 1988. History of Soil Penetration Testing. *Proceedings of the 1st International Symposium on Penetration Testing*, Vol. 1, pp. 157–220.

Brooker, E. and Ireland, H., 1965. Earth Pressure at Rest Related to Stress History. *Canadian Geotechnical Journal*, Vol. 1, No. 1, pp. 1–15.

Brown, R.E., 1977. Drill Rod Influence on Standard Penetration Test. *Journal of the Geotechnical Engineering Division, ASCE*, Vol. 103, No. GT11, pp. 1332–1336.

Bureau of Reclamation, U.S. Department of Interior, 1974. Earth Manual.

Burland, J.B. and Burbidge, M.C., 1985. Settlement of Foundations on Sand and Gravel. *Proceedings of the Institute of Civil Engineers*, Part 1, Vol. 78, pp. 1325–1381.

Cangir, B. and Dipova, N., 2017. Estimation of Undrained Shear Strength of Konyaalti Silty Clays. *Indian Journal of Geo Marine Sciences*, Vol. 46, No. 3, pp. 513–520.

Chang, M.F. and Broms, B.B., 1991. Design of Bored Piles in Residual Soils Based on Field Performance Data. *Canadian Geotechnical Journal*, Vol. 28, pp. 200–209.

Clayton, C.R.I., Hababa, M.B., and Simons, N.E., 1985. Dynamic Penetration Resistance and the Prediction of the Compressibility of a Fine-Grained Sand – A Laboratory Study. *Geotechnique*, Vol. 35, No. 1, pp. 19–31.

Crova, R., Jamiolkowski, M., Lanellotta, R., and Presti, D.C.F., 1993. Geotechnical Characterization of Gravelly Soils at Massena Site: Selected Topics. *Predictive Soil Mechanics*, pp. 199–218.

Cubrinovski, M. and Ishihara, K., 2001. Correlation Between Penetration Resistance and Relative Density of Sandy Soils. *Proceedings of the 15th International Conference on Soil Mechanics and Geotechnical Engineering*, Vol. 1, pp. 393–396.

Daniel, C., Howie, J., and Sy, A., 2003. A Method for Correlating Large Penetration Test (LPT) to Standard Penetration Test (SPT) Blow Counts. *Canadian Geotechnical Journal*, Vol. 40, No. 1, pp. 66–77.

Daniel, C., Howie, J., Campanella, R., and Sy, A., 2004. Characterization of SPT Grain Size Effects in Gravels. *Proceedings of the 2nd International Symposium on Geotechnical and Geophysical Site Characterization*, Vol. 1, pp. 299–303.

D'Appolonia, D.J., D'Appolonia, E., and Brissette, R.F., 1970. Closure of Settlement of Spread Footings on Sand. *Journal of the Soil Mechanics and Foundation Division, ASCE*, Vol. 96, No. SM2, pp. 754–761.

Das, U., 2014. A Study on the Effect of Distorted Sampler Shoe on Standard Penetration Test Result in Cohesionless Soil. *International Journal of Innovative Research in Science, Engineering and Technology*, Vol. 3, No. 10, pp. 16654–16558.

Davidson, J., Maultsby, J., and Spoor, K., 1999. Standard Penetration Test Energy Calibrations. Final Report Florida DOT Contract No. BB261, Department of Civil Engineering, University of Florida.

Decourt, L., 1982. Prediction of the Bearing Capacity of Piles Based Exclusively on N Values of the SPT. *Proceedings of the 2nd European Symposium on Penetration Testing*, Vol. 1, pp. 292–234.

Decourt, L., 1989. The Standard Penetration Test – State of the Art Report. *Proceedings of the 12th International Conference on Soil Mechanics and Foundation Engineering*. Vol. 4, pp. 2405–2416.

Decourt, L., 1994. The Behavior of a Building with Shallow Foundations on a Stiff Lateritic Clay. *Vertical and Horizontal Deformations of Foundations and Embankments, ASCE*, Vol. 2, pp. 1505–1515.

Decourt, L. and Quarsma, A.R, 1994. Practical Applications of thje Standard Penetration Test Complemented by Torque Measurements, SPT-T; Present Stage and Future Trends. *Proceedings of the 13th International Conference on Soil Mechanics and Geotechnical Engineering*, pp. 143–146.

DeGodoy, N.S., 1971. Discussion of the Standard Penetration Test. *Proceedings of the 4th Pan American Conference on Soil Mechanics and Foundation Engineering*, Vol. 3, pp. 100–103.

DeMello, V., 1971. The Standard Penetration Test. *Proceedings of the 4th Panamerican Conference on Soil Mechanics*, Vol. 1, pp. 1–86.

Denver, H., 1982. Modulus of Elasticity for Sand Determined by SPT and CPT. *Proceedings of the 2nd European Symposium on Penetration Testing*, Vol. 1, pp. 35–40.

Dikmen, U., 2009. Statistical Correlations of Shear Wave Velocity and Penetration Resistance for Soils. *Journal of Geophysical Engineering*, Vol. 6, pp. 61–72.

Dung, N. and Chung, G., 2013. Behavior of the Standard Penetration Test (SPT) in Sandy Deposits. *Proceedings of the 3rd International Symposium on Geotechnical and Geophysical Site Characterization*, Vol. 1, pp. 365–373.

Esfehanizadeh, M., Nabizadeh, F., and Yazarloo, R., 2015. Correlation between Standard Penetration (N_{SPT}) and Shear Wave Velocity (V_S) for Young Coastal Sands of the Caspian Sea. *Arab Journal of Geoscience*, Vol. 8, pp. 7333–7341.

Fabbrocino, S., Lanzano, G., Forte, G., de Magistris, F., and Fabbrocino, G., 2015. SPT Blow Count vs. Shear Wave Velocity Relationship in the Structurally Complex Formations of the Molise Region (Italy). *Engineering Geology*, Vol. 187, pp. 94–97.

Fatehnia, M., Hayden, M., and Landschoot, M., 2015. Correlation between Shear Wave Velocity and SPT-N Values for North Florida Soil. *Electronic Journal of Geotechnical Engineering*, Vol. 20, pp. 12421–12430.

Fletcher, G.F.A., 1965. Standard Penetration Test: It's Uses and Abuses. *Journal of the Soil Mechanics and Foundation Division, ASCE*, Vol. 91, No. SM4, pp. 67–75.

Frost, D.J., 1992. Evaluation of the Repeatability and Efficiency of Energy Delivered with a Diedrich Automatic SPT Hammer System. Report prepared for Diedrich Drill Inc., La Porte, IN.

Gautam, D., 2017. Empirical Correlation between Uncorrected Standard Penetration Resistance (N) and Shear Wave Velocity (V_S) for Kathmandu Valley, Nepal. *Geomatics, Natural Hazards and Risk*, Vol. 8, No. 2, pp. 496–508.

Giacheti, H.L., Pedrini, R., and Rocha, B., 2013. The Seismic SPT Test in a Tropical Soil and the G_o/N Ratio. *Proceedings of the 18th International Conference on Soil Mechanics and Geotechnical Engineering*, 4 pp.

Gibbs, H.J. and Holtz, W.G., 1957. Research on Determining the Density of Sands by Spoon Penetration Testing. *Proceedings of the 4th International Conference on Soil Mechanics and Foundation Engineering*, Vol. 1, pp. 35–39.

Golder, H.Q., 1961. Discussion to Session IV. *Proceedings of the 5th International Conference on Soil Mechanics and Foundation Engineering.* Vol. 3, p. 163.

Granger, V.L., 1963. The Standard Penetration Test in Central Africa. *Proceedings of the 3rd Regional Conference for Africa on Soil Mechanics and Foundation Engineering*, Vol. 1, pp. 153–156.

Hall, J.R., 1982. Drill Rod Energy as a Basis for Correlation of SPT Data. *Proceedings of the 2nd European Symposium on Penetration Testing*, Vol. 1, pp. 57–60.

Hara, A., Ohta, T., Niwa, M., Tanaka, S., and Banno, T., 1974. Shear Modulus and Shear Strength of Cohesive Soils. *Soils and Foundations*, Vol. 14, No. 3, pp. 1–12.

Harder, L.F. and Seed, H.B., 1986. Determination of Penetration Resistance for Coarse-Grained Soils Using the Becker Hammer Drill. Report UCB/EERC-86/06, Earthquake Engineering Research Center, University of California, Berkeley.

Harrison, C., Bonin, G., and Esford, F., 2017. Summary of Drilling and Large Penetration Test Program to Confirm the Density of Soils Subjected to Dynamic Compaction for the Bay-Goose Dike at the Meadowbank Gold Projects in Nunavut, Canada. *Proceedings of GeoOttawa 2017*, 9 pp.

Hasancebi, N and Ulusay, R., 2007. Empirical Correlations between Shear Wave Velocity and Penetration Resistance for Ground Shaking Assessments. *Bulletin of Engineering Geology and the Environmnet*, Vol. 66, pp. 203–213.

Hatanaka, M. and Feng, L., 2006. Estimating Relative Density of Sandy Soils. *Soils and Foundations*, Vol. 46, No. 3, pp. 299–313.

Hatanaka, M. and Uchida, A., 1996. Empirical Correlation Between Cone Resistance and Internal Friction Angle of Sandy Soils. *Soils and Foundations*, Vol. 36, No. 4, pp. 1–10.

Hettiarachchi, H. and Brown, T., 2009. Use of SPT Blow Counts to Estimate Shear Strength Properties of Soils: Energy Balance Approach. *Journal of Geotechnical and Geoenvironmental Engineering, ASCE*, Vol. 135, No. 6, pp. 830–834.

Heydarzadeh, A., Fahker, A., and Moradi, M., 2013. A Feasibility Study of Standard Penetration Test with Torque Measurement (SPT-T) in Iran. *Proceedings of the 4th International Symposium on Geotechnical and Geophysical Site Characterization*, Vol. 1, pp. 553–559.

Hirayama, H., 1994. Secant Young's Modulus from N-Value or c_u Considering Strain Levels. *Proceedings of the International Symposium on Pre-Failure Deformation of Geomaterials*, Vol. 1, pp. 247–252.

Honeycutt, J., Kiser, S., and Anderson, J., 2014. Database Evaluation of Energy Transfer for Central Mine Equipment Automatic Hammer Standard Penetration Tests. *Journal of Geotechnical and Geoenvironmental Engineering, ASCE*, Vol. 140, No. 1, pp. 194–200.

Imai, T., 1977. P- and S-Wave Velocities of the Ground in Japan. *Proceedings of the 9th International Conference on Soil Mechanics and Foundation Engineering*, Vol. 2, pp. 257–260.

Imai, T. and Tonouchi, K., 1982. Correlations of N Value with S-wave Velocity and Shear Modulus. *Proceedings of the 2nd European Symposium on Penetration Testing*, Vol. 1, pp. 67–72.

Ireland, H.O., Moretto, O., and Vargas, M., 1970. The Dynamic Penetration Test: A Standard That Is Not Standardized. *Geotechnique*, Vol. 20, No. 2, pp. 185–192.

Jafari, M., Shafiee, A., and Razmkhan, A., 2002. Dynamic Properties in Fine-Grained Soils in the South Tehran. *Journal of Seismology and Earthquake Engineering*, Vol. 4, pp. 25–35.

Jain, P.K. and Handa, S.C., 1979. In-Situ Penetration Resistance of Cohesionless Soils. *Proceedings of the International Symposium on In Situ Testing of Soils and Rocks and Performance of Structures*, pp. 203–206.

Jamiolkowski, M., Ghionna, V.N., Lancellotta, R., and Pasqualini, E., 1988. New Correlations of Penetration Tests for Design Practice. *Proceedings of the 1st International Symposium on Penetration Testing*, Vol.1, pp. 263–296.

Jinan, Z., 1985. Correlation between Seismic Wave Velocity and the Number of Blow of SPT and Depths. *Selected Papers from the Chinese Journal of Geotechnical Engineering, ASCE*, pp. 92–100.

Jones, G.A. and Rust, E., 1989. Foundations on Residual Soil Using Pressuremeter Moduli. *Proceedings of the 12th International Conference on Soil Mechanics and Foundations Engineering*, Vol. 1, pp. 519–523.

Kaito, T., Sakaguchi, S., Nishigaki, Y., Miki, K., and Yukami, H., 1971. Large Penetration Test. *Tsuchi-to-Kiso*, Vol. 629, pp. 15–21.

Kalteziotis, N., Sabatakakis, N., and Vasiliou, I., 1992. Evaluation of Dynamic Characteristics of Greek Soil Formations. *Proceedings of the 2nd Hellenic Conference on Geotechnical Engineering*, Vol. 2, pp. 239–246.

Kanai, S. and Yabuuchi, S., 1989. Bearing Capacity of Nodular Piles. *Proceedings of the International Conference on Piling and Deep Foundations*, Vol. 1, pp. 73–79.

Kayabasi, A., 2012. Prediction of Pressuremeter Modulus and Limit Pressure of Clayey Soils by Simple and Non-Linear Multiple Regression Techniques: A Case Study from Mersin, Turkey. *Environmental Earth Sciences*, Vol. 66, pp. 2171–2183.

Kelley, S.P. and Lens, J., 2010. Evaluation of SPT Hammer Energy Variability. Report to Vermont Department of Transportation, Geodesign.

Kelley, S.P. and Lutenegger, A.J., 2004. Unit Skin Friction from the Standard Penetration Test Supplemented with the Measurement of Torque. *Journal of Geotechnical and Geoenvironmental Engineering, ASCE*, Vol. 130, No. 4, pp. 540–543.

Kim, D.-S., Bang, E.-S., and Seo, W.-S., 2004. Evaluation of Shear Wave Velocity Profile Using SPT Based Uphole Test. *Proceedings of the 2nd International Site Characterization Conference*, Vol. 1, pp. 707–712.

Kirar, B., Maheshawri, B., and Muley, P., 2016. Correlation between Shear Wave Velocity (V_s) and SPT Resistance (N) for Roorkee Region. *International Journal of Geosynthetics and Ground Engineering*, Vol. 2, No. 1, p. 9.

Kishida, H., 1967. Ultimate Bearing Capacity of Piles Driven into Loose Sand. *Soils and Foundations*, Vol. 7, No. 3, pp. 20–29.

Koester, J., Daniel, C., and Anderson, M., 2000. In Situ Investigation of Liquefiable Gravels. Transportation Research Record No. 1714, pp. 75–82.

Koike, M., Matsui, T., and Matsui, K., 1988. Vertical Loading Tests of Large Bored Piles and Their Estimation. *Proceedings of the 1st International Geotechnical Seminar on Bored and Augered Piles*, pp. 531–536.

Kovacs, W.D., 1975. On Dynamic Shear Moduli and Poisson's Ratios of Soil Deposits. *Soils and Foundations*, Vol. 15, No. 1, pp. 93–96.

Kovacs, W.D. and Salomone, L.A., 1982. SPT Hammer Energy Measurement. *Journal of the Geotechnical Engineering Division, ASCE*, Vol. 108, No. GT4, pp. 599–621.

Kovacs, W.D., Salomone, L.A., and Yokel, F.Y., 1981. Energy Measurement in the Standard Penetration Test. *National Bureau of Standards Building Science Series 135*, 99 pp.

Kraemer, S., 1996. *Geotechnical Earthquake Engineering*. Prentice Hall.

Kulhawy, F.H. and Mayne, P., 1990. Manual on Estimating Soil Properties for Foundation Design. EPPRI.

Kulhawy, F.H. and Trautmann, C.H., 1996. Estimation of In Situ Test Uncertainty. *Uncertainty in the Geologic Environment: From Theory to Practice, ASCE*, Vol. 1, pp. 269–286.

Kulhawy, F.H., Jackson, C., and Mayne, P.W., 1989. First-Order Estimation of K_o in Sands and Clays. *Foundation Engineering: Current Principles and Practices, ASCE*, pp. 121–134.

Kuwambara, F. and Tanaka, M., 1998. Statistical Analysis of Shaft Friction of Vertically Loaded Bored Piles. *Proceedings of the 3rd International Geotechnical Seminar on Bored and Augered Piles*, pp. 293–297.

Lake, L.M., 1974. Discussion on Session 1, Settlement of Structures, BGS, p. 663.

Lamb, R., 1997. SPT Energy Measurements with the PDA. *Proceedings of the 45th Annual Geotechnical Engineering Conference*, University of Minnesota, St. Paul, MN.

Lee, S.H.H., 1992. Analysis of the Multicollinearity of Regression Equations of Shear Wave Velocity. *Soils and Foundations*, Vol. 32, No. 1, pp. 205–214.

Liao, S.S.C. and Whitman, R.V., 1986. Overburden Correction Factors of SPT in Sand. *Journal of Geotechnical Engineering, ASCE*, Vol. 112, No. 3, pp. 373–377.

Liebich, B., 2005. Standard Penetration Test Energy Testing and Hammer Efficiency Measurements. California Department of Transportation Geotechnical Manual.

Lutenegger, A.J., 2008. The Standard Penetration Test – More Than Just a One Number Test. *Proceedings of the 3rd International Symposium on Site Characterization*, pp. 481–485.

Lutenegger, A.J., 2009. Estimating Shaft Resistance of Driven Piles from SPT-Torque Tests. *Contemporary Topics in In Situ Testing, Analysis, and Reliability of Foundations, ASCE GSP 186*, pp. 9–17.

Lutenegger, A.J. and Adams, M.T., 1998. Bearing Capacity of Footings on Compacted Sand. *Proceedings of the 4th International Conference on Case Histories in Geotechnical Engineering*, pp. 1216–1224.

Lutenegger, A.J. and Adams, M.T., 2003. Characteristic Load-Displacement Curves of Shallow Foundations. *Proceedings of the International Conference on Shallow Foundations*, Paris, France, Vol. 2, pp. 381–393.

Lutenegger, A.J. and Kelley, S.P., 1998. Standard Penetration Tests with Torque Measurement. *Proceedings of the International Symposium on Site Characterization*, Vol. 2, pp. 939–945.

Marcuson, W.F., 1978. Determination of In Situ Density of Sands. ASTM Special Technical Publication 654, pp. 318–340.

Marcuson, W.F. and Bieganousky, W.A., 1977a. Laboratory Standard Penetration Tests on Fine Sands. *Journal of the Geotechnical Engineering Division, ASCE*, Vol. 103, No. GT6, pp. 565–588.

Marcuson, W.F. and Bieganousky, W.A., 1977b. SPT and Relative Density in Coarse Sands. *Journal of the Geotechnical Engineering Division, ASCE*, Vol. 103, No. GT11, pp. 1295–1309.

Martin, R.E., 1977. Estimating Foundation Settlements in Residual Soils. *Journal of the Geotechnical Engineering Division, ASCE*, Vol. 103, No. GT3, pp. 197–212.

Martin, R.E., 1987. Settlement of Residual Soils. *Foundations and Excavation in Decomposed Rocks of the Piedmont Province, ASCE*, pp. 1–14.

Matsui, T., 1993. Case Studies on Cast-In-Place Bored Piles and Some Considerations for Design. *Proceedings of the 2nd International Geotechnical Seminar on Bored and Augered Piles*, pp. 77–101.

Matsumoto, K. and Matsubara, M., 1982. Effects of Rod Diameter in the Standard Penetration Test. *Proceedings of the 2nd European Symposium on Penetration Testing*, Vol. 1, pp. 107–112.

Matsumoto, T., Sekiguchi, H., Yoshida, H., and Kita, K., 1992. Significance of Two-Point Strain Measurement in SPT. *Soils and Foundations*, Vol. 32, No. 2, pp. 67–82.

Matsumoto, T., Phan, L., Oshima, A., and Shimono, S., 2015. Measurements of Driving Energy in SPT and Various Dynamic Cone Penetration Tests. *Soils and Foundations*, Vol. 55, No. 1, pp. 201–212.

Mayne, P.W., 1995. Profiling Yield Stress in Clays by In Situ Tests. Transportation Research Record No. 1479, pp. 43–50.

Mayne, P.W. and Kemper, J.B., 1988. Profiling OCR in Stiff Clays by CPT and SPT. *Geotechnical Testing Journal, ASTM*, Vol. 11, No. 2, pp. 139–147.

Mayne, P.W. and Kulhawy, F.H., 1982. K_o – OCR Relationships in Soil. *Journal of the Geotechnical Engineering Division, ASCE*, Vol. 108, No. GT6, pp. 851–872.

Meyerhof, G.G., 1956. Penetration Tests and Bearing Capacity of Cohesionless Soils. *Journal of the Soil Mechanics Division, ASCE*, Vol. 82, SM1, pp. 1–12.

Meyerhof, G.G., 1976. Bearing Capacity and Settlement of Pile Foundations. *Journal of the Geotechnical Engineering Division, ASCE*, Vol. 102, No. GT3, pp. 197–228.

Morgano, C.M. and Liang, R., 1992. Energy Transfer in SPT – Rod Length Effect. *Application of Stress-Wave Theory to Piles*, pp. 121–127.

Muromachi, T., Aguro I., and Miyashita, T., 1974. Penetration Testing in Japan. *Proceedings of the European Symposium on Penetration Testing*, Vol. 1, pp. 193–200.

Nassaji, F. and Kalantari, B., 2011. SPT Capability to Estimate Undrained Shear Strength of Fine-Grained Soils of Tehran, Iran. *Electronic Journal of Geotechnical Engineering*, Vol. 16, pp. 1229–1238.

National Highway Institute (NHI), 2002. Course 132031 Subsurface Investigations – Geotechnical Site Characterization Reference Manual; Publication No. FHWA NHI-01-031.

Nayak, N.V., 1979. Use of the Pressuremeter in Geotechnical Design Practice. *Proceedings of the International Symposium on In Situ Testing of Soils and Rocks and Performance of Structures*, Roorkee pp. 432–436.

Nevels, J. and Laguros, J., 1993. Correlation of Engineering Properties of the Hennessey Formation Clays and Shales. *Proceedings of the International Symposium on Geotechnical Properties of Hard Soils and Soft Rock*, Vol. 1, pp. 215–221.

Nixon, I.K., 1982. Standard Penetration Test: State of Art Report. *Proceedings of the 2nd European Symposium on Penetration Testing*, Vol. 1, pp. 3–24.

Odebrecht, E., Schnaid, F., Rocha, M., and Bernardes, G., 2004. Energy Measurements for Standard Penetration Tests and the Effects of the Length of Rods. *Proceedings of the 2nd International Symposium on Geotechnical and Geophysical Site Characterization*, Vol. 1, pp. 351–358.

Ohsaki, Y. and Iwasaki, R., 1973. On the Dynamic Shear Moduli and Poisson's Ratio of Soil Deposits. *Soils and Foundations*, Vol. 13, No. 4, pp. 61–73.

Ohta, Y. and Goto, N., 1978. Empirical Shear Wave Velocity Equations in Terms of Characteristic Soil Indexes. *Earthquake Engineering and Structural Dynamics*, Vol. 6, pp. 167–187.

Ohya, S., Imai, T., and Nagura, M., 1982. Relationships between N-Value by SPT and LLT Measurement Results. *Proceedings of the 2nd European Symposium on Penetration Testing*, Vol. 1, pp. 125–130.

Oskorouchi, A.M. and Mehdibeigi, A., 1988. Effect of Overburden Pressure on SPT N-Values in Cohesive Soils of North Iran. *Proceedings of the 1st International Symposium on Penetration Testing*, Vol. 1, pp. 363–367.

Palmer, D.J. and Stuart, J.G., 1957. Some Observations on the Standard Penetration Test and a Correlation of the Test with a New Penetrometer. *Proceedings of the 4th International Conference on Soil Mechanics and Foundation Engineering*, Vol. 1, pp. 231–236.

Papadopoulos, B.P., 1992. Settlement of Shallow Foundations on Cohesionless Soils. *Journal of Geotechnical Engineering, ASCE*, Vol. 118, No. 3, pp. 377–393.

Parsons, J.D., 1966. discussion of Standard Penetration Test: Its Uses an Abuses. *Journal of the Soil Mechanics and Foundation Division, ASCE*, Vol. 92, No. SM3, pp. 103–105.

Parry, R.H.G., 1977. Estimating Bearing Capacity in Sand from SPT Values. *Journal of the Geotechnical Engineering Division, ASCE*, Vol. 103, No. GT9, pp. 1112–1116.

Peck, R.B., Hanson, W.E., and Thornburn, T.H., 1953. *Foundation Engineering.* John Wiley & Sons Inc., New York.

Peck, R.B., Hansen, W.E., and Thornburn, T.H., 1974. *Foundation Engineering.* John Wiley & Sons, Inc., New York.

Pedrini, R.A.A. and Giacheti, H.L., 2013. The Seismic SPT to Determine the Maximum Shear Modulus. *Proceedings of the 4th International Geophysical and Geotechnical Site Characterization Conference, ISC-4*, pp. 337–342.

Pedrini, R.A.A., Rocha, B., and Giacheti, H.L., 2018. The Up-Hole Seismic Test Together with the SPT: Description of the System and Method. *Soils and Rocks, Sao Paulo*, Vol. 41, No. 2, pp. 133–148.

Peixoto, A., Carvalho, D., and Giacheti, H., 2004. SPT-T: Test Procedure and Applications. *Proceedings of the 2nd International Symposium on Geotechnical and Geophysical Site Characterization*, Vol. 1, pp. 359–366.

Pinto, C. and Abramento, M., 1997. Pressuremeter Tests on Gneissic Residual Soil in Sao Paulo, Brazil. *Proceedings of the 14th International Conference on Soil Mechanics and Foundation Engineering*, Vol. 1, pp. 175–176.

Pitilakis, K., Lontzetidid, K., Raptakis, D., and Tika, T., 1998. Geotechnical and Seismic Survey for Site Characterization. *Proceedings of the 1st International Conference on Site Characterization*, Vol. 2, pp. 1339–1344.

Pitilakis, K., Raptakis, D., Lontzetidis, K., Tika-Vassilikou, T., and Jongmans, D. 1999. Geotechnical and Geophysical Descriptions of Euro-Seistests Using Field and Laboratory Tests and Moderate Strong Ground Motions. *Journal of Earthquake Engineering*, Vol. 3, No. 3, pp. 381–409.

Plytas, C., Baltzoglou, A., Chlimintzas, G., Anagnostopoulos, G., Kozompolis, A. and Koutalia, C., 2011. Empirical determination of the undrained shear strength of very stiff to (very) hard cohesive soils from SPT tests. Proceedings of the 15the European Conference on Soil Mechanics and Foundation Engineering, pp. 61–66.

Poulos, H., 1989. Pile Behaviour – Theory and Applications. *Geotechnique*, Vol. 39, No. 3, pp. 365–415.

Ranzine, S., 1988. SPTF Technical Note. *Solos e Rochas*, Vol. 11, pp. 29–30.

Raptakis, D., Anastasiadis, A., Pitilakis, K., and Lontzetidis, K., 1994. Shear Wave Velocity and Damping of Greek Natural Soils. *Proceedings of the 10th European Conference on Earthquake Engineering.*

Reese, L.C., Touma, F.T., and O'Neill, M., 1976. Behavior of Drilled Piers under Axial Loading. *Journal of the Geotechnical Engineering Division, ASCE*, Vol. 102, No. GT5, pp. 493–510.

Riggs, C.O., Schmidt, N.O., and Rassieur, C.L., 1983. Reproducible SPT Hammer Impact Force with an Automatic Free Fall SPT Hammer System. *Geotechnical Testing Journal, ASTM*, Vol. 6, No. 4, pp. 165–172.

Riggs, C.O., Mathes, G.M., and Rassieur, C.L., 1984. A Field Study of an Automatic SPT Hammer System. *Geotechnical Testing Journal, ASTM*, Vol. 7, No. 3, pp. 158–163.

Rocha, B., Pedrini, R., and Giachetti, H., 2015. G_0/N Ratio in Tropical Soils from Brazil. *Electronic Journal of Geotechnical Engineering*, Vol. 20, No. 7, pp. 1915–1933.

Robert, Y., 1997. A Few Comments on Pile Design. *Canadian Geotechnical Journal*, Vol. 34, pp. 560–567.

Robertson, P.K., Campanella, R.G., and Wightman, A., 1983. SPT-CPT Correlations. *Journal of the Geotechnical Engineering Division, ASCE*, Vol. 109, No. 11, pp. 1454–1459.

Robertson, P.K., Woeller, D.J., and Addo, K.O., 1992. Standard Penetration Test Energy Measurements Using a System Based on the Personal Computer. *Canadian Geotechnical Journal*, Vol. 29, No. 4, pp. 551–557.

Rocha Filho, P., Antunes, F., and Falcao, M.F.Q., 1985. Qualitative Influence of the Weathering Degree upon the Mechanical Properties of Young Gneiss Residual Soil. *Proceedings of the 1st International Conference on Geomechanics in Tropical Lateritic and Saprolitic Soils*, Vol. 1, pp. 281–294.

Rogers, J., 2006. Subsurface Exploration Using the Standard Penetration Test and the Cone Penetration Test. *Environmental & Engineering Geoscience*, Vol. 12, No. 2, pp. 161–179.

Rollins, K.M., Evans, M.D., Diehl, N.B., and Daily, W.D. III, 1998. Shear Modulus and Damping Relationships for Gravels. *Journal of Geotechnical and Geoenvironmental Engineering, ASCE*, Vol. 124, No. 5, pp. 396–405.

Ruge, J., Mendoza, C., Colmenares, J., Cunha, R., and Otalvaro, I., 2018. Analysis of the Undrained Shear Strength through the Standard Penetration Test with Torque (SPT-T). *International Journal of GEOMATE*, Vol. 14, No. 41, pp. 102–110.

Saiki, K., 1983. Methods of Analysis for Correcting the Standard Penetration Resistance Measured in Normally Consolidated Clays. *Recent Developments in Laboratory and Field Tests and Analysis of Geotechnical Problems*, pp. 375–380.

Sambhandharaksa, S. and Pitupakorn, W., 1985. Predictions of Prestressed Concrete Pile Capacity in Bangkok Stiff Clay and Clayey Sand. *Proceedings of the 8th Southeast Asian Geotechnical Conference*, pp. 3.58–3.65.

Sanglerat, G., 1972. *The Penetrometer and Soil Exploration*. Elsevier, Amsterdam, 464 pp.

Sanglerat, G. and Sanglerat, T.R., 1982. Pitfalls of the SPT. *Proceedings of the 2nd European Symposium Penetration Testing*, Vol. 1, pp. 143–145.

Sarac, D. and Popovic, M., 1982. Penetration Tests for Determination of Characteristics of Flood Dike Materials. *Proceedings of the 2nd European Symposium on Penetration Testing*, Vol. 1, pp. 147–152.

Schmertmann, J.H., 1978. Use of the SPT to Measure Dynamic Soil Properties? – Yes, But...! ASTM Special Technical Publication 654, pp. 341–355.

Schmertmann, J.H., 1979. Statics of SPT. *Journal of the Geotechnical Division, ASCE*, Vol. 105, No. GT5, pp. 655–670.

Schmertmann, J.H., 1984. Discussion of Reproducible SPT Hammer Force with an Automatic Free Fall SPT Hammer System. *Geotechnical Testing Journal, ASTM*, Vol. 7, No. 3, pp. 167–168.

Schmertmann, J.H. and Palacios, A., 1979. Energy Dynamics of SPT. *Journal of the Geotechnical Engineering Division, ASCE*, Vol. 105, No. GT8, pp. 909–926.

Schnaid, F., Odebrecht, E., Rocha, M, and Bernardes, G., 2009. Prediction of Soil Properties from the Concepts of Energy Transfer in Dynamic Penetration Tests. *Journal of Geotechnical and Geoenvironmental Engineering, ASCE*, Vol. 135, No. 8, pp. 1092–1100.

Schultze, E. and Melzer, K.J., 1965. The Determination of the Density of Modulus of Compressibility of Non-Cohesive Soils by Soundings. *Proceedings of the 6th International Conference on Soil Mechanics and Foundation Engineering*, Vol. 1, pp. 354–358.

Seed, H.B., 1976. Evaluation of Soil Liquefaction Effects on Level Ground during Earthquakes. *ASCE Specialty Session, Liquefaction Problems in Geotechnical Engineering, ASCE*.

Seed, H.B. and De Alba, P., 1986. Use of SPT and CPT Tests for Evaluating the Liquefaction Resistance of Sands. *Use of In Situ Tests in Geotechnical Engineering, ASCE*, pp. 281–302.

Seed, H.B., Idriss, I.M., and Arango, I., 1983. Evaluation of Liquefaction Potential Using Field Performance Data. *Journal of the Geotechnical Engineering Division, ASCE*, Vol. 109. No. GT3, pp. 458–482.

Seed, H.B., Tokimatsu, K., Hardner, L.F., and Chung, R.M., 1985. Influence of SPT Procedures in Soil Liquefaction Resistance Evaluations. *Journal of Geotechnical Engineering, ASCE*, Vol. 111, No. GT12, pp. 1425–1445.

Seed, R.B., Harder, L.F. Jr., and Youd, T.L., 1988. Effects of Borehole Fluid on Standard Penetration Test Results. *Geotechnical Testing Journal, ASTM*, Vol. 11, No. 4, pp. 248–256.

Serajuddin, M. and Chowdhury, M., 1996. Correlation between Standard Penetration Resistance and Unconfined Compression Strength of Bangladesh Cohesive Deposits. *Journal of Civil Engineering, The Institution of Engineers, Bangladesh*, Vol. CE24, No. 1, pp. 69–83.

Shi-Ming, H., 1982. Experience on a Standard Penetration Test. *Proceedings of the 2nd European Symposium on Penetration Testing*, Vol. 1, pp. 61–66.

Shioi, Y. and Fukui, J., 1982. Application of N-Value to Design of Foundations in Japan. *Proceedings of the 2nd European Symposium on Penetration Testing*, Vol. 1, pp. 159–164.

Sivrikaya, O. and Togrol. E., 2002. Relations between SPT-N and q_u. *Proceedings of the 5th International Congress on Advances in Civil Engineering*, Istanbul, pp. 943–952.

Sivrikaya, O. and Togrol. E., 2006. Determination of Undrained Strength of Fine-Grained Soils by Means of SPT and Its Application in Turkey. *Engineering Geology*, Vol. 86, pp. 52–69.

Skempton, A.W., 1986. Standard Penetration Test Procedures and the Effects in Sands of Overburden Pressure, Relative Density, Particle Size, Aging Overconsolidation. *Geotechnique*, Vol. 36, No. 3, pp. 425–447.

Sowers, G., 1979. *Introductory Soil Mechanics and Foundations*. Macmillan, New York, 621 pp.

Sowers, G.B. and Sowers, G.F., 1961. *Introductory Soil Mechanics and Foundations*. The Macmillan Co., 2nd Edition, New York, 386 pp.

Spangler, M.G., 1960. *Soil Engineering*. International Textbook Co., 2nd Edition, Scranton, PA, 483 pp.

Stamatopoulos, A.C. and Kotzias, P.C., 1974. The Use of the Standard Penetration Test in Classifying Rocks. *Proceedings of the 3rd International Congress of the International Society for Rock Mechanics*, p. 85.

Stamatopoulos, A.C. and Kotzias, P.C., 1993. Refusal to the SPT and Penetrability. *Geotechnical Engineering of Hard Soils – Soft Rock*, Vol. 1, pp. 301–306.

Stark, T., Long, J., and Assem, P., 2013. Improvement for Determining the Axial Capacity of Drilled Shafts in Illinois. Research Report No. FHWA-ICT-13-017, Illinois Center for Transportation.

Stark, T., Long, J., Baghdady, A., and Osouli, A., 2017. Modified Standard Penetration Test-Based Drilled Shaft Design Method for Weak Rocks. Research Report No. FHWA-ICT-17-018, Illinois Center for Transportation.

Stroud, M.A., 1974. The Standard Penetration Test in Insensitive Clays and Soft Rocks. *Proceedings of the 2nd European Symposium on Penetration Testing*, Vol. 2, pp. 367–375.

Stroud, M.A., 1989. The Standard Penetration Test – Its Application and Interpretation. *Proceedings of the Conference on Penetration Testing in the U.K.*, pp. 29–46.

Stroud, M.A. and Butler, F.G., 1975. The Standard Penetration Test and the Engineering Properties of Glacial Materials. *Proceedings of the Symposium on Engineering Behavior of Glacial Materials*, pp. 117–128.

Suzuki, Y., Goto, S., Hatanaka, M., and Tokimatsu, K., 1993. Correlation between Strengths and Penetration Resistances in Gravelly Soils. *Soils and Foundations*, Vol. 33, No. 1, pp. 92–101.

Sy, A. and Campanella, R.G., 1993. Dynamic Performance of the Becker Hammer Drill and Penetration Test. *Canadian Geotechnical Journal*, Vol. 30, No. 4, pp. 607–619.

Sy, A. and Campanella, R.G., 1994. Becker and Standard Penetration Tests (BPT-SPT) Correlations with Consideration of Casing Friction. *Canadian Geotechnical Journal*, Vol. 31, No. 3, pp. 343–356.

Sykora, D.W. and Stokoe, K.H., 1983. Correlations of In Situ Measurements in Sands with Shear Wave Velocity. Geotechnical Engineering Report GR83-33. University of Texas at Austin, Austin, TX.

Sykora, D. and Koester, J., 1988. Correlations between Dynamic Shear Resistance and Standard Penetration Resistance in Soils. *Earthquake Engineering and Soil Mechanics II, ASCE*, pp. 389–404.

Takesue, K., Sasao, H., and Makihara, Y., 1996. Cone Penetration Testing in Volcanic Soil Deposits. *Advances in Site Investigation Practice, Institution of Civil Engineers, Thomas Telford*, London, England, pp. 452–463

Tavares, A.X., 1988. Bearing Capacity of Footings on Guabirotuba Clay Based on SPT N-Values. *Proceedings of the 1st International Symposium on Penetration Testing*, Vol. 1, pp. 375–379.

Terzagi, K. and Peck, R.B., 1948. *Soil Mechanics in Engineering Practice*. John Wiley.

Terzaghi, K. and Peck, R., 1967. *Soil Mechanics in Engineering Practice*. 2nd Edition. John Wiley & Sons, New York, 729 pp.

Terzaghi, K., Peck, R.B., and Mezri, G., 1996. *Soil Mechanics in Engineering Practice*. John Wiley & Sons, New York.

Toh, C.t., Ting, W.H., and Ooi, T.A., 1987. Allowable Bearing Pressure of the Kenney Hill Formation. *Proceedings of the 9th Southeast Asian Geotechnical Conference*, pp. 6/55–6/66.

Tokimatsu, K., 1988. Penetration Tests for Dynamic Problems. *Proceedings of the International Symposium on Penetration Testing*, Vol. 1, pp. 117–136.

Tokimatsu, K. and Yoshimi, Y., 1983. Empirical Correlation of Soil Liquefaction Based on SPT N-Value and Fines Content. *Soils and Foundations*, Vol. 23, No. 4, pp. 56–74.

Toledo, R.D., 1986. Field Study of the Stiffness of a Gneissic Residual Soil Using Pressuremeter and Instrumented Pile Load Test. MS Thesis, Civil Engineering Department, Pontifical Catholic University, Rio de Janeiro.

Towhata, I. and Ronteix, S., 1988. Probabilistic Prediction of Shear Wave Velocity from SPT Blowcounts and Its Application to Seismic Microzonation. *Proceedings of the Symposium on Geotechnical Aspects of Restoration and Maintenance of Infra-Structures and Historical Monuments*, pp. 423–439.

Tsiambaos, G. and Sabtakakis, N., 2011. Empirical Estimation of Shear Wave Velocity from In Situ Tests on Soil Formations in Greece. *Bulletin of Engineering Geology and the Environment*, Vol. 70, pp. 291–297.

Tsuchiya, H. and Toyooka, Y. 1982. Comparison between N-Value and Pressuremeter Parameters. *Proceedings of the 2nd European Symposium on Penetration Testing*, Vol. 1, pp. 169–174.

Uma Maheswari, R., Boominathan, A., and Dodagoudar, G., 2010. Use of Surface Waves in Statistical Correlations of Shear Wave Velocity and Penetration Resistance of Chennai Soils. *Geotechnical and Geological Engineering*, Vol. 28, pp. 119–137.

Valiquette, M., Robinson, B., and Borden, R., 2010. Energy Efficiency and Rod Length Effects in SPT Hammers. *Proceedings of the 89th Annual Meeting of the Transportation Research Board*, Washington, DC.

Vallee, R.P. and Skryness, R.S., 1979. Sampling and In-Situ Density of a Saturated Gravel Deposit. *Geotechnical Testing Journal, ASTM*, Vol. 2, No. 3, pp. 136–142.

Van der Graaf, H.J. and Van den Heuvel, M.H.J.P., 1992. Determination of the Penetration Energy in the Standard Penetration Test. *Application of Stress-Wave Theory to Piles*, pp. 253–257.

Veijayaratnam, M., Poh, K., and Tan, S., 1993. Seismic Velocities in Singapore Soils and Some Geotechnical Applications. *Proceedings of the 11th Southeast Asian Geotechnical Conference*, pp. 813–818.

Viana da Fonseca, A., Fernandes, M., and Cardoso, A., 1998. Characterization of a Saprolitic Soil from Porto Granite Using In Situ Testing. *Proceedings of the 1st International Conference on Site Characterization*, Vol. 2, pp. 1381–1387.

Viswanath, M. and Mayne, P., 2013. Direct SPT Method for Footing Response in Sands Using a Database Approach. *Proceedings of the 4th International Symposium on Geotechnical and Geophysical Site Characterization*, Vol. 2, pp. 1131–1136.

Wada, A., 2003. Development Mechanism of Pile Skin Friction. *Proceedings of the BGA International Conference on Foundations*, pp. 921–929.

Webb, D.L., 1969. Settlement of Structures on Deep Alluvial Sandy Sediments in Durban, South Africa. *Proceedings of the Conference on In Situ Investigations in Soils and Rocks, BGS*, pp. 181–187.

White, F., Ingram, P., Nicholson, D., Stroud, M. and Betru, M., 2019. An update of the SPT-cu relationship proposed by M. Stroud in 1974. Proceedings of the 17th European Conference on Soil Mechanics and Foundation Engineering, 8 pp.

Whited, G.C. and Edil, T.B., 1986. Influence of Borehole Stabilization Techniques on Standard Penetration Test Results. *Geotechnical Testing Journal, ASTM*, Vol. 9, No. 5, pp. 180–188.

Wightman, A., Yan, L., and Diggle, D.A., 1993. Improvements to the Becker Penetration Test for Estimate of SPT Resistance in Gravelly Soils. *Proceedings of the 46th Canadian Geotechnical Conference*, pp. 379–388.

Winter, E., Nordmark, T.S., and Burns, T., 1989. Skin Friction Load Tests for Caissons in Residual Soils. *Foundation Engineering: Current Principles and Practices, ASCE*, Vol. 2, pp. 1070–1075.

Winter, C., Wagner, A., and Komurka, V., 2005. Investigation of Standard Penetration Torque Testing (SPT-T) to Predict Pile Performance. Report No. 0092-04-09, Wisconsin Highway Research Program.

Wrench, B.P. and Nowatzki, E.A., 1986. A Relationship between Deformation Modulus and SPT N for Gravels. *Use of In Situ Tests in Geotechnical Engineering, ASCE*, pp. 1163–1177.

Wroth, C.P., 1984. The Interpretation of In Situ Tests. *Geotechnique*, Vol. 34, No. 4, pp. 449–489.

Wroth, C.P., Randolph, M.F., Houslby, G. and Fahey, M., 1979. *A Review of Engineering Properties of Soils, with Particular Reference to the Shear Modulus.* CUED/D Soils TR75, Cambridge University Engineering Department.

Yagiz, S., Akyol, E., and Sen, G., 2008. Relationship between the Standard Penetration Test and the Pressuremeter Test on Sandy Silty Clays: A Case Study from Denizli. *Bulletin of Engineering Geology and Environment*, Vol. 67, pp. 405–410.

Yokel, F.Y., 1982. Energy Transfer in Standard Penetration Test. *Journal of the Geotechnical Engineering Division, ASCE*, Vol. 108, No. GT9, pp. 1197–1202.

Yoshida, Y., Ikemi, M., and Kokusho, T., 1988. Empirical Formulas of SPT Blow-Counts for Gravelly Soils. *Proceedings of the 1st International Symposium on Penetration Testing*, Vol. 1, pp. 381–387.

Chapter 3

Dynamic Cone Penetration Test (DCP)

3.1 INTRODUCTION

As discussed in Chapter 2, the Standard Penetration Test (SPT) is typically performed at depth intervals of about 1.5 m (5 ft). A rapid and inexpensive alternative for obtaining more continuous data is to use a Dynamic Cone Penetrometer Test (DCP). These tests are also often referred to as drive cones, dynamic probing, or dynamic sounding Tests. Reviews of the DCP are given by Melzer & Smoltczyk (1982) and Stefanoff et al. (1988). Broms & Flodin (1988) presented an historical review of the development of DCP tests and reported that there are records tracing this type of test to Germany as early as 1699. The use of DCPs in the U.S. was described at least as early as 1948 by Terzaghi & Peck (1948).

DCPs may be performed using a drill rig, using essentially the same equipment used to conduct the SPT; they may be performed using portable automatic rigs, or they may be performed by hand using simple light weight equipment. In addition to their use in routine site investigation work and for estimating individual soil properties, DCPs are often used for shallow site work to evaluate footing subgrade characteristics, as a quality control tool for backfill around buried pipes or compacted fill for shallow foundations or retaining walls, to locate depth to shallow bedrock, and for a variety of other applications. Test results have been presented in *clays* (e.g., Butcher et al 1995; Lawson et al. 2018), *sands* (e.g., Palmer & Stuart 1957; Mohan et al. 1970; Singh & Sharma 1973; Coyle & Bartoskewitz 1980; Muromachi & Kobayashi 1982), and *residual soils* (Chang & Wong 1986; Kelley & Lutenegger 1999) as well as in *gravelly* materials (e.g., Rao et al. 1982; Hanna et al. 1986; Talbot 2017).

The DCP may be used to supplement other *in situ* tests, or it may be used in situations where site accessibility is very limited or mobilization costs for a drill rig or other equipment may be very high. They may also be used for shallow investigations where the use of a drill rig is not practical.

3.2 MECHANICS

DCPs are conducted by attaching a cone on the bottom of a set of drive rods and attaching an anvil on the top of the rods. The drive rods are marked off in the specified driving intervals and then driven with a simple drop hammer system. The cone may be sacrificial or may be threaded or pinned onto the end of the rods. Figure 3.1 illustrates the principle of the DCP and shows the different components of the test. All that is needed to perform the test

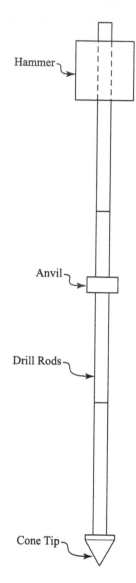

Figure 3.1 Principle of Drive Cone Penetration Tests.

are a drop hammer, an anvil, a set of drive rods, and a conical tip. DCPs have some potential advantages over other types of penetration tests:

1. They are simple to perform;
2. Most drilling companies have the necessary equipment to perform the test (e.g., SPT hammer and drill rods) and can obtain a suitable cone tip;
3. They may be used in a wide range of soil types;
4. They provide more continuous test results than the SPT;
5. They are rapid and inexpensive;
6. They may be used to provide a good indication of relative differences in density or stiffness of the soil being penetrated;

7. The tests do not require a large static reaction that is needed for the CPT/CPTU and DMT;
8. They generally do not require a borehole; therefore, there are no cuttings to dispose of;
9. There are essentially no moving parts to the equipment; and
10. The test does not require power or any electronic components.

However, this type of test also has some limitations, which include the following:

1. All forms of the test are not currently standardized;
2. There can be problems with unknown and nonuniform energy input during the test;
3. No soil sample is obtained;
4. Rod friction may cause errors in the test results; and
5. Interpretation of results is almost entirely empirical.

Even with these limitations, DCPs may be used in many site investigations to supplement test borings and other *in situ* tests. They provide a large amount of data quickly at very low cost. Compared to a CPT/CPTU deployed off the back of a drill rig, DCPs are much faster and easier to perform since no reaction is needed. Three or four DCP profiles can easily be conducted to a depth of 6.1 m (20 ft) in about an hour using a drill rig with an automatic SPT hammer.

An early attempt to standardize the DCP was presented by ASTM D18 Subcommittee 2 (ASTM 1970) in which the use of a 57.2 mm (2.25 in.) diameter 60° cone with flush joint rods having an outside diameter of 44.4 mm (1.75 in.) was recommended. The cone was to be driven with a 63.5 kg (140 lbs) SPT hammer having a drop height of 76.2 cm (30 in.). The number of blows for each 30.5 cm (12 in.) were to be recorded. There is no record that this suggested method was ever adopted by ASTM as a standard; however, a light duty DCP procedure is described in ASTM D6951. The DCP is also covered by the International Organization for Standardization (ISO) under Dynamic Probing ISO 22476-2, which describes four classes of DCP.

3.3 EQUIPMENT

The DCP is generally classified according to the mass of the hammer as light, medium, heavy, or super heavy, as indicated in Table 3.1. There are several variables that can be altered for the different configurations and categories of the DCPs including: (1) the hammer mass, (2) the hammer drop height, (3) the cone diameter, (4) the cone apex angle, (5) the drive rod diameter, and (6) the driving distance.

In a survey of about 50 countries taken in 1988 by Stefanoff et al. (1988), the overwhelming responses indicated that the most common types of drive cones in use were the light and super heavy; the light probably because it is simple, inexpensive, and portable and the super

Table 3.1 Classification of DCP by Hammer Mass.

Classification	Hammer mass (kg)	Abbr.
Light	≤ 10	DPL
Medium	>10<40	DPM
Heavy	≥40<60	DPH
Super heavy	>60	DPSH

heavy probably because it uses the same driving equipment as the SPT. The results of the survey presented by Stefanoff et al. (1988) clearly showed that there was extensive use of the DCP throughout the world at the time, a trend which has likely continued.

Test equipment and procedures for different classes of DCP have been presented as part of the International Reference Test Procedures. Table 3.2 gives a summary of the proposed characteristics of the different categories of the DCP (Stefanoff et al.1988). Note that all of the cones for these proposed international standards have an apex angle of 90°. In the past, some countries have made use of the DIN 4094 specifications developed in Germany for DCP, given in Table 3.3.

Table 3.2 Proposed international DCP reference test specifications

Factor	Reference test procedure			
	DPL	DPM	DPH	DPSH
Hammer mass, kg	10±0.1	30±0.3	50±0.5	63.5±0.5
Height of fall, m	0.5±0.01	0.5±0.01	0.5±0.01	0.75±0.02
Mass of anvil and guide rod	6	18	18	30
Rebound (max), %	50	50	50	50
Length to diameter (D) ratio (hammer)	>1<2	>1<2	>1<2	>1<2
Diameter of anvil (d), mm	100<d<0.5D	100<d<0.5D	100<d<0.5D	100<d<0.5D
Rod length, m	1±0.1%	1–2±0.1%	1–2±0.1%	1–2±0.1%
Maximum mass of rod, kg/m	3	6	6	8
Rod deviation (max), first 5m, %	1.0	1.0	1.0	1.0
Rod deviation (max), mm	2.0	2.0	2.0	2.0
Below 5m, %	0.2	0.2	0.2	0.2
Rod eccentricity (max), mm	22±0.2	32±0.3	32±0.3	32±0.3
Rod OD, mm	6±0.2	9±0.2	9±0.2	9±0.2
Rod ID, mm				
Apex angle, deg.	90	90	90	90
Nominal area of cone, cm²	10	10	15	20
Cone diameter, new, mm	35.7±0.3	35.7±0.3	43.7±0.3	50.5±0.5
Cone diam. (min), worn, mm	34	34	42	49
Mantle length of cone, mm	35.7±1	35.7±1	43.7±1	50.5±2
Cone taper angle, upper, deg.	11	11	11	11
Length of cone tip, mm	17.9±0.1	17.9±0.1	21.9±0.1	25.3±0.4
Max wear of cone tip length, mm	3	3	4	5
Number of blows per x cm penetration	10 cm; N_{10}	10 cm; N_{10}	10 cm; N_{10}	20 cm; N_{20}
Standard range of blows	3–50	3–50	3–50	5–100
Kinetic energy, kJ	0.050	0.150	0.250	0.4734
Energy per blow: (Mgh/A), kJ/m²	50	150	167	238

Table 3.3 Outline of DIN 4094 dynamic penetrometers

Class	Name	Hammer mass (kg)	Drop height (cm)	Cone diameter (mm)	Cone area (cm²)
Light	LR 5	10	50	25.2	5
	LR 10	10	50	35.6	10
Medium	MRSA	30	20	35.6	10
	MRSB	30	50	35.6	10
Heavy	SRS 10	50	50	35.6	10
	SRS 15	50	50	43.7	15

3.4 TEST PROCEDURES

3.4.1 Light DCP

Light DCPs (DPL) have a hammer mass in the range of 4.5–10 kg (10–22 lbs) and are often used for rapid, low-cost investigation of shallow soil conditions. In most cases, light DCPs are performed by hand using portable equipment. This makes the test easy for a one or two-person crew to conduct. In this configuration, however, the test is labor intensive and is typically only applicable for depths of 3.0–6.0 m (10–20 ft) depending on site conditions. A wide range of combinations of hammer mass, fall height, cone size, cone apex angle, and drive distance have been used to conduct the DPL as given in Table 3.4.

3.4.1.1 Sowers Cone

Sowers & Hedges (1966) described the use of a light DCP for footing inspection and light field exploration. The test is shown in Figure 3.2 and is performed in conjunction with a small diameter hand auger hole. A 38.1 mm (1.5 in.) diameter cone with an apex angle of 45° is driven with a hammer having a mass of 6.8 kg (15 lbs) and a drop height of 508 mm (20 in.). Sowers & Hedges (1966) recommended that the cone point first be seated (50.8 mm) 2 in. into the bottom of the borehole. The number of hammer blows required to then drive

Table 3.4 Summary of light DCP equipment

Hammer mass (kg)	Drop height (cm)	Cone diameter (mm)	Apex angle (°)	Drive distance (mm)	References
2.5	30.5	12/18	45	pen./blow	DeGaridele-Thoron & Javor (1983)
4.5	45.7	15.9	-	25.4	Rostron et al. (1969)
4.5	30	29	20	300	Chan & Chin (1972) Ooi & Ting (1975)
5	28	25	60	300	Ooi & Ting (1975)
5	50	25	60	100	Sugiyama et al. (1998)
5/10	33	20/50	60	50	Acar et al. (1991)
6.8	50.8	38.1	45	44.5	Sowers & Hedges (1966) Robinson (1988) Elton (1989)
8	57.5	20	60	pen./blow	Kleyn et al. (1982) Kleyn & Van Zyl (1988) Chua (1988) Ayers & Thompson (1989) Chua & Lytton (1992) Ford & Eliason (1993) Ampadu & Arthur (2006)
8	57.5	20	90	pen./blow	Livneh (1987a)
9	60	16	0	150	Fityus (1998)
9.1	50.8	10	30	25	Scala (1956) Smith (1988)
10	50	35	60	N/A	Singh & Sharma (1973)
10	50	36	90	100	Borowczyk & Frankowski (1981)
10	50	25.2	60	pen./blow	DeHenau (1982)
10	50	35.7	90	100	Paunescu & Gruia (1982)
10	50	44.5	60	100	Author

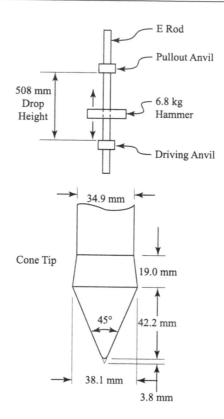

Figure 3.2 Sowers Cone.

the cone a further distance of 44.5 mm (1.75 in.) is recorded as N_S. The current supplier of the equipment actually recommends that the initial seating distance be 44.5 mm (1.75 in.).

This configuration of the DCP is often referred to as the "Sowers Cone Penetrometer" (e.g., Robinson 1988; Elton 1989) and appears to be used in the U.S. for footing inspection work and other shallow investigations, especially in the southeastern part of the U.S. Sowers & Hedges (1966) suggested some general correlations between N_S and SPT N values for different soil types; however, there were no details on the SPT equipment used. Hajduk et al. (2007) evaluated correlations between the Sowers Cone and CPT and DMT results for soils in the Charleston, S.C. area. Figure 3.3 shows some typical results obtained by the author at two sites using the Sowers Cone.

3.4.1.2 ASTM Light "Pavement" DCP

Kleyn et al. (1982) described a light DCP for subgrade investigations in South Africa. As shown in Figure 3.4, the penetrometer consists of a sliding hammer with a mass of 8 kg (17.6 lbs) with a drop height of 575 mm (22.6 in.). The cone has a diameter of 20 mm (0.8 in.) and an apex angle of 60°. Rods with a diameter of 16 mm (0.625 in.) are used. The device is essentially the same as the US Army Corps of Engineers Waterways Experiment Station (WES) Light DCP. In this arrangement, the rods are marked off so that the penetration distance achieved with each hammer drop (mm/blow) can be recorded. This configuration of the test has been used extensively for very shallow work and appears to be gaining popularity for pavement design.

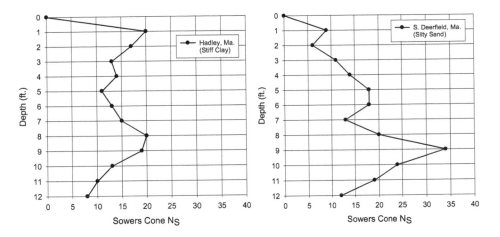

Figure 3.3 Typical results from Sowers Cone.

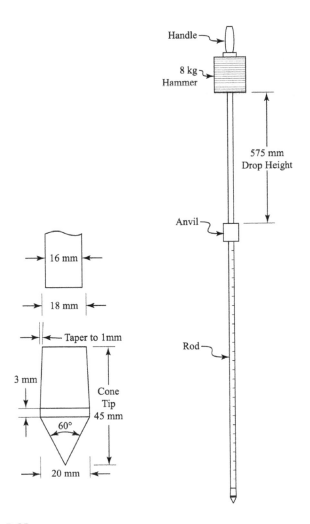

Figure 3.4 ASTM Light DCP.

This configuration of light DCP is available commercially and was adopted by ASTM, which developed a standard test method for a Light DCP (ASTM D6951 *Standard Test Method for Use of the Dynamic Cone Penetrometer in Shallow Pavement Applications*), which also allows for a lighter 4.6 kg (10.1 lbs) hammer if the penetration achieved using the 8 kg (17.6 lbs) hammer is too large.

3.4.1.3 Mackintosh & JKR Probe

According to Nixon (1989), one of the earliest forms of a light DCP was developed by F.H. Mackintosh in 1922 for "shallow testing" by "detecting changes in the hardness of strata from the speed of driving". The "Mackintosh Probe" consists of a 27.4 mm (1.1 in.) diameter probe with an apex angle of 30° connected to 12.7 mm (0.5 in.) diameter rods as shown in Figure 3.5. Some diagrams of the probe (e.g., Fakher & Khodaparasat 2004) actually show the probe as "torpedo"-shaped rather than cylindrical. The probe is driven at the bottom of a borehole using a 4.5 kg (10 lbs) hammer with a drop height of 0.30 m (1 ft).

Some authors define the results obtained from the Mackintosh Probe as the number of blows (M) for 100 mm (4 in.) penetration (e.g., Fakher et al. 2006), while others have defined M as the number of blows for 30 cm (1 ft) penetration (e.g., Fatt & Kee 1972; Chan & Chin 1972; Kong 1983; Hossain & Ali 1990). The Mackintosh Probe has been used extensively in Southeast Asia and parts of the Middle East for relatively shallow (< 8 m (26 ft)) investigations (e.g., Chan & Chin 1972; Fatt & Kee 1972; Ooi & Ting 1975; Kong 1983). Several correlations between Mackintosh Probe results and SPT results have been presented (e.g., Chan & Chin 1972; Fatt & Kee 1972; Sabtan & Sherbata 1994; Fakher et al. 2006) and suggest that the value of M (blows/ft) is nearly the same as N from the SPT.

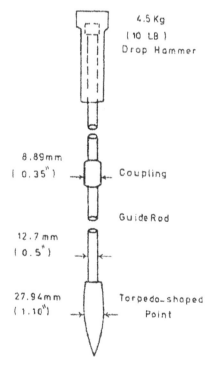

Figure 3.5 Schematic of Mackintosh Probe. (From Sabtan & Sherbata 1994.)

Ooi & Ting (1975) described the use of a Light DCP used in Malaysia similar to the Mackintosh Probe for evaluating the allowable bearing capacity of spread footings. The "JKR" probe consists of a 60^0 cone with a diameter of 25 mm (1 in.) connected to drive rods of 12 mm (0.5 in.) diameter. The cone is driven with a 5 kg (11 lbs) hammer using a drop height of 280 mm (11 in.). The number of hammer blows required for a penetration distance of 0.30 m (1 ft) is recorded.

3.4.1.4 Lutenegger Drive Cone

For several years, the author has used a light DCP that is a variation of the DPL given in Table 3.2. As shown in Figure 3.6, the test uses a 10 kg (22 lbs) hammer with a drop height of 500 mm (19.7 in.). Standard EW drill rods (34.9 mm (1.37 in.) diameter) are used, and a $60°$ cone tip with a diameter of 44.5 mm (1.75 in.) giving an area of 15.5 cm² is used. This gives a cone/rod diameter ratio of 1.28. EW rods in lengths of 0.9 m (3 ft) are readily available and are easy to transport and handle. The system is meant for continuous driving and uses either disposable cones that fit into the end of the rods or cones that are pinned in the lead rod to allow retrieval and reuse. The number of hammer blows for a penetration of 10 cm (4 in.) are recorded. Tests have typically been used for shallow site characterization, footing inspection and evaluation of backfill compaction. The larger diameter cone gives more sensitivity than the small diameter cone and rods used in the ASTM D6951. Figure 3.7 shows typical results obtained at four sites.

3.4.2 Medium DCP

The medium DCP (DPM) has a hammer mass that is too large to be comfortably operated by hand for efficient use but too small for use with a drill rig and a conventional SPT hammer. Several configurations of this test have been reported as summarized in Table 3.5. Figure 3.8 shows a schematic of a portable DCP rig mounted on a trailer.

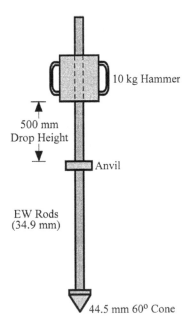

Figure 3.6 LDCP used by the author.

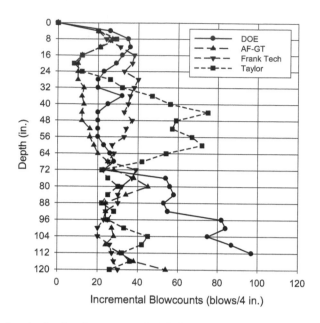

Figure 3.7 Typical results obtained at four sites using LDCP.

Table 3.5 Summary of Different Medium DCP Equipment.

Hammer mass (kg)	Drop height (cm)	Cone diameter (mm)	Apex angle (°)	Drive distance (mm)	References
15.5	61	33.3	60	102	Lacroix & Horn (1973)
30	25	35.7	30	200	Schmid (1975)
30	20	35.7	60	100	Triggs & Liang (1988)

Triggs & Liang (1988) presented results using a small commercially available penetration rig with a 30 kg (66 lbs) hammer that can be transported in components and assembled at a site by a two-person crew. The equipment is manufactured in Italy and can be used for both static and dynamic penetration. In the dynamic mode, the hammer is operated by a small gasoline engine and an automatic lift/drop mechanism. The cone may be either fixed to the rods or may be expendable.

In the UK, a medium DCP (DPM15) with a cone diameter of 43.7 mm (1.72 in.) (giving a cone area of 15 cm²) is sometimes used (Butcher et al. 1995). Because of the larger area of the cone, the DPM15 has a specific work per blow twice that of the DPL.

3.4.3 Heavy DCP

A heavy DCP (DPH) with a hammer mass of 50 kg (110 lbs) was described by Card & Roche (1989) and Scarff (1989). The equipment was developed in the UK and is mounted on a small self-contained trailer pulled by a vehicle to the site. Table 3.6 gives a summary of some different heavy DCP equipment used.

In most cases, the hammer is raised by a chain drive mechanism operated by a small gasoline engine and released automatically to give a drop height of 50 cm (19.7 in.) with a driving rate typically between 25 and 30 blows/min. This is similar in design to the mechanism used with some automatic SPT hammers. The rods are guided between rollers at the base of the

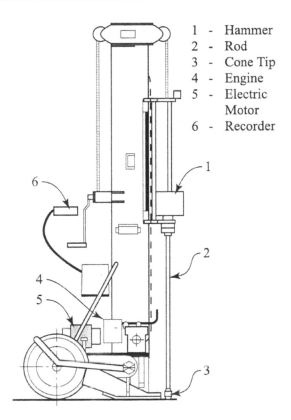

1 - Hammer
2 - Rod
3 - Cone Tip
4 - Engine
5 - Electric Motor
6 - Recorder

Figure 3.8 Trailer-mounted automatic medium DCP.

Table 3.6 Summary of heavy DCP equipment

Hammer mass (kg)	Drop height (cm)	Cone diameter (mm)	Apex angle (°)	Drive distance (mm)	References
50	50	43.7	60	200	Teferra (1983) Kayalar (1988)
50	50	43.7	90	100	Paunescu & Gruia (1982) Card & Roche (1989) Scarff (1989) Tonks & Whyte (1989) Butcher et al. (1995)
50	50	43.7	90	200	Kayalar (1988)

mast and at the drive head to resist bending. This equipment also conforms to the German Standard DIN 4094 Part I (Card & Roche 1989) and has been used in Eastern Europe (e.g., Paunescu & Gruia 1982).

3.4.4 Super Heavy DCP

For routine site investigation work in which a drill rig will be on site, the use of the super heavy DCP (DPSH) appears to be the most practical and useful configuration of the test. The most common technique for conducting super heavy DCPs is to use the 63.5 kg (140 lbs)

Table 3.7 Reported use of hammers larger than 63.5 kg (140 lbs) to conduct super heavy DCPs

Hammer mass (kg)	Drop height (cm)	Cone diameter (mm)	Apex angle (°)	Drive distance (mm)	References
75.6	61	76.2	60	305	Coyle & Bartoskewitz (1980)
155	30	70	60	300	Chan & Chin (1972)
159	60	63.5	60	150	Rodin (1961) Ergun (1982)
470	100	92	30	N/A	Hanna et al. (1986)
120	100	74	60	100	Talbot (2017)

SPT hammer although other hammer masses have also been used. AW rods are used for driving, and normally a cone with a 60° apex and a diameter of 63.5 mm (2.5 in) is used, although in some cases, a 76.2 mm cone (3.0 in.) may also be used. The number of hammer blows to advance the cone 150 mm (6 in.) is typically recorded, and the results are reported as N_C blows/30 cm (1 ft).

Borros produces a small portable rig that provides equipment in the "SPT" configuration, i.e., 63.5 kg (140 lbs) hammer, 76 cm (30 in.) drop height, for driving a 50.5 mm (2 in.) diameter cone with an apex angle of 90°. The driving is automatic, being operated by a small gasoline engine, and is conducted at a rate of about 20 blows/min. The number of blows required to drive the cone a distance of 10 cm (3.9 in.) is recorded.

In some cases, it may be desirable to use a hammer with a mass larger than 63.5 kg (140 lbs), especially in very dense or gravelly materials. For example, Hanna et al. (1986) used a 470 kg (1034 lbs) hammer with a 92 mm (3.6 in.) diameter cone to evaluate ground improvement in sands and sandy gravel. Table 3.7 presents a summary of reported use of some heavy hammers for conducting super heavy DCPs.

3.5 TEXAS CONE PENETROMETER

According to Lawson et al. (2018), the Texas Cone Penetrometer (TCP) was used as early as 1949 by the Texas Highway Department. The test uses a 75.6 kg (170 lbs) hammer falling 61 cm (24 in.) to drive a 76 mm (3 in.) diameter cone with a 60° apex angle a distance of 30 cm (12 in.). The test is typically performed in the bottom of a nominal 100 mm (4 in.) borehole. The test appears to be more or less routinely used by the Texas Highway Department for site investigations and has been used to correlate to undrained shear strength in clays (Hamoudi et al. 1974; Gudavalli et al. 2008) and shear strength in sands (Coyle & Bartoskewitz 1980).

Results from the test were traditionally presented as the number of hammer drops to drive the cone 30 cm (12 in.) (Touma & Reese 1972); however, in very strong materials, an alternative is to use the penetration (mm) for 100 hammer drops. More recent use of the TCP has been described by Gudavalli et al. (2008) and Nam & Vipulanandan (2010). The test has been used for estimating compressive strength of clay shale (Cavusoglu et al. 2004) and undrained shear strength of low plasticity clay (Vipulanandan et al. 2008). Early correlation between TCP and SPT showed that SPT N=0.5 TCP (Coyle & Bartoskewitz 1980). Updated correlations between the TCP and SPT have recently been presented by Lawson et al. (2018) for different soils. Maghaddam et al. (2017) have recently presented measurements of hammer efficiency and correction factors for interpreting TCP data.

3.6 SWEDISH RAM SOUNDING TEST

In the Scandinavian countries, the Swedish Ram Sounding Test (SRS) is often used (e.g. Dahlberg & Bergdahl 1974). This test consists of an automatic apparatus with a 63.5 kg (140 lbs) hammer with a drop height of 50 cm (19.7 in.) and uses a 45 mm (1.8 in.) diameter cone with a 90° apex angle. The cone has a mantle with a length of 90 mm (3.5 in.) with the same diameter as the cone. The number of hammer blows to drive the cone a distance of 20 cm (7.9 in.) is designated as N_{20}. As a part of the apparatus, a torque wrench is provided so that some evaluation of rod friction can be obtained at intervals during probing. This is often done after each 1 m rod section is attached.

Broms & Flodin (1988) indicate that the ram sounding test can actually have other configurations depending on the diameter of the cone and rods. In the DPA, a 62 mm (2.4 in.) diameter cone is used with 40–45 mm (1.6–1.8 in.) rods, while in the DPB, the cone diameter is 51 mm (2.0 in.), and 32 mm (1.3 in.) diameter rods are used. In both cases, the same hammer mass and drop height as the SRS are used, and the number of blows to drive the cone a distance of 20 cm is recorded as N_{20}, as previously described.

3.7 FACTORS AFFECTING TEST RESULTS

Many of the factors that can affect the results of the DCP are similar to those that can affect SPT results. This is especially true for super heavy DCPs that use SPT equipment to perform the test. While it is true that different hammer masses, drop heights, cone diameter, and drive distances can be used, in some respects, the DCP may be considered to be less variable than the SPT and subject to fewer errors, especially associated with issues such as borehole diameter, use of drilling fluid, spoon geometry, etc. The principal variables of concern for the DCP are as follows:

1. Hammer mass,
2. Hammer drop height,
3. Cone diameter,
4. Cone apex angle, and
5. Drive distance.

Like the SPT, an important factor influencing DCP results using the SPT equipment is the energy of the system. Since different hammers and drop systems produce different test results, the author recommends that, like the SPT, all super heavy DCPs be performed using a calibrated automatic SPT hammer, and results standardized to an energy level of 60% in order to provide a reference level of comparison.

Since side resistance along the rods will affect the test results and is one of the major concerns about this test, some attempts should be made to reduce skin friction. This may be accomplished using a variety of techniques; however, the simplest approach is to use driving rods smaller in diameter than the cone. Experience suggests that if the cone/rod diameter ratio is on the order of about 1.3, there will be little or no significant effect of rod friction on the measured values of penetration resistance in most granular soils. A practical upper limit of this ratio of about 1.5 is suggested so that the rod diameter will be compatible with the driving energy and reduce the potential for bending during the test.

3.8 PRESENTATION OF TESTS RESULTS

Results from DCPs are typically presented in one of the following ways:

1. Incremental number of blows for a given driving distance;
2. Cumulative number of blows for the entire depth;
3. Penetration distance achieved with each drop of the hammer; or
4. Dynamic penetration resistance, r_d or q_d, calculated from different driving formulae.

3.8.1 Incremental Penetration Resistance

Results of DCPs are often simply presented as a plot of incremental penetration resistance versus depth, as previously shown in Figure 3.7. The penetration resistance is defined over a specific penetration interval, typically 10–20 cm (4–12 in.). An example using results from a super heavy DCP profile conducted at a medium dense fine sand site is shown in Figure 3.9. These results clearly show changes in driving penetration resistance at several depths, indicating some changes in stratigraphy.

In this case, a 63.5 mm (2.5 in.) diameter 60° cone was used with AW rods in lengths of 1.5 m (5 ft.) and an automatic SPT hammer to perform the tests. Since the tests were performed above the water table, there is little possibility of the sand collapsing back around the drive rods, and therefore, almost no side resistance builds up on the rods. As a check in the field, the engineer checked the rod friction by simply rotating the rods by hand after each section of rod was driven. Additionally, the penetration record indicates lower blow counts after penetrating through a stiff layer where blow counts were high, again indicating little buildup of rod friction.

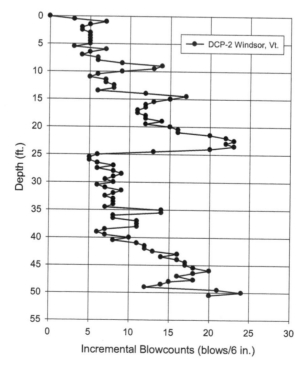

Figure 3.9 Incremental driving resistance from super heavy DCP (blows/6 in.).

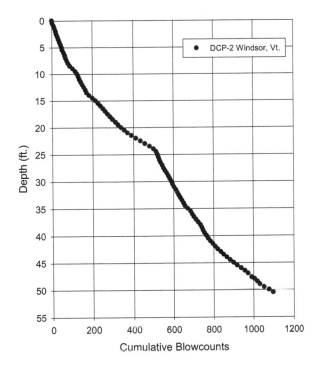

Figure 3.10 Cumulative penetration resistance.

3.8.2 Cumulative Penetration Resistance

DCP results can also be presented as cumulative penetration resistance or total driving record as a function of depth. This is done by numerically adding the results for each consecutive driving interval. The incremental results from Figure 3.9 are shown as cumulative driving resistance in Figure 3.10 as an example. These results are typically interpreted by identifying changes in the slope of the penetration record. Figure 3.10 indicates changes at depths of about 15, 23, and 41 ft. In this case, the changes may indicate differences in grain-size distribution, relative density, or a different stratigraphic unit.

3.8.3 Penetration Distance per Hammer Blow

It may be useful to record the actual penetration distance achieved with each drop of the hammer. This requires a measurement of the advance of the drive rods after each hammer blow. This is usually done by measuring the distance between a mark on the rods and some stationary reference point, usually the top of the ground or casing or using an electronic system. The record of penetration distance per hammer blow may give a more refined look at changes in penetration resistance that might be missed using the incremental penetration resistance, which tends to "average" out the resistance over the interval.

A slight modification to this procedure is to choose a specific number of hammer blows, say ten, and measure the resulting penetration produced. The average penetration per blow over the measured interval is obtained by dividing the distance by the number of blows. Data from the penetration record may be used to calculate the incremental and cumulative resistance by simply counting the number of hammer blows that were required to produce the set driving increment.

3.8.4 Dynamic Penetration Resistance

Bolomey (1974) suggested that it may be useful to express the penetration resistance in terms of the resistance values, r_d or q_d, using different pile driving formulae. The unit dynamic driving resistance is obtained as

$$r_d = (MgH)/(Ae) \tag{3.1}$$

or

$$q_d = \left[(MgH)/(Ae)\right]\left[M/(M+M')\right] = r_d\left[M/(M+M')\right] \tag{3.2}$$

where:

r_d and q_d=dynamic penetration resistance (Pa)
M=mass of the hammer (kg)
M'=total mass of the rods, anvil, etc. (kg)
H=hammer drop height (m)
A=cross sectional area of the cone point (m^2)
e=average penetration distance (m/drop)=D/N
N=number of blows/10 cm
D=drive interval (mm)
g=acceleration of gravity (9.81 m/s^2)

Dynamic penetration resistance is particularly useful when comparing DCP results obtained from different equipment, such as of different size cones, hammer mass, or height at the same site.

3.9 INTERPRETATION OF TEST RESULTS

The interpretation and use of DCPs may be highly varied depending on the application, the soil, the need, and the experience of the engineer. For example, Stefanoff et al. (1988) found that the predominant use of DCPs was *qualitative*, i.e., to distinguish between different soil layers. However, DCPs have also been used to interpret individual soil properties in both coarse-grained and fine-grained soils and for the design of deep and shallow foundations. Because of the wide variation in equipment and definition of N_C obtained from any particular set of equipment, caution should be used when applying any empirical correlations.

3.9.1 Correlations to SPT

In the past, a common method of using DCP results has been to correlate the test results to the SPT (e.g., Palmer & Stuart 1957; Rodin 1961; Mohan et al.1970; Solymar 1984; Robinson 1988; Cearns & McKenzie 1989; Elton 1989; Butcher et al. 1995). In this way, correlations or design methods available for the SPT could be applied. This is one approach that may have some merit, especially if the DCP is being used to supplement SPT data in an investigation.

In some cases, considerable scatter can exist in these correlations and may be related to variations in both the SPT and DCP equipment and test procedures. Results from the DCP generally relate to end resistance to driving, while results from the SPT relate to both end and side resistance along the spoon. As previously discussed in Chapter 2, the distribution of end and side resistance in the SPT depends on soil type. Therefore, it should not be too surprising that correlations between the DCP and SPT may be highly variable and probably should be based on local correlations to specific geologic deposits.

It is possible to replace the driving shoe of the SPT split spoon with a 60° solid point of the same diameter as the spoon barrel (i.e., 50.8 mm (2 in.)) and then use the spoon as a drive probe. Results presented by Palmer & Stuart (1957), Palmer (1957), and Gawad (1976) in granular soil materials indicated that in this case, the blow counts are essentially the same, i.e., $N_{SPT} = N_C$. In each of these cases, however, it should be noted that the split spoon used did not have any internal relief. This suggests that either the spoon plugged during most of the tests or that the combined internal and external wall friction along the spoon compensated for the increased end area using the conical tip.

Table 3.8 gives a summary of some reported correlations between super heavy DCP and SPT. Results often show considerable scatter, most likely because of variability in both tests and natural variability between adjacent boreholes. However, it has generally been shown that when a casing, sleeve, drilling fluid, or sufficient cone/rod diameter ratio has been used to eliminate the rod friction, $N_C = N_{SPT}$ (e.g., Gadsby 1971; Mohan et al. 1971; Meardi 1971; Goel 1982) with a range of N_C/N_{SPT} from about 0.5 to 2.

Results from light and medium DCPs have also been used to correlate to SPT N values (e.g., Fatt & Kee 1972; Sabtan & Sherbata 1994; Fakher et al. 2006; Opuni et al 2017). Lacroix & Horn (1973) had suggested that it would be possible to estimate the SPT N value from any penetrometer with either an open end or solid conical tip accounting for not just differences in geometry but also dive energy and drive distance as follows:

$$N = N_1 (2 \text{ in.}/D_1)^2 (12 \text{ in.}/L_1)(w_1/140 \text{ lb})(H_1/30 \text{ in.}) = (2N_1W_1H_1)/(175D_1^2L_1) \qquad (3.3)$$

Table 3.8 Reported correlations between DPSH and SPT

Cone diameter (mm)	Cone apex angle (°)	Hammer Mass (kg)	Drop Height (cm)	Drive Distance (cm)	Correlation N_C/N_{SPT}	References
50.8	60	63.5	76.2	30	2	Meyerhof (1956)
50.8	60	63.5	76.2	30	1	Palmer & Stuart (1957)
50.8	60	63.5	76.2	30	1	Rodin (1961)
63.5	60	63.5	76.2	30	1.5–4	Mohan et al. (1970)
63.5	60	63.5	76.2	30	1	
57.1	60	63.5	76.2	30	1.5 (sleeved)	Gadsby (1971)
35.6	60	63.5	76.2	30	0.9	Gawad (1976)
51.0	60	63.5	76.2	30	0.6	
76.0	60	63.5	76.2	30	0.5	
51	60	63.5	76.2	30	1.6	Goel (1982)
63	60	63.5	76.2	30	0.8–3.5	
50.8	60	63.5	75	10	1.15	Muromachi &
45	90	63.5	50	20	1	Kobayashi (1982)
62.5	60	63.5	76.2	30	1.6	Rao et al. (1982)
45	90	63.5	50	20	0.5–0.67	Chang & Wong (1986)
45	90	63.5	50	20	0.66 (sandy) 0.5 (gravelly)	McGrath et al. (1989)
51	90	63.5	76	20	0.83–1.1	Cabrera & Carcole (2007)
51	90	63.5	76	30	0.6–2	Macrobert (2017)

where
  N=SPT blow counts
  N_1=measured blow counts from another penetrometer
  D_1=outside diameter of the nonstandard spoon or conical point (in.)
  L_1=depth of penetration (in.)
  W_1=weight of hammer (lbs)
  H_1=height of hammer drop (in.)

In all cases where the DPC blow count value is converted to SPT N values via some correlation, the resulting N values should be stated as "equivalent" or "comparable" N values so that there is no confusion as to how they were obtained.

3.9.2 Correlations to CPT

DCP results have also been correlated to CPT tip resistance. For example, Swann (1982) presented a correlation between light DCPs (DIN 4094) and static cone resistance values as

$$q_c = 0.5 N_C \tag{3.4}$$

where
  q_c=static cone tip resistance (MN/m^2)
  N_{10}=penetration resistance/10 cm

A comparison between DCP results (converted to static resistance using an energy balance equation) and CPT results was presented by Triggs & Liang (1988) using the medium DCP described in Section 3.4.2. They found that for a wide range of soils, results from the two tests were very similar.

Butcher et al. (1995) showed that the dynamic penetration resistance, q_d, can be related to the static cone penetration resistance in both soft and stiff clays. This is probably the most desirable approach to correlating DCP results to other *in situ* tests, since any configuration of the test should give the same value of q_d.

3.9.3 Direct Correlations to Soil Properties

In addition to the use of DCPs as a qualitative indication of changes in stratigraphy, the test results may also be used to provide estimates of individual soil properties. As with the SPT, caution should be used since most of the reported correlations are empirical and may be site-specific.

3.9.3.1 Relative Density of Sands

A number of correlations have been reported for estimating relative density of coarse-grained soils from results of different DCPs as summarized in Table 3.9.

3.9.3.2 Undrained Shear Strength of Clays

Tonks & Whyte (1989) showed a correlation between SRS 15 and undrained shear strength at three different sites as

$$s_u(kPa) = (K_1)(N_{10}) \tag{3.5}$$

Table 3.9 Summary of correlations between relative density and DCP

Equation	Class of DCP	References
$D_r = 0.429 \log N_{10} + 0.071$	DPL	Borowczyk & Frankowski (1981)
$D_r = 0.441 \log N_{20} + 0.196$	DPSH	
(D_r in decimal)		
$\log D_r = 0.554 \log N_{10} + 0.980$	DPL	Paunescu & Gruia (1982)
$D_r = 189.9/(DPI)^{0.53}$	DPL	Mohammadi et al. (2008)
(DPI in mm/drop)		
$D_r = 0.71 \log N'_c - 0.035$	DPSH	Hanna et al. (1986)
(D_r in %)		
(N'_c = blow count corrected for overburden using CPT correction factors of Seed & Idriss 1981)		
$D_r = 0.385 \log N_{20} - 0.385 \log \sigma_{vo} - 0.145$	DPH	Teferra (1983)
(above limiting depth)		
$D_r = 0.270 \log N'_{20} + 0.340$		
(below limiting depth)		
(D_r in decimal)		
(σ_{vo} in kg/cm^2)		
($N'_{20} = N_{20}$ at limiting depth)		
$D_R(\%) = -52\log(DPI \times D_{50})^{0.3} + 150$	DPL	MacRobert et al. (2019)

Values of K_1 ranged from about 5 to 9, typically averaging about 7. A comparison between the results of light DCPs and the undrained shear strength of silty clays was also reported by Waschkowski (1982); unfortunately, no details on the DCP equipment were given. Several correlations have also been suggested to estimate undrained shear strength from Mackintosh Probe results (Hossain & Ali 1990; Fakher et al. 2006; Khodaparasat 2010; Khodaparasat et al. 2015). McGrath et al. (1989) suggested a correlation between the results of the Swedish ram sounding tests and undrained shear strength of clays.

Butcher et al. (1995) suggested a general correlation between results from the DCP and undrained shear strength of clays and found that the undrained shear strength could be correlated directly to driving resistance q_d as

$$s_u(kPa) = (q_d/22) \quad \text{(for stiff clays)} \tag{3.6a}$$

$$s_u(kPa) = (q_d/170) + 20 \quad \text{(for soft clays)} \tag{3.6b}$$

Alshkan et al. (2020) suggested a correlation between the unconfined compressive strength(q_u) of fine-grained soils in Iraq using a ght DCP (8 kg drop weight, 575 mm drop height, 20 mm cone tip) as:

$$q_u(kPa) = 1033.6(DPI)^{-0.968} \tag{3.7}$$

with DPI in mm/blow

3.9.3.3 California Bearing Ratio

A common use of light DCPs has been in the area of subgrade support characteristics. Because only shallow tests are needed, the equipment is particularly suitable. Results of DCPs have been used extensively for estimating CBR values for pavements and subgrades (e.g., Scala 1956; Kleyn et al. 1982; Smith & Pratt 1983; DeGaridele-Thoron & Javor 1983;

Table 3.10 Different DCPs Used for Evaluating CBR.

Hammer mass (kg)	Drop height (cm)	Cone apex angle (°)	Cone diameter (mm)	References
2.5	30.5	45	12 & 18	DeGaridele-Thoron & Javor (1983)
8	57.5	90	20	Livneh (1987b)
8	57.5	60	20	Kleyn (1975) Harrison (1987) Chua (1988) Ayers & Thompson (1989) Webster et al. (1992) Karunaprema & Edirisinghe (2002) Abu-Farsakh et al. (2005) Misra et al. (2006)
8	57.5	30	20	Kleyn (1975) Livneh & Ishai (1988)
9.1	50.8	30	20.2	Scala (1956) Smith (1983)
63	50	90	45	McGrath et al. (1989)

Harrison 1987; Livneh 1987b; Kleyn & Van Zyl 1988; McGrath et al. 1989). A comparison of different light DCPs used for evaluating the CBR of subgrades is given in Table 3.10.

Correlations between CBR and DCP naturally depend on the DCP used. Several correlations have been suggested between DCP results and CBR, especially for pavement design, as given in Table 3.11. All of these correlations have the general form

$$\log CBR = a + b \log (DCP)^c \qquad (3.8)$$

where

DCP = ratio between the penetration and the number of blows (mm/blow)

3.9.3.4 Resilient Modulus

Several correlations have been suggested for estimating the resilient modulus (M_R) of sub-grades for pavement design from results obtained using light duty DCP equipment. Some reported correlations are given in Table 3.12.

3.9.3.5 Compaction Control

DCPs may also be useful as a quality control tool for evaluating the effects of compaction or ground improvement (e.g. Acar et al. 1991; Gabr et al. 2001) For example, Acar et al. (1991) showed that the relative density or relative compaction of granular fills could be evaluated

Table 3.11 Reported correlations between light DCP results and CBR

Correlation	References
log CBR = 2.55 − 1.14 log (DCP)	Smith (1983)
log CBR = 2.81 − 1.32 (log DCP)	Harrison (1987)
log CBR = 2.20 − 0.71 (log DCP)$^{1.5}$	Livneh & Ishai (1988)
log CBR = 2.46 − 1.12 (log DCP)	Webster et al. (1992)
log CBR = 1.97 − 0.67 (log DCP)	Karunaprema & Edirisinghe (2002)

Table 3.12 Reported correlations between light DCP results and resilient modulus

Correlation	Units	References
M_R (MPa) $= 235$ (DCPI)$^{-0.48}$	mm/blow	George & Uddin (2000)
M_R (MPa) $= 532$ (DCPI)$^{-0.492}$ (for fine-grained soils)	mm/blow	Rahim & George (2004)
M_R (MPa) $= 285$ (DCPI)$^{-0.475}$ (for coarse-grained soils)	mm/blow	Rahim & George (2004)
M_R (MPa) $= 338$ (PR)$^{-0.39}$	mm/blow	Abu-Farsakh et al. (2005)
M_R (MPa) $= 16.25 + 928.2/$DCPI	mm/blow	Herath et al. (2005)
M_R (MPa) $= 152/$(DCPI)$^{1.096}$	mm/blow	Mohammadi et al. (2008)

using a light DCP. Details of the DCP used are given in Table 3.4. The number of blows for 50 mm drive distance was directly proportional to the relative density. Tests performed before and after ground improvement work can be used to indicate the depth and relative degree of improvement and the uniformity of the work.

3.10 SUMMARY OF DCP

DCPs are a simple test for performing site investigations or to supplement other *in situ* tests or test borings. Table 3.13 presents a summary of reported application of DCP results.

Table 3.13 Summary of reported applications of DCP results

Application	Class of DCP	References
Stratigraphy	DPL	author
	DPSH	Kelley & Lutenegger (1999)
Liquefaction potential	DPSH	Ashfield et al. (2013)
		Talbot (2017)
Compaction control	DPL	Lacroix & Horn (1973)
		Hanna et al. (1986)
		Acar et al. (1991)
		Ford & Eliason (1993)
		Gabr et al. (2001)
		Karunaprema & Edirisinghe (2002)
		Ampadu & Arthur (2006)
		Mousavi et al. (2018)
	DPM	Schmid (1975)
	DPH	Schmid (1975)
Footing inspection	DPL	Sowers & Hedges (1966)
Design of shallow foundations	DPL	Ooi & Ting (1975)
		Swann (1982)
		Formazin & Hausner (1985)
		Ampadu (2005)
	DPH	Kayalar (1988)
		Bohdan & Tomasz (2001)
	DPSH	Khanna et al. (1953)
Design of deep foundations	DPL	Engel et al. (1994)
		Nilsson & Cunha (2004)
		Silva & Miguel (2008)
	DPM	Van Leijden et al. (1982)

The author recommends that engineers give consideration to the use of both light DCPs and super heavy DCPs performed in the following manner:

1. For super heavy DCPs, use a 63.5 kg (140 lbs) automatic hammer, a drop height of 76 cm (30 in.), AW drill rods, a 63.5 mm (2.50 in.) diameter cone ((area = 31.7 cm²) with an apex angle of 60°, and a drive interval of 15 cm (6 in.).

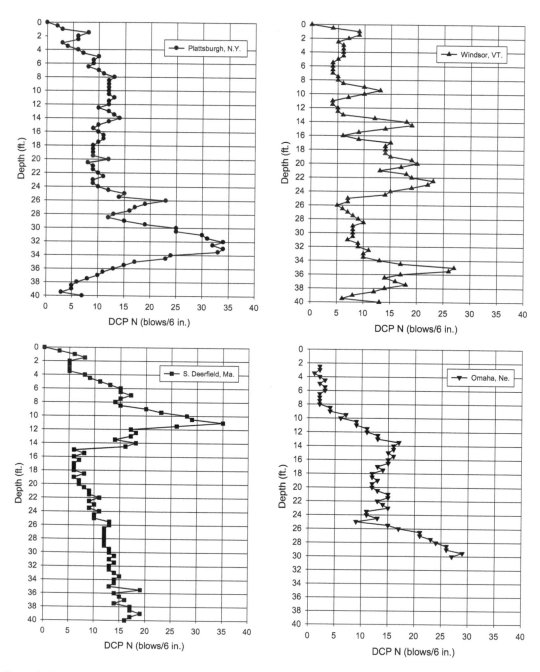

Figure 3.11 Super heavy DCP results obtained at four sites.

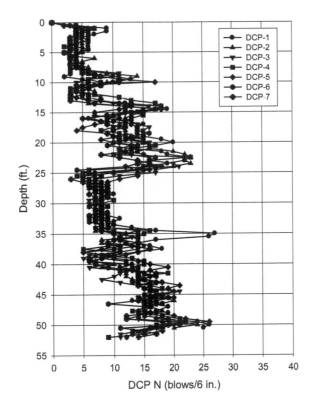

Figure 3.12 Variability of results from seven super heavy DCP profiles at a site.

2. For light DCPs, select a suitable hammer mass and drop height, and a suitable cone diameter and drive interval to match expected soil conditions. Use a cone with an apex angle of 60°;
3. Test results should be presented graphically as incremental and cumulative penetration resistance versus depth and as dynamic penetration resistance, r_d or q_d, versus depth.

Figure 3.11 shows the results of super heavy DCPs obtained by the author at four sites. Each profile was completed in about 25 min. To illustrate the reproducibility of results, the results from seven super heavy DCP profiles performed at a site in a medium dense sand by one operator using an automatic hammer and AW rods are shown in Figure 3.12. The variation in test results can essentially be attributed to natural soil variability.

REFERENCES

Abu-Farsakh, M.Y., Nazzal, M.D., Alshibli, K., and Seyman, E., 2005. Application of Dynamic Penetrometer in Pavement Construction Control. *Transportation Research Record No. 1913*, pp. 53–61.
Acar, Y.B., Puppala, A.J., and Seals, R.K., 1991. Calibration of a Dynamic Penetrometer for Compaction Quality Control of Boiler Slag. *Geotechnical Testing Journal, ASTM*, Vol. 14, No. 1, pp. 56–63.
Ampadu, S., 2005. A Correlation Between the Dynamic Cone Penetrometer and Bearing Capacity of a Local Soil Formation. *Proceedings of the 16th International Conference on Soil Mechanics and Geotechnical Engineering*, Vol. 2, pp. 355–358.

Ampadu, S.I.K and Arthur, T.D., 2006. The Dynamic Cone Penetrometer in Compaction Verification on a Model Road Pavement. *Geotechnical Testing Journal, STM,* Vol. 29, No. 1, pp. 70–79.

Ashfield, D., Ashby, G., Power, P., Fitch, N., Ladley, E., and Smith, T., 2013. Use of the Dynamic Cone Penetrometer to Assess the Liquefaction Potential of Christchurch Alluvial Soils. *Proceedings of the 19th New Zealand Geotechnical Symposium,* 8 pp.

Alshkane, Y.M., Rashed, K.A. and Daoud, H.S., 2020. Unconfined compression strength (UCS) and compressibility indices prediction from dynamic cone penetration index (DCP) for cohesive soils in Kurdistan Region/Iraq. Geotechnical and Geological Engineering, Vol. 38, pp. 3683–3695.

ASTM, 1970. Suggested Method for Dynamic Cone Soil Penetration Test. ASTM STP No. 479, pp. 62–63.

ASTM Standard D6951-03. Test Method for Use of the Dynamic Cone Penetrometer in Shallow Pavement Applications. Annual Book of ASTM Standards.

Ayers, M.E. and Thompson, M.R., 1989. Evaluation Techniques for Determining the Strength Characteristics of Ballast and Subgrade Materials. University of Illinois, Department of Civil Engineering Report.

Bohdan, Z. and Tomasz, G., 2001. Estimation Method of the Bearing Capacity of Shallow and Pile Foundations from Dynamic Probings. *Proceedings of the 15th International Conference on Soil Mechanics and Geotechnical Engineering,* Vol. 1, pp. 551–554.Bolomey, H., 1974. Dynamic Penetration Resistance Formulae. *Proceedings of the European Symposium on Penetration Testing,* Vol. 2.2, pp. 35–46.

Borowczyk, M. and Frankowski, Z.B., 1981. Dynamic and Static Sounding Results Interpretation. *Proceedings of the 10th International Conference on Soil Mechanics and Foundation Engineering,* Vol. 2, pp. 451–454.

Broms, B.B. and Flodin, N., 1988. History of Soil Penetration Testing. *Proceedings of the 1st International Symposium on Penetration Testing,* Vol. 1, pp. 157–220.

Butcher, A.P., McElmeel, K., and Powell, J.J.M., 1995. Dynamic Probing and its use in Clay Soils. *Proceedings of the International Conference on Advances in Site Investigation Practice,* pp. 383–395.

Cabrera, M. and Carcole, A., 2007. Relationship Between Standard Penetration Test (SPT) and Dynamic Penetration Super Heavy Test (DPSH) for Natural Sand Deposits. *Proceedings of the 14th European Conference on Soil Mechanics and Geotechnical Engineering,* Vol. 3, pp. 1691–1695.

Card, G.B. and Roche, D.P., 1989. The Use of Continuous Dynamic Probing in Ground Investigation. *Proceedings of the Conference on Penetration Testing in the U.K.,* pp. 119–122.

Cavusoglu, E., Nam, M., O'Neill, M., and McClelland, M., 2004. Multi-Method Strength Characterization for Soft Cretaceous Rocks in Texas. *Geo-Support 2004, ASCE,* pp. 199–210.

Cearns, P.J. and McKenzie, A., 1989. Application of Dynamic Cone Penetrometer Testing in East Anglia. Penetration Testing in the U.K., pp. 123–127.

Chan, S. and Chin, F., 1972. Engineering Characteristics of the Soils along the Federal Highway in Kuala Lumpur. *Proceedings of the 3rd Southeast Asian Conference on Soil Engineering,* pp. 41–45.

Chang, M.F. and Wong, I.H., 1986. Penetration Testing in the Residual Soils of Singapore. Proceedings of the Specialty Geomechanics Symposium on Interpretation of Field Testing for Design Parameters.

Chua, K.M., 1988. Determination of CBR and Elastic Modulus of Soils Using a Portable Pavement Dynamic Cone. *Proceedings of the 1st International Symposium on Penetration Testing,* Vol. 1, pp. 407–414.

Chua, K.M. and Lytton, R.L, 1992. Dynamic Analysis Using the Portable Pavement Dynamic Cone Penetrometer. Transportation Research Record No. 1192, pp. 27–38.

Coyle, H.M. and Bartoskewitz, R.E., 1980. Prediction of Shear Strength of Sand by Use of Dynamic Penetration Tests. Transportation Research Record No. 749, pp. 46–54.

Dahlberg, R. and Bergdahl, U., 1974. Investigation on the Swedish Ram Sounding Method. *Proceedings of the European Symposium on Penetration Testing,* Vol. 2.2, pp. 93–102.

DeGaridele-Thoron, R. and Javor, E., 1983. A Small Dynamic Penetrometer to Estimate CBR Value. *Proceedings of the International Symposium on In Situ Testing of Soil and Rock*, Vol. 2, pp. 43–47.

DeHenau, A., 1982. The Use of a Light Percussion Sounding Apparatus in Road Engineering. *Proceedings of the 2nd European Symposium on Penetration Testing*, Vol. 1, pp. 271–276.

Elton, D., 1989. Hand Held Dynamic Cone Correction Factors. *Symposium on Engineering Geology and Geotechnical Engineering*, pp. 39–44.

Engel, R., Riaz, M., Dalah, V., Hanhilammi, D., Husein, A.I., and Lang, R.Y., 1994. Dynamic Probing Test to Determine Driven Pile Capacity. *Proceedings of the International Conference Design and Construction of Deep Foundations*, Vol. 3, pp. 1321–1336.

Ergun, M.U., 1982. A Site Investigation Through Penetration Tests. *Proceedings of the 2nd European Symposium on Penetration Testing*, Vol. 1, pp. 257–262.

Fakher, A. and Khodaparasat, M., 2004. The Repeatability of the Mackintosh Probe Test. *Proceedings of the 2nd International Symposium on Site Characterization*, Vol. 1, pp. 325–330.

Fakher, A., Khodaparasat, M., and Jones, C.J.F.P., 2006. The use of the Mackintosh Probe for site investigation in soft soils. *Quarterly Journal of Engineering Geology and Hydrogeology*, Vol. 39, pp. 189–196.

Fatt, C. and Kee, C., 1972. Engineering Characteristics of the Soils Along the Federal Highway in Kuala Lumpur. *Proceedings of the 3rd Southeast Asian Conference on Soil Engineering*, pp. 41–45.

Fityus, S., 1998. Calibration of the Blunt Tipped Dynamic Penetrometer for Silica Sands. *Proceedings of the 1st International Conference on Site Characterization*, Vol. 2, pp. 927–932.

Ford, G.R. and Eliason, B.E., 1993. Comparison of Compaction Methods in Narrow Subsurface Drainage Trenches. *Paper Presented at the Transportation Research Board Annual Meeting*, Washington, D.C.

Formazin, J. and Hausner, H., 1985. Correlations Between Soil Parameters and Penetration Testing. *Proceedings of the 11th International Conference on Soil Mechanics and Foundation Engineering*, Vol. 2, pp. 459–463.

Gabr, M.A., Coonse, J. and Lambe, P.C., 2001. A Potential Model for Compaction Evaluation of Piedmont Soils Using Dynamic Cone Penetrometer (DCP). *Geotechnical Testing Journal, ASTM*, Vol. 24, No. 3, pp. 308–313.

Gadsby, J.W., 1971. Discussion of The Correlation of Cone Size in the Dynamic Cone Penetration Test with the Standard Penetration Test. *Geotechnique*, Vol. 21, No. 2, pp. 188–189.

Gawad, T., 1976. Standard Penetration Resistance in Cohesionless Soils. *Soils and Foundations*, Vol. 16, No. 4, pp. 47–60.

George, K. and Uddin, W., 2000. Subgrade Characterization for Highway Subgrade Design. Final Report FHWA/MS-DOT-RD-00-131, Mississippi Deportment of Transportation.

Goel, M.C., 1982. Various Types of Penetrometers, Their Correlations and Field Applicability. *Proceedings of the 2nd European Symposium on Penetration Testing*, Vol. 1, pp. 263–270.

Gudavalli, S., Vipulanandan, C., Puppala, A.J., Pala, N., Fang, X., Jao, M., Yin, S., and Wang, M.C., 2008. Development of Correlation Between TCP Blow Count and Undrained Shear Strength of Low Plasticity Clay. *Proceedings of the International Symposium on Geotechnical and Geophysical Site Characterization*, Vol. 2, pp. 1401–1404.

Hajduk, E., Shiner, B., and Meng, J., 2007. A Comparison of the Sowers Dynamic Cone Penetrometer Test with Cone Penetration and Flat Blade Dilatometer Testing. *Proceedings of GeoDenver, ASCE*.

Hamoudi, M.M., Coyle, H.M., and Bartoskewitz, R.E., 1974. Correlation of the Texas Highway Department Cone Penetrometer Test with Unconsolidated-Undrained Strength of Cohesive Soils. Research Report No. 10-1, TTI, Texas A&M University, 133 pp.

Hanna, A.W., Ambrosii, G., and McConnell, A.D., 1986. Investigation of a Coarse Alluvial Foundation for an Embankment Dam. *Canadian Geotechnical Journal*, Vol. 23, pp. 203–215.

Harrison, J.A., 1987. Correlation Between California Bearing Ratio and Dynamic Cone Penetration Strength Measurement of Soils. *Proceedings of the Institution of Civil Engineers*, Part 2, No. 83, Technical Note No. 463.

Herath, A., Mohammed, L., Gaspard, K., Gudishala, R., and Abu-Farsakh, M., 2005. The Use of Dynamic Cone Penetrometer to Predict Resilient Modulus of Subgrade Soils. *Geofrontiers Congress 2005, ASCE.*

Hossain, D. and Ali, K., 1990. Mackintosh vs. Vane Estimation of Undrained Shear Strength Correlation for a Sabkha Clay of Saudi Arabia. *Quarterly Journal of Engineering Geology*, Vol. 23, pp. 269–272.

Karunaprema, K. and Edirisinghe, A., 2002. A laboratory Study to Establish Some Useful Relationships for the Use of Dynamic Cone Penetrometer. *Electronic Journal of Geotechnical Engineering*, paper 2002-0228, 18 pp.

Kayalar, A.S., 1988. Statistical Evaluation of Dynamic Cone Penetration Test Data for Design of Shallow Foundations in Cohesionless Soils. *Proceedings of the 1st International Symposium on Penetration Testing*, Vol. 1, pp. 429–434.

Kelley, S.P. and Lutenegger, A.J., 1999. Enhanced Site Characterization in Residual Soils Using the SPT-T and Drive Cone Tests. *Behavioral Characteristics of Residual Soils. ASCE*, pp. 88–100.

Khanna, P.L., Varghese, P.C., and Hoon, R.C., 1953. Bearing Pressure and Penetration Tests on Typical Soil Strata in the Region of the Hirakud Dam Project. *Proceedings of the 3rd International Conference on Soil Mechanics and Foundation Engineering*, Vol. 1, pp. 246–252.

Khodaparasat, M., 2010. The Repeatability in Results of Mackintosh Probe Test in Soft Soils. *Proceedings of the International Conference on Geotechnical Engineering*, Lahore, pp. 237–243.

Khodaparasat, M., Rajabi, A.M., and Mohammadi, M., 2015. The New Empirical Formula Based on Dynamic Probing Test Results in Fine Cohesive Soils. *International Journal of Civil Engineering – Geotechnique*, Vol. 13, No. 2, pp. 105–113.

Kleyn, E.G., 1975. Use of the Dynamic Cone Penetrometer (DCP). Report No. 2/74, Transvaal Road Department, South Africa.

Kleyn, E.G. and Van Zyl, G.D., 1988. Application of Dynamic Cone Penetrometer (DCP) to Light Pavement Design. *Proceedings of the 1st International Symposium on Penetration Testing*, Vol. 1, pp. 435–444.

Kleyn, E.G., Maree, J.H., and Savage, P.F., 1982. The Application of a Portable Dynamic Cone Penetrometer to Determine In Situ Bearing Properties of Road Pavement Layers and Subgrades in South Africa. *Proceedings of the 2nd European Symposium on Penetration Testing*, Vol. 1, pp. 277–282.

Kong, T.B., 1983. In Situ Soil Testing at the Bekok Dam Site, Johor, Peninsular Malaysia. *Proceedings of the International Symposium on In Situ Testing*, Vol. 2, pp. 403–409.

Lacroix, Y. and Horn, H.M., 1973. Direct Determination and Indirect Evaluation of Relative Density and It's Use on Earthwork Construction Projects. ASTM Special Technical Publication 523, pp. 251–280.

Lawson, W.D., Terrel, E.O., Surles, J.G., Moghaddam, R.B., Seo, H., and Payawickrama, P.W., 2018. Side-by-Side Correlation of Texas Cone Penetration and Standard Penetration Test Blowcount Values. *Geotechnical & Geological Engineering*, Vol. 36, No. 5, pp. 2769–2787.

Livneh, M., 1987a. The Use of Dynamic Cone Penetrometer for Determining the Strength of Existing Pavements and Subgrades. *Proceedings of the 9th Southeast Asian Geotechnical Conference*, pp. 9-1–9-10.

Livneh, M., 1987b. The Correlation Between Dynamic Cone Penetrometer Values (DCP) and CBR Values. Transportation Research Institute, Technion-Israel Institute of Technology, Publication No. 87-303.

Livneh, M. and Ishai, I., 1988. The Relationship Between In-Situ CBR Test and Various Penetration Tests. *Proceedings of the 1st International Symposium on Penetration Testing*, Vol. 1, pp. 445–452.

MacRobert, C., 2017. Interpreting DPSH Penetration Values in Sand Soil. *Journal of the South African Institute of Civil Engineering*, Vol. 59, No. 3, pp. 11–15.

MacRobert, C.J., Bernstein, G.S. and Nchabeleng, M.M., 2019. Dynamic Cone Penetrometer (DCP) Relative Density Correlations for Sand. *Soils and Rocks, Sao Paulo*, Vol. 43, No. 2, pp. 201-207.

Maghaddam, R., Lawson, W., Surles, J., Seo, H., and Jayawickrama, P., 2017. Hammer Efficiency and Correction Factors for the TxDOT Texas Cone Penetration Test. *Geotechnical and Geological Engineering*, Vol. 35, pp. 2147–2162.

McGrath, P.G., Motherway, F.K., and Quinn, W.J., 1989. Development of Dynamic Cone Penetration Testing in Ireland. *Proceedings of the 12th International Conference on Soil Mechanics and Foundation Engineering*, Vol. 1, pp. 271–275.

Meardi, G., 1971. Discussion of The Correlation of Cone Size in the Dynamic Cone Penetration Test with the Standard Penetration Test. *Geotechnique*, Vol. 21, No. 2, pp. 184–188.

Melzer, K.J. and Smoltczyk, U., 1982. Dynamic Penetration Testing: State-of-the-Art Paper. *Proceedings of the 2nd European Symposium on Penetration Testing*, Vol. 1, pp. 191–202.

Meyerhof, G.G., 1956. Penetration Tests and Bearing Capacity of Cohesionless Soils. *Journal of the Soil Mechanics Division, ASCE*, Vol. 82, No. SM1, pp. 1–12.

Misra, A., Upadhyaya, S., Horn, C., Kondagari, S., and Gustin, F., 2006. CBR and DCP Correlation for Class C Fly-Ash Stabilized Soil. *Geotechnical Testing Journal, ASTM*, Vol. 29, No. 1, pp. 1–7.

Mohammadi, S., Nikoudel, M., Rahimi, H., and Khamehchiyan, M., 2008. Application of the Dynamic Cone Penetrometer (DCP) for Determination of the Engineering Parameters of Sandy Soils. *Engineering Geology*, Vol. 101, pp. 195–203.

Mohan, D., Aggarwal, V.S., and Tolia, D.S., 1970. The Correlation of Cone Size in the Dynamic Cone Penetration Test with the Standard Penetration Test. *Geotechnique*, Vol. 20, No. 3, pp. 315–319.

Mohan, D., Aggarwal, V.S., and Tolia, D.S., 1971. Closure to The Correlation of Cone Size in the Dynamic Cone Penetration Test with The Standard Penetration Test. *Geotechnique*, Vol. 21, No. 2, pp. 189–190.

Mousavi, S., Gabr, M., and Borden, R., 2018. Correlation of Dynamic Cone Penetrometer Index to Proof Roller Test to Assess Subgrade Soils Stabilization Criterion. *International Journal of Geotechnical Engineering*, Vol. 12, No. 3, pp. 284–92.

Muromachi, T. and Kobayashi, S., 1982. Comparative Study of Static and Dynamic Penetration Tests Currently in Use in Japan. *Proceedings of the 2nd European Symposium on Penetration Testing*, Vol. 1, pp. 297–302.

Nam, M.S. and Vipulanandan, C., 2010. Relationship Between Texas Cone Penetrometer Tests and Axial Resistances of Drilled Shafts Socketed in Clay Shale and Limestone. *Journal of Geotechnical and Geoenvironmental Engineering, ASCE*, Vol. 136, No. 8, pp. 1161–1165.

Nilsson, T. and Cunha, R., 2004. Advantages and Equations for Pile Design in Brazil via DPL Tests. *Proceedings of the 2nd International Symposium on Site Characterization*, Vol. 2, pp. 1519–123.

Nixon, I.K., 1989. Introduction to Papers 10–13. *Proceedings of the Conference on Penetration Testing in the U.K.* pp. 105–110.

Ooi, T.A. and Ting, W.H., 1975. The Use of a Light Dynamic Cone Penetrometer in Malaysia. *Proceedings of the 4th Southeast Asian Conference on Soil Engineering*, pp. 3-62–3-79.

Opuni, K., Nyako, S., Ofosu, B., Mensah, F., and Sarpong, K., 2017. Correlations of SPT and DCPT Data for Sandy Soils in Ghana. *Lowland Technology International*, Vol. 19, No. 2, pp. 145–150.

Palmer, D.J., 1957. Discussion. *Proceedings of the 4th International Conference on Soil Mechanics and Foundation Engineering*, Vol. 3, p. 125.

Palmer, D.J. and Stuart, J.G., 1957. Some Observations on the Standard Penetration Test and a Correlation of the Test with a New Penetrometer. *Proceedings of the 4th International Conference on Soil Mechanics and Foundation Engineering*, Vol. 1, pp. 231–236.

Paunescu, M. and Gruia, A., 1982. Some Aspects Concerning the Study of Foundation Soils, Using Cone Penetration Tests. *Proceedings of the 2nd European Symposium on Penetration Testing*, Vol. 1, pp. 317–322.

Rahim, A. and George, K., 2004. Dynamic Cone Penetrometer to Estimate Subgrade Resilient Modulus for Low Volume Roads Design. *Proceedings of the 2nd International Symposium on Site Characterization*, Vol. 1, pp. 367–371.

Rao, B.G., Narahari, D.R., and Balodhi, G.R., 1982. Dynamic Cone Probing Tests in Gravelly Soils. *Proceedings of the 2nd European Symposium on Penetration Testing*, Vol. 1, pp. 337–344.

Robinson, L., 1988. Dynamic CPT for Test Pit Field Investigations: Experiences with Sowers' Cone Penetrometer. *Proceedings of the 24th Symposium on Engineering Geology and Soils Engineering*, pp. 31–39.

Rodin, S., 1961. Experiences with Penetrometers with Particular Reference to the Standard Penetration Test. *Proceedings of the 4th International Conference on Soil Mechanics and Foundation Engineering*, Vol. 1, pp. 517–521.

Rostron, J.P., Schwartz, A.E., and Gioiosa, T.E., 1969. A Drop Hammer Penetrometer for Determining the Density of Soils and Granular Materials. Highway Research Record No. 284, pp. 70–73.

Sabtan, A.A. and Sherbata, W.M., 1994. Mackintosh Probe as an Exploration Tool. *Bulletin of the International Association of Engineering Geologists*, No. 50, pp. 89–94.

Scala, A.J., 1956. Simple Methods of Flexible Pavement Design Using Cone Penetrometers. *New Zealand Engineer*, Vol. 11, No. 2, pp. 34–44.

Scarff, R.D., 1989. Factors Governing the Use of Continuous Dynamic Probing in U.K. Ground Investigation. *Proceedings of the Conference on Penetration Testing in the U.K.*, pp. 129–132.

Schmid, W.E., 1975. Evaluation of Vibratory Compaction Field Tests. *In Situ Measurement of Soil Properties, ASCE*, Vol. 1, pp. 373–394.

Seed, H.B. and Idriss, I.M., 1981. Evaluation of Liquefaction Potential of Sand Deposits Based on Observations of Performance in Previous Earthquakes. *In Situ Testing to Evaluate Liquefaction Susceptibility*, ASCE Preprint 81–544, 21 pp.

Silva, D. and Miguel, M., 2008. Estimation of Pile Bearing Capacity Using Dynamic Probing in Tropical Soils. *Proceedings of the Symposium on Geotechnical and Geophysical Site Characterization*, Vol. 1, pp. 523–527.

Singh, A. and Sharma D., 1973. Development of a Light Weight Dynamic Penetrometer. *Journal of the Indian Geotechnical Society*, Vol. 3, No. 1, pp. 182–188.

Smith, R.B., 1983. In Situ CBR and Dynamic Cone Penetrometer Testing. *Proceedings of the International Symposium on In Situ Testing of Soil and Rock*, Vol. 2, pp. 149–154.

Smith, R.B., 1988. Cone Penetrometer and In Situ CBR Testing of an Active Clay. *Proceedings of the 1st International Symposium on Penetration Testing*, Vol. 1, pp. 459–465.

Smith, R.B. and Pratt, D.N., 1983. A Field Study of In-Situ California Bearing Ratio and Dynamic Penetrometer Testing for Road Subgrade Investigation. *Australian Road Research Board*, Vol. 13, No. 4, pp. 285–294.

Solymar, Z.V., 1984. Compaction of Alluvial Sands by Deep Blasting. *Canadian Geotechnical Journal*, Vol. 21, No. 2, pp. 305–321.

Sowers, G.F. and Hedges, C.S., 1966. Dynamic Cone for Shallow In-Situ Penetration Testing. ASTM Special Technical Publication 399, pp. 29–37.

Stefanoff, G., Sanglerat, G., Bergdahl, U., and Melzer, K.J., 1988. Dynamic Probing (DP): International Reference Test Procedure. *Proceedings of the 1st International Symposium on Penetration Testing*, Vol. 1, pp. 53–70.

Sugiyama, T., Muraishi, H., Noguchi, T., Okada, K., and Kakio, T., 1998. Spatial Distribution Characteristics of Soil Strength in Embankment Estimated by Portable Dynamic Cone Penetration Test. *Proceedings of the International Conference on Geotechnical Site Characterization*, Vol. 2, pp. 953–958.

Swann, L.H., 1982. The Use of Dynamic Soundings on Evaluating Settlements. *Proceedings of the 2nd European Symposium on Penetration Testing*, Vol. 1, pp. 345–349.

Talbot, M., 2017. Dynamic Cone penetration Tests for Liquefaction Evaluation of Gravelly Soils. Report ST-2017-7102, U.S. Bureau of Reclamation.

Teferra, A., 1983. Estimation of the Angle of Internal Friction of Non-Cohesive Soils from Sounding Tests. *Indian Geotechnical Journal*, Vol. 13, No. 4, pp. 211–221.

Terzaghi, K. and Peck, R.B., 1948. Soil Mechanics in Engineering Practice. John Wiley and Sons, New York.

Tonks, D.M. and Whyte, I.L., 1989. Dynamic Soundings in Site Investigations: Some Observations and Correlations. *Proceedings of the Conference on Penetration Testing in the U.K.*, pp. 113–117.

Touma, F. and Reese, L., 1972. The Behavior of Axially Loaded Drilled Shafts in Sand. Report No. CFHR 3-5-72-176-1. Center for Highway Research, University of Texas.

Triggs, J.K. and Liang, R.Y.K., 1988. Development of and Experiences from a Light-Weight, Portable Penetrometer Able to Combine Dynamic and Static Cone Tests. *Proceedings of the 1st International Symposium on Penetration Testing*, Vol. 1, pp. 467–473.

Van Leijden, W., Pachen, H.M.A., and VadenBerg, J., 1982. Dynamic Probing Research for Pile Driving Predictions in the Netherlands. *Proceedings of the 2nd European Symposium on Penetration Testing*, Vol. 1, pp. 285–290.

Vipulanandan, C., Puppala, A.J., Jao, M., Kim, M.S., Vasundevan, H., Kumar, P., and Mo, Y.L., 2008. *Correlation of Texas Cone Penetrometer Test Values and Shear Strength of Texas Soils.* University of Houston, Houston, TX.

Waschkowski, E., 1982. Dynamic Probing and Practice. *Proceedings of the 2nd European Symposium on Penetration Testing*, Vol. 1, pp. 357–362.

Webster, S.L., Grau, R.H., and Williams, T.P., 1992. Description and Application of Dual Mass Dynamic Cone Penetrometer. Instruction Report GL-92-3, US Army Engineers Waterways Experiment Station.

Chapter 4

Cone Penetration (CPT) and Piezocone (CPTU) Tests

4.1 INTRODUCTION

Cone Penetration Tests have been in use in many parts of the world for over 70 years for determining site stratigraphy, evaluating strength characteristics and other soil properties, designing foundations, and a wide array of other applications. Cone tests are highly versatile and possess many of the attributes desirable in an *in situ* test previously described in Chapter 1. The test represents a simple concept and can be relatively simple to perform. Cone tests can be used in a wide range of soil types and have a wide range of applications. In this chapter, the use and various aspects of the cone penetrometer (CPT) and piezocone (CPTU) are discussed. The focus in this chapter is on electric cones, and mechanical cones are only briefly addressed.

A large volume of work has been published about CPTs, cone testing, data interpretation, and cone design applications. Two European cone testing symposia were held in 1974 and 1982. In addition, since 1988, the International Symposia on Penetration Testing (now Geotechnical and Geophysical Site Characterization) have been devoted in part to CPT and CPTU testing. Since 1995, several specialty international symposia on cone penetration testing have been held. Also at least three books have been prepared on cone testing (Sanglerat 1972; Meigh 1987; Lunne et al. 1997). An historical perspective of cone penetration testing was presented by Massarsch (2014). Since the CPT/CPTU provides a near continuous record of the stratigraphy, it is useful for rapid and more complete evaluation of detailed soil layering that can often be missed during conventional test drilling and sampling.

In general, the empirical interpretation of results from the CPT/CPTU has been at an advanced state of maturity for the last 10–15 years in the author's opinion. New publications often provide some additional data sets to an existing correlation between measured values and soil parameters, but correlations do not change significantly.

4.2 MECHANICS OF THE TEST – CPT/CPTU

The CPT is an intrusive, full displacement cylindrical probe, usually machined from stainless steel, with a diameter of about 35.7 mm (1.405 in.) that is attached to either conventional drill rods or special CPT rods and pushed from the ground surface, with or without a borehole. The cone has a tip apex angle of 60° and is advanced at a rate of 2 cm/s (about 15 s/ft) using the static thrust provided by the hydraulics of a conventional drill rig or special hydraulic pushing rig. This concept is shown schematically in Figure 4.1. During the advance, forces ore pressures acting on the cone tip are measured.

The dimensions of the cone and recommended test procedures are described in detail by ASTM Test Method D3441 *Standard Test Method for Deep Quasi-Static, Cone and*

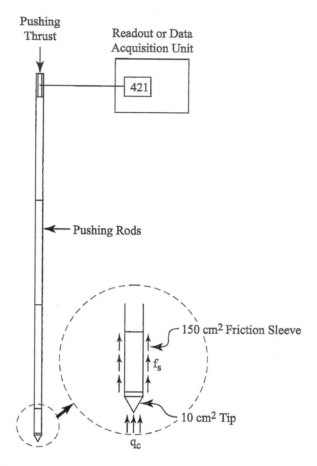

Figure 4.1 Principle of cone penetrometer testing.

Friction-Cone Penetration Tests in Soil, and ASTM D5778 *Standard Test Method for Electronic Friction Cone and Piezocone Testing of Soils.* The CPT and CPTU are also standardized by the International Organization for Standardization in ISO 22476-1 Geotechnical Investigation and Testing-Field Testing-Part 1: Electrical Cone and Piezocone Penetration Test.

4.2.1 Mechanical Cones

Early static CPTs were developed around simple mechanical systems consisting of a cone, push rods, and an external load cell. The "Dutch" CPT was apparently first used around 1930 to determine the thickness and bearing capacity of hydraulic fill in the Netherlands. The cone had an end area of $10\,cm^2$ and an apex angle of $60°$ and was pushed by one or two men. This limited the depth of exploration to a maximum of about 3 m (10 ft). Cone resistance was measured with a pressure gauge or mechanical load cell at the ground surface. Most early cones only measured the cone tip resistance and used a double rod system to first advance the cone by pushing on outer rods and then pushing on only the cone using a set of inner rods.

A modification to mechanical cones was made by Begemann (1953) who introduced the idea of attaching a sleeve behind the cone tip to evaluate the local friction. Although Begemann (1953) actually referred to this design as the "adhesion jacket-cone", the term "friction cone" is more commonly used. The surface area of Begemann's sleeve was $150\,cm^2$.

Again, a double rod system is required, with the cone first being advanced to the test depth by pushing on the outer rods. At the test depth, only the inner rod is advanced about 40 mm (1.5 in.) to measure the tip resistance. After the 40 mm (1.5 in.) push, the inner rod is continued to be pushed to engage the friction sleeve. An additional push of about 40 mm (1.5 in.) gives both the tip and sleeve resistance. The sleeve resistance is then obtained by subtraction. Figure 4.2 shows the sequence of advancing a mechanical cone with a friction sleeve.

Mechanical cones have a number of drawbacks, which makes their use somewhat impractical. The double rod system is cumbersome and usually must be fabricated as a special set of rods; i.e., conventional drill rods cannot usually be used. Because of the design and construction, soil particles may enter or adhere to some of the sliding components, and the cone may become jammed. There may also be frictional losses in the double rod system. The mantle cone and friction jacket cone can only be used to provide test results at intervals of about 0.15 m (6 in.) and therefore provide discontinuous data rather than a continuous profile. It is sometimes difficult for the operator to accurately read the load cell, especially in highly stratified soils or where the cone travels through alternating soft and stiff layers. In situations where the load may change dramatically over relatively short distances, some individual layering may be missed.

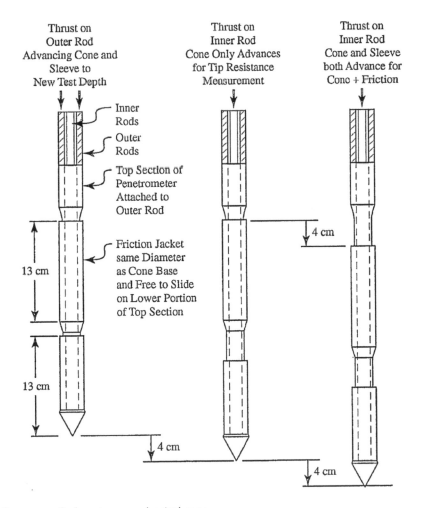

Figure 4.2 Sequence of advancing a mechanical cone.

Because of these difficulties and other problems, the results obtained with mechanical cones tend to be less reliable than those from electric cones. Additionally, there is considerable evidence which shows that the results between mechanical and electric cones are not the same. With available electronics and low cost/simple electric cones, electric cones have essentially replaced mechanical cones.

4.2.2 Electric Cones

Electric cones have been in more common use since about the 1970s and are designed with load cells located within the cone body to measure the tip and sleeve force. The load cells are usually equipped with strain gages. Advanced technology using electronic components to measure the load and transmit data to the surface have been introduced. Even cones with a down-hole memory module have been used, eliminating the need for a cable. A section through a typical electric CPT cone body is shown in Figure 4.3. Most electric cones for routine work have a tip area of $10 \, cm^2$, and the friction sleeve area is $150 \, cm^2$ although other sizes are available.

A computer is used to record the data automatically as the test proceeds. The results are displayed in real time as the cone is advanced so that the operator has an immediate indication of soil conditions. Normally, the data are presented in terms of unit tip and sleeve resistance versus depth. These two measurements may be used to give an indication of site stratigraphy, to estimate soil properties, and for direct foundation design.

Electric cones have a number of clear advantages over mechanical cones including the following: the ability to use almost any rod system, since only a single push rod is required;

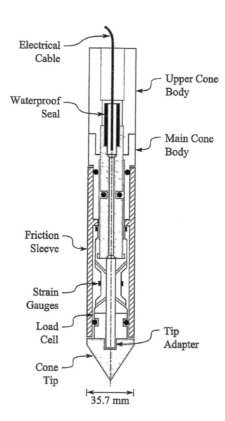

Figure 4.3 Section through electric cone penetrometer.

the data may be automatically recorded for more easier and more rapid reporting; the testing interval is closer so that there is more enhanced delineation of stratigraphy; and the results are usually more reliable since the test is essentially operator-independent, i.e., the results are usually obtained automatically. The design of all electric cones is not the same. For the most part, electric cones can be divided into two types: (1) those with the tip and sleeve load cells that are designed to be totally independent and (2) those with the tip and sleeve load cells more-or-less in series. The latter design is often called a "subtraction" cone.

The friction sleeve must be free to move in order to provide an accurate response. If sufficient clearance is not provided at the ends of the sleeve in the design, there may be an error in either the tip or sleeve resistance or both. Also, there must be a sufficient gap between the cone tip and friction sleeve so that tip force transfer is not impeded. Normally, rubber "O" rings are used to keep water out of the load cells and often some type of soil seal is used at the ends of the friction sleeve to keep soil from entering the area between the sleeve and tip or sleeve and body.

Some cones are also equipped with an internal inclinometer to monitor the deviation of the cone from vertical. This measurement is not necessarily used to provide any correction to the test data but is used more to provide advance warning of a problem. Van de Graaf & Jenkel (1982) discussed correcting the CPT depth using results from an internal inclinometer. They illustrated that depth errors of as much as 1.2 m can occur in a CPT sounding of 30 m. Typically, if the cone deviates more than about 5° from vertical, the probability of damage or loss of the cones starts to become high. A sharp deviation may mean that the cone has encountered an obstruction such as a cobble or random uncontrolled fill, while a gradual deviation may mean that the rods were simply not vertical at the beginning of the test or the pushing is not vertical.

For several years, the author has used a simple electric cone designed with a single high-capacity load cell to measure only the tip resistance in very dense and coarse granular soils and in other situations where the potential for damage to the cone or even cone loss is high. A photo of the cone is shown in Figure 4.4. The cone body and load cell are fabricated from a single piece of stainless steel, and a protective sleeve is used to prevent damage to the strain gages. These cones are very inexpensive to fabricate since there are only three parts: body, sleeve, and tip. A similar design has been described by Treen et al. (1992). Other cone designs using different tip and sleeve areas have also been used for special testing.

Figure 4.4 High capacity tip-only electric cone.

4.2.3 Electric Piezocone

The CPTU (sometimes denoted as PCPT or CPTu) is an electric cone that is built essentially the same as a standard electric friction cone except that it also has a pressure transducer mounted inside the cone body to measure soil pore water pressure as the cone advances. The measurement of pore water pressure provides an additional means to characterize the subsurface soil conditions. An alternative design for a CPTU is to eliminate the friction sleeve and measure only pore water pressure and tip resistance.

According to Vlasblom (1985), the first piezocone was constructed by the Delft Soil Mechanics Laboratory in 1962 although the first measurements of pore water pressure during cone penetration were reported by Janbu & Sennesset (1974). A year later, two variations of a pore pressure probe were introduced simultaneously by Torstenson (1975) and Wissa et al. (1975). Probes described by both Torstenson (1975) and Wissa et al. (1975) measured only pore water pressure although the filter element location was different. A standard test method that describes the equipment and procedures for conducting piezocone tests is given in ASTM Test Method D5778.

In order to measure pore water pressures, a porous element is mounted on the cone body and is connected to an internal pressure transducer. In saturated soils, since the cone must force soil out of the way in order to be advanced, there is usually a tendency to generate pore water pressures adjacent to the cone. The pressures may be sensed with the transducer. In order to do this, the porous element must be deaired and saturated with a fluid to communicate the pressure from the soil to the transducer. Typically, water, silicone oil, or glycerin are used as the fluid. Piezocone filter elements for measuring pore water pressure are typically made from porous plastic (HDPE), sintered stainless steel, sintered bronze, or porous ceramic. In general, it has been shown that the filter material has little or no influence on the measured pore water pressure (e.g., Jacobs & Coutts 1992).

There are a number of possible locations for the porous element on a CPTU. As shown in Figure 4.5, the element may be located at the cone tip (u_1), along the cone face (u_1), behind the cone base (u_2), or along the cone shaft (u_3). The first three positions are the most commonly used. There are both advantages and disadvantages of having the pore pressure element at different locations.

Tip elements generally provide the most reliable measurements and almost always give positive pore water pressures. A tip element also appears to give the most detailed stratigraphic results. However, the element tends to clog more easily than other locations and can

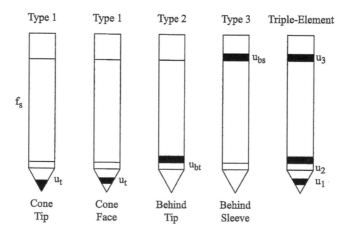

Figure 4.5 Different locations for pore pressure measurement in CPTU.

be easily damaged. Elements along the face of the cone also give good results but require special machining for the cone tip to be made in two parts. The filter element is also sometimes difficult to fabricate. In order to correct the tip resistance for pore pressure effects, a filter element is needed at the cone base (u_2 – Type 2). This position is less sensitive to fine changes in stratigraphy and can give both positive and negative pore water pressure values. However, the filter is less prone to clogging and is easy to fabricate.

Since a number of useful correlations exist that make use of the pore water pressure measurement, it is essential that the CPTU provide reliable pore water pressures during the test. This requires careful deairing and calibration as well as careful handling in the field. The filter element of the CPTU is usually deaired in a vacuum chamber after the element and cone tip are assembled onto the cone. The same chamber may be used to calibrate the cone, provided that some mechanism is available for holding the cone in the chamber while the chamber is pressurized. Saturation fluid (usually water, glycerin, or silicone oil) is kept at a level above the filter element and a vacuum is applied to remove the air. After deairing, the cone is kept in the chamber or transformed to another container and kept immersed in fluid to maintain saturation. In some cases, a rubber membrane or fluid filled plastic bag is placed over the cone for handling. It is imperative that the filter element and the system remain saturated during handling.

In many cases, a pilot hole will be drilled or punched with a dummy cone to the water table and then backfilled with water to the ground surface. The cone is then lowered through the water column without allowing the filter element to pass through air. In situations where the CPTU will be advanced through an unsaturated zone, such as near the ground surface, it is preferable to use a more viscous fluid and one with a higher surface tension, such as glycerin or silicone oil, to prevent loss of saturation of the pore water pressure system through cavitation.

The CPTU is able to help identify soil stratigraphy in more detail than the CPT through the use of the measured pore water pressure simply because different soils produce different pore pressure responses when penetrated by the cone. In addition to providing information on the stratigraphy to supplement the tip and sleeve measurements, the CPTU can also help with environmental site investigations in at least two more areas of interests: (1) identification of the depth to the water table and (2) estimation of the saturated hydraulic conductivity of the soil.

There are a number of factors that may influence the actual pore water pressure measurement. These include the exact location of the filter element, the thickness of the filter element, the stiffness of the fluid system, the reliability of the pressure transducer, and the deairing of the system. Figure 4.6 shows a typical deairing chamber used to saturate the porous element.

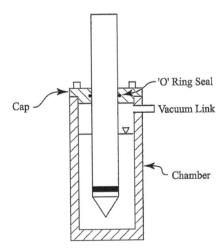

Figure 4.6 Deairing chamber used to saturate the porous element of CPTU.

4.3 DEPLOYING CONE PENETROMETERS

The deployment of the cone and other environmental *in situ* equipment represents a problem in practicality in that the various instruments must be installed in order to be of any value. There are a number of different methods for deploying the various instruments described, and the final choice of which system to use will depend on a number of factors, including depth of testing required, site geology, site mobility, equipment availability, budget, time constraints, personnel availability, etc. Figure 4.7 illustrates different common methods of deploying the CPT/CPTU.

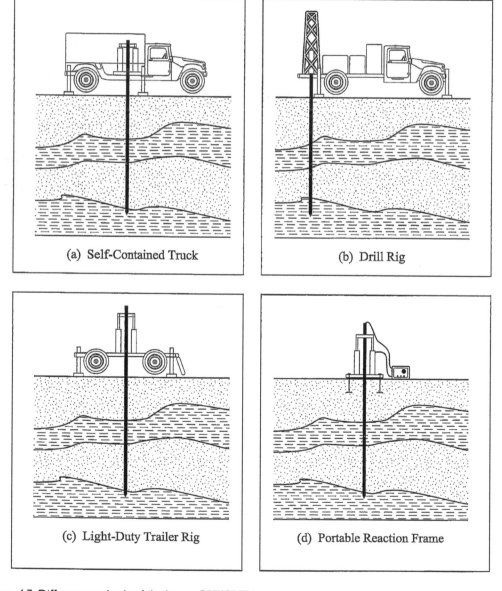

<table>
<tr><td>(a) Self-Contained Truck</td><td>(b) Drill Rig</td></tr>
<tr><td>(c) Light-Duty Trailer Rig</td><td>(d) Portable Reaction Frame</td></tr>
</table>

Figure 4.7 Different methods of deploying CPT/CPTU.

4.3.1 Self-Contained Truck

A number of specialty contracting firms have outfitted large trucks equipped with the necessary equipment and tools for deploying cones and other *in situ* tests. The box is usually weatherproof and can be heated or air conditioned so that testing can proceed in almost any weather conditions day or night. Pushing thrust is limited to the full weight of the truck since pushing usually is through the center. The data acquisition system is also contained in the truck. Cone rods typically of 1 m (3.28 ft) length are gripped with a hydraulic clamp on the outside of the rod. This means that as one rod is advanced, another rod may be threaded on at the top, speeding up the sounding. A thick-walled rod guide tube is usually extended out the bottom of the truck and provides lateral support to the rods to prevent buckling when pushing in very stiff materials.

4.3.2 Drill Rig

A conventional drill rig can be used to advance cones by using the hydraulics off the back of the rig. Because pushing is from the back of the rig, only about half of the rig weight is available, and therefore, the thrust is severely limited. One way to overcome this problem is to push as far as possible, pull out the tool, advance a hole to the test depth, using hollow stem or continuous flight augers, and then reinsert the tool to advance further. Another possibility is to install a section of auger to act as an anchor and attach a chain to the auger and to the truck. Sometimes, the unsupported rod length out of the ground allows rod buckling. This can be remedied by installing a temporary rod guide on the deck of the drill rig, near the base of the mast. Pushing of the rods must take place from a rigid location, usually on the head of the drill rig.

4.3.3 Light-Duty Trailer

A small light-duty pushing frame mounted on a trailer or self-propelled vehicle can be used in areas which do not require much thrust. Even then, the rig will most likely require some type of anchoring to keep the rig stable and provide sufficient thrust. These rigs have the advantage of low cost, low cost of mobilization, and ease of operation.

4.3.4 Portable Reaction Frame

In special cases, where mobility is limited or where the soil conditions are extremely soft, a small portable hydraulic frame can be used in conjunction with either a 12V battery powered hydraulic system or a gasoline engine powered system. Clearly, the reaction and depth of testing may have limitations for many projects, however, the author has successfully used this type of set up to advance up to 24 m (80 ft) in soft clay.

The depth of penetration for CPTs and CPTUs can be significantly increased if the frictional resistance between the cone rods and the soil above the cone can be eliminated or reduced. A friction reducer is often used for this purpose and typically consists of an enlarged section behind the cone such as a ring welded onto the first push rod.

4.4 TEST PROCEDURES

The CPT and CPTU are standardized by ASTM and ISO. Many of the recommendations in these standards relate to recording the test data and standardization of the cone geometry (i.e., apex angle, tip and sleeve area, etc.). The rate of advance is set at 20 mm/s (0.8 in./s), and because there are very few external components required to perform the test, the procedure is relatively simple.

In addition to recording the cone tip resistance and sleeve friction, a measurement of depth is also needed. There are a number of electro-mechanical systems that may be used to measure depth, many of which use a rotary potentiometer attached to the cone rods and some reference point. For advanced systems, such as self-contained cone trucks, an electrical system is typically used. Whatever technique is used, it is important that a permanent, stable reference point be used.

4.5 FACTORS AFFECTING TEST RESULTS

Even though the test equipment and procedure are generally specified, there are a number of factors that can influence the test results obtained from the CPT. Unlike the SPT, however, the CPT does not suffer from as many problems or uncertainties, simply because of the configuration of the test and the manner in which data are collected. In the following sections, a brief discussion is given on a number of the different variables that may affect CPT/CPTU results. In some cases, the discussion is intentionally brief since it is assumed that most CPT work will be conducted using a $10 \, cm^2$ ($1.55 \, in.^2$) cone with a tip apex angle of $60°$.

Errors can also occur in the use of electric cones and primarily relate to either calibration error or zero load error as pointed out by DeRuiter (1982). The load cells need to be calibrated frequently to check for errors or calibration changes and readings with zero load need to be obtained before and after each sounding to check for any mechanical or electrical problems.

4.5.1 Cone Design

Cone design may affect test results simply because of the way that the load is developed and transferred on the various parts of the cone. As previously indicated, mechanical and electrical cones do not always give the same results at the same site. Even within a given type of cone (e.g., electric), the test results may not be the same. It would be expected that in sands, different electrical cones would give similar results but in clays, where excess pore water pressures generate during penetration may actually reduce the measured tip resistance, the design of different electric cones may be more important. Additionally, subtraction cones may not always give the same results as cones with independent load cells.

4.5.2 Cone Diameter

The diameter of the CPT cone tip is specified to be $35.7 \pm 0.4 \, mm$ ($1.406 \pm 0.016 \, in.$) to give a projected end area of $10 \, cm^2$ ($1.55 \, in^2$). However, ASTM D3441 provides a note indicating that "cone tips with larger end areas may be used to increase measurement sensitivity in weak soils". The use of different size cones (in the range of $5-20 \, cm^2$ is allowed by ASTM "provided the cone tip and friction sleeve (if any) area is noted". The use of different size cones in the same soils has been considered by a number of investigators and is sometimes referred to as the "scale effect". The bulk of available data in both sands and clays suggests that there is generally no significant difference in cone tip resistance with cone tip areas ranging from 2.5 to $20 \, cm^2$. However, DeLima & Tumay (1991) found that using a $1.27 \, cm^2$ cone in comparison with a 10 and $15 \, cm^2$ cone at several sites did produce a scale effect especially for the measured sleeve resistance and hence friction ratio. It also appeared that the scale effect increased with an increasing cone resistance.

It seems logical to expect a difference in cone tip resistance with different size cones in sands and perhaps for the difference to be more pronounced as the grain-size increases simply because of size scaling of the cone relative to the size of the soil grains. Similar observations have been noted for model/prototype/full-scale piles in sand. At the present time, there are no real data to support this, however, and it is likely that within the range of cone sizes typically used, it may be difficult to distinguish clear trends in most natural sands.

4.5.3 Rate of Penetration

Variations in the rate of cone penetration can also produce variations in both tip resistance and local friction. Most investigations indicate that lower cone resistance and lower local friction is obtained at lower penetration rates in both sand and clay (e.g., Kok 1974; Dayal & Allen 1975; Tekamp 1982; Powell & Quarterman 1988; Chung et al. 2006; Kim et al. 2008; Lehane et al. 2009; Oliveira et al. 2011). However, Campanella & Robertson (1981) noted that the rate of penetration had only a minor influence on tip resistance but significantly affected sleeve friction in a clayey silt. They showed that sleeve friction doubled when the penetration rate was reduced by one order of magnitude.

ASTM requires a penetration rate of 1–2 cm/s (2–4 ft/min) ± 2.5%, which actually gives a wide range of allowable advance rates. It is also noted by ASTM that "Rates of penetration either slower or faster than the standard rate may be used for special circumstances, such as pore pressure measurements". In clays, the rate of penetration may be more important, too slow a rate producing some dissipation of pore water pressure, while too fast a rate producing a noticeable increase resulting from increased shear strength. The two may actually offset each other, for different reasons, so that provided the test is conducted within the specified rate, there appear to be no significant effects on the results.

4.5.4 Surface Roughness of Friction Sleeve

The surface roughness of the friction sleeve may influence the measured sleeve resistance in some soils as pointed out by Jekel (1988). A new sleeve or tip may not give the same test results as a worn sleeve or tip.

4.6 DATA REDUCTION AND PRESENTATION OF RESULTS

Data reduction of results obtained from the CPT/CPTU is relatively straightforward. Unit cone tip resistance is obtained by dividing the measured tip load (force) by the cone tip projected end area as

$$q_c = F_T/A_T \qquad (4.1)$$

where
  q_c = tip resistance (end bearing)
  F_T = tip force
  A_T = tip area (normally $10\,cm^2$)

Unit sleeve friction or skin friction is obtained by dividing the measured sleeve force by the sleeve area as

$$f_s = F_s/A_s \qquad (4.2)$$

where

f_s = sleeve friction
F_s = sleeve force
A_s = sleeve area (normally $150\,\text{cm}^2$)

Normal units for q_c and f_s are either kg/cm^2 or tons/ft^2. An additional parameter that combines the tip and sleeve measurements is called the friction ratio and is defined as

$$\text{FR or } R_f = f_s/q_c \times 100\% \tag{4.3}$$

Results are presented as q_c, f_s, and R_f versus depth so that an indication of the vertical variation in these parameters at the site may be obtained. An example of a typical CPT profile obtained from an electric cone is shown in Figure 4.8.

It is also useful to have an indication of any changes in *relative* cone resistance which may be made by calculating the normalized net tip resistance as

$$Q_c = (q_c - \sigma_{vo})/\sigma'_{vo} \tag{4.4}$$

and the normalized sleeve resistance as

$$F_r = f_s/(q_c - \sigma_{vo}) \tag{4.5}$$

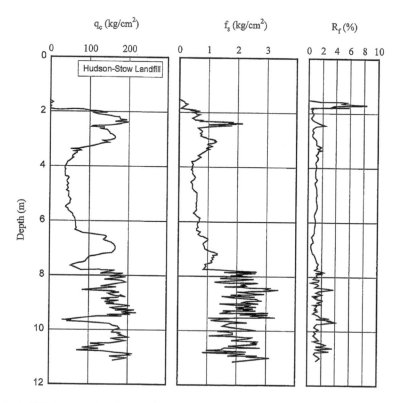

Figure 4.8 Typical CPT data obtained in sand.

where

σ_{vo} = total vertical stress

σ'_{vo} = effective vertical stress

The additional measurement of pore water pressure in the CPTU allows for alternative interpretation of the test results. Depending on the design of the cone, water pressures can act on the exposed parts behind the cone tip and on the ends of the friction sleeve. These water pressures can result in measured tip resistance and sleeve friction values that do not reflect the true resistance of the soil thereby introducing errors in the test results. These errors can be overcome by correcting the measured values for what are called "unequal end area effects".

As the cone is advanced, soil must be displaced in order to make room for the cone to pass. This movement of soil is accompanied by a change in stress conditions next to the soil and is composed of a combination of changes in the soil normal stress acting on the cone and changes in shear stress. The degree of each component is controlled in part by the composition of the soil, the original stress state, the degree of overconsolidation and other variables such as rate of advance, cone tip angle, and location along the cone body.

For a standard 60° apex cone, most of the stress changes acting at the cone tip and along the cone face result from compressional stresses which produce increases in pore water pressure, much like compressing a soil in a consolidation test. At the cone shoulder and along the friction sleeve, the stress changes are largely produced by shear stresses, since the cone apex and face have already passed, displacing most of the soil outward. Along these points, the pore water pressures may be positive or negative, depending on the specific soil behavior. This means that it is important to know the exact location of the pore water pressure measuring point, relative to either the cone tip or cone base in order to make correct interpretations of the test results.

Pore water pressures will not be measured with the CPTU unless the soil is saturated or near saturated. In fine-grained soils that have low hydraulic conductivity, the excess pore water pressure does not have time to dissipate in the time while the cone is advancing. In freely draining coarse-grained soils with high hydraulic conductivity, the generated pore water pressure dissipates about as quickly as it is developed so that the measured pore water pressure during the test is close to the *in situ* value. This means that in saturated sands and sand and gravel deposits, the CPTU can be used to construct the *in situ* pore water pressure profile.

The measured cone tip resistance, q_c, can be corrected for pore pressure effects to give the corrected tip resistance, q_t as

$$q_t = q_c + (1-a)u_2 \tag{4.6}$$

where

u = pore water pressure generated behind the cone tip

a = net area ratio = A_N/A_T

The areas A_N and A_T are defined in Figure 4.9. In order to determine q_t, the porous element must be located at the base of the cone (Type 2).

The value of the net area ratio, a, in Equation 4.6 can be calculated by measuring A_n and A_T; however, a more accurate technique is to place the cone in a water pressure chamber and apply various levels of known fluid pressure and measure the corresponding cone tip resistance. It is desirable to have a cone design with as large a value of a as possible, e.g., on the order of 0.85–0.95; however, most cone designs probably have a value of a in the range of 0.5–0.7 This correction is especially significant in soft saturated clays where the measured tip resistance is low and the measured pore water pressure is high. Figure 4.10 shows a

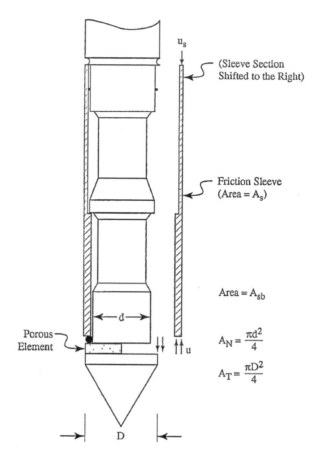

Figure 4.9 Pore water pressures acting on cone tip and sleeve.

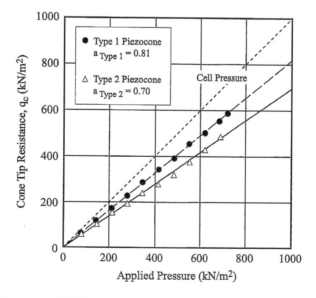

Figure 4.10 Typical calibration of CPTU to determine net area ratio.

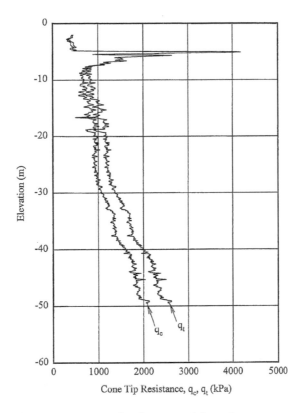

Figure 4.11 CPTU profile showing uncorrected and corrected tip resistance.

typical calibration to obtain the net area ratio. Figure 4.11 shows a CPTU profile performed in soft clay and the difference between q_c and q_t.

A similar pore pressure correction is necessary to obtain the correct value of sleeve friction; however, this would require a measurement of the pore water pressure at both ends of the friction sleeve. Campanella & Robertson (1981) and Konrad (1987) demonstrated that the distribution of f_s along the friction sleeve is nonuniform and increases as the sleeve is moved further back behind the cone tip. This suggests that pore water pressure is also nonuniform along the friction sleeve. The friction ratio R_f may be redefined in terms of the corrected tip resistance as

$$R_f = f_s/q_t \tag{4.7}$$

Some cone designs use equal end area friction sleeves. If it is assumed that the pore pressure distribution along the entire length of the sleeve is equal, the two forces will cancel. However, this will not be the case in all soils.

The measured pore water pressure obtained from the CPTU is a *total* pore water pressure, u_T, which is defined as

$$u_T = u_e + u_o \tag{4.8}$$

where
  u_e = excess pore water pressure
  u_o = *in situ* pore water pressure at the test depth

The magnitude of u_T depends on the measuring location since the stress conditions acting along the cone length vary. The value of the *in situ* pore water pressure u_o is primarily a function of geologic conditions and is generally a positive value.

An additional cone parameter may be defined using the results of the CPTU, incorporating both the corrected tip resistance and the measured pore water pressure and is denoted as B_q, the pore pressure parameter, defined as

$$B_q = (u_T - u_o)/(q_t - \sigma_{vo})$$

(4.9)

Wroth (1988) and others have suggested that the corrected tip resistance could be expressed as a normalized (and therefore nondimensional) parameter using the expression:

$$Q = (q_t - \sigma_{vo})/\sigma'_{vo}$$

(4.10)

The normalized excess pore water pressure may also be defined as

$$U = (u_T - u_o)/\sigma'_{vo}$$

(4.11)

Since B_q has already been defined by Equation 4.9, the normalized pore water pressure is given as

$$U = Q\, B_q$$

(4.12)

The normalized tip resistance and normalized pore water pressure can be useful for estimating soil properties in clays. The parameter B_q should always be defined in terms of the corrected tip resistance, q_t, and not simply q_c.

Wroth (1988) suggested that the friction ratio should also be expressed in nondimensional form as

$$F = f_s/(q_t - \sigma_{vo})$$

(4.13)

As with the CPT, the value of cone tip resistance and sleeve friction from a CPTU (i.e, q_t and f_s) should be presented versus depth. Additionally, the measured pore water pressure should also be presented, and an estimate of the *in situ* pore water pressure, u_o, should be shown as a reference. The calculated values of Q, U, and F may also be shown. Figure 4.12 shows the typical examples of CPTU results in a soft clay.

4.7 INTERPRETATION OF RESULTS FOR STRATIGRAPHY

The results from the CPT/CPTU may be used to evaluate site stratigraphy as well as provide estimates of a number of specific soil properties for both coarse-grained and fine-grained soils. In the past 10–15 years, the interpretation of individual soil properties from CPT/CPTU has more or less reached a stage of maturity. Additional observations have added to the existing database but have not substantially changed many empirical correlations that already exist.

CPT/CPTU results can often provide an indication of subsurface soil conditions by showing differences in penetration resistance. The simplest way to get an initial indication of changes in stratigraphy is to look at the penetration records of q_c or q_t and f_s versus depth. The added measurement of pore water pressure also can be used to identify major changes in soil stratigraphy. The friction ratio may help as an indicator of soil type.

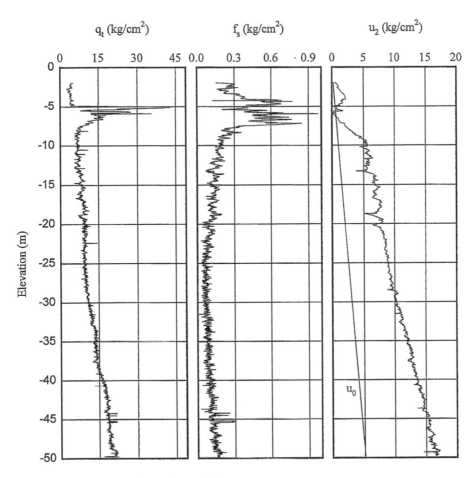

Figure 4.12 Typical CPTU data obtained in soft clay.

The results from CPT/CPTU may be used to identify soil conditions as an alternative to obtaining samples from test borings. This is an indirect approach but is based on many years of experience and well documented investigations but can sometimes be misleading. There is no reliable universal soil identification scheme that works in all cases. It should also be remembered that the response of a CPT (i.e., q_c and f_s) is an average response, influenced by a relatively large volume of soil. This makes the detection of very thin layers almost impossible. In this discussion, the term *identification* is used rather than classification for evaluating soil conditions from CPT/CPTU data since the identification is based on soil behavior or response to the test.

4.7.1 Soil Identification from q_c, f_s, and R_f

Electric cones provide independent measurements of cone tip resistance, q_c, and sleeve "friction", f_s. These values are uncorrected for pore pressure effects but can be combined to define the friction ratio, FR, as

$$FR = R_f = (f_s/q_c) \times 100\% \tag{4.14}$$

Begemann (1965) suggested that the sleeve friction could be used in conjunction with the cone tip resistance to develop a soil profile and suggested that values of FR less than about 2.5%

would indicate sand, greater than 3.5% would indicate clays, and between 2% and 4% as mixed composition soils.

Some investigations (e.g., Muromachi 1981; Zervogiannis & Kalterziotis 1988) have shown that the friction ratio, R_f, could be correlated to the mean grain size, D_{50}, and there is also some evidence indicating that in some deposits, the friction ratio is related to the content of fines (e.g., Suzuki et al. 1995). However, other results show very large scatter and an apparent lack of correlation between friction ratio and content of fines (e.g., Arango 1997). Several charts have been suggested for identifying soil type by combining the friction ratio and cone tip resistance obtained from electric cones (e.g., Douglas & Olsen 1981; Douglas 1984; Robertson et al. 1986). Most of these charts are of the general form shown in Figure 4.13, where tip resistance and friction ratio are plotted, and specific zones of soil behavior are suggested.

Other soil identification charts have been suggested using normalized parameters of tip resistance and sleeve friction:

$$Q = (q_c - \sigma_{vo})/\sigma'_{vo} \tag{4.15}$$

$$F = f_s/(q_c - \sigma_{vo}) \tag{4.16}$$

4.7.2 Soil Identification from q_t, B_q, and R_f

The additional measurement of pore water pressure obtained with a CPTU allows for enhanced interpretation of soil conditions since the measured pore water pressure during penetration is a function of soil behavior for a fixed porous element location. As a result, a number of soil

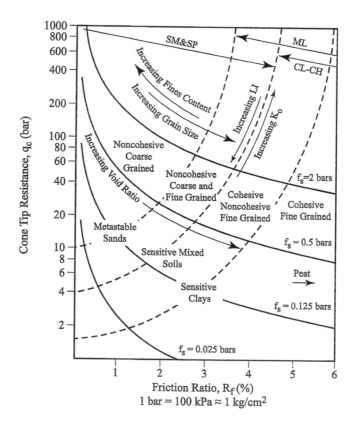

Figure 4.13 Soil identification chart based on CPT. (After Douglas & Olsen 1981.)

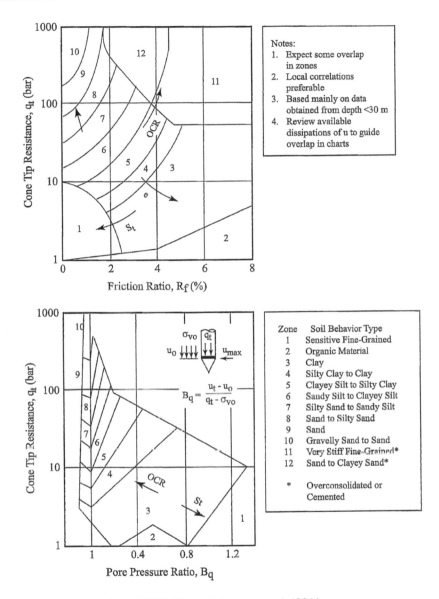

Figure 4.14 Soil identification from CPTU. (From Robertson et al. 1986.)

identification charts have been proposed for use with the CPTU using the combined measurements of corrected tip resistance, q_t, corrected sleeve friction, f_t, and pore pressure, u_2 (e.g., Jones & Rust 1982; Senneset & Janbu 1985; Robertson et al. 1986; Campanella & Robertson 1988; Robertson 2009). Figure 4.14 shows typical soil identification charts for CPTU.

4.7.3 Soil Identification from Q_t, B_q, and F_r

Normalized CPTU measurements may also be used for soil identification, as shown in Figure 4.15, in which

$$Q_t = (q_t - \sigma_{vo})/\sigma'_{vo}$$
(4.17)

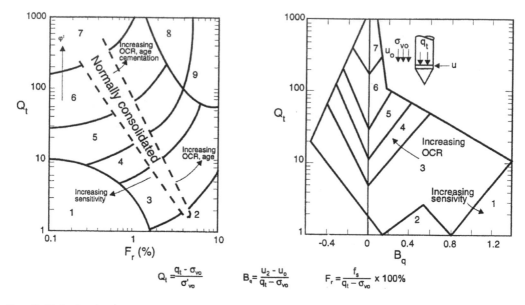

$$Q_t = \frac{q_t - \sigma_{vo}}{\sigma'_{vo}} \qquad B_q = \frac{u_2 - u_o}{q_t - \sigma_{vo}} \qquad F_r = \frac{f_s}{q_t - \sigma_{vo}} \times 100\%$$

Zone	Soil behaviour type	Zone	Soil behaviour type	Zone	Soil behaviour type
1.	Sensitive, fine grained;	4.	Silt mixtures clayey silt to silty clay	7.	Gravelly sand to sand;
2.	Organic soils-peats;	5.	Sand mixtures; silty sand to sand silty	8.	Very stiff sand to clayey sand
3.	Clays-clay to silty clay;	6.	Sands; clean sands to silty sands	9.	Very stiff fine grained

Figure 4.15 Soil identification from normalized CPTU parameters. (From Robertson 1990.)

$$F_r = \left[f_t / (q_t - \sigma_{vo}) \right] \times 100\% \tag{4.18}$$

$$B_q = (u_2 - u_0) / (q_t - \sigma_{vo}) \tag{4.19}$$

4.7.4 Soil Behavioral Type from CPTU, I_C, and I_{CRW}

The tip resistance, sleeve resistance, and pore water pressure obtained from the CPTU may be combined to obtain a CPTU soil behavioral type. Robertson & Wride (1998) suggested using normalized parameters of CPTU tip resistance and sleeve friction to give

$$I_{CRW} = \left[\left\{ 3.47 - \log(Q_{t1}) \right\}^2 + \left\{ 1.22 + \log(F) \right\}^2 \right]^{0.5} \tag{4.20}$$

Jefferies & Been (2006) suggested I_c as

$$I_C = \left[\left\{ 3 - \log\left(Q_t \left[1 - B_q\right] + 1\right) \right\}^2 + \left\{ 1.5 + 1.3 \log(F_r) \right\}^2 \right]^{0.5} \tag{4.21}$$

Values of soil behavioral type based on I_C and I_{CRW} are given in Table 4.1. Figure 4.16 shows a chart developed based on CPTU soil behavioral type I_C.

It is also possible to estimate the per cent fines (% < No. 200 sieve) using the CPT index I_C. The content of fines may be estimated as shown in Table 4.2 (after Mayne et al. 2009).

Eslami & Fellenius (1997) presented a very simplified chart for identifying soil type from a Type-2 CPTU based on effective cone resistance ($q_E = q_t - u_2$) and sleeve friction (f_s). The chart is shown in Figure 4.17 and identifies five basic soil types. Even though this

Table 4.1 Soil behavioral type from CPTU.

CPTU I_c	CPTU I_{CRW}	Soil behavioral type zone	Soil identification
<1.25	<1.31	7	Gravelly sands
1.25–1.80	1.31–2.05	6	Clean to silty sands
1.80–2.40	2.05–2.60	5	Sandy mixtures
2.40–2.76	2.60–2.95	4	Silty mixtures
2.76–3.22	2.95–3.60	3	Clays
>3.22	>3.60	2	Organic soils
N/D	N/D	1	Sensitive clays

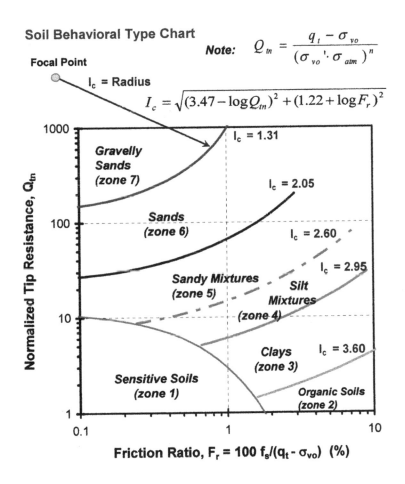

Figure 4.16 Soil behavioral type chart based on CPTU. (From Mayne 2014; after Robertson 2009.)

approach was developed for use in a method to use the cone to estimate capacity of driven piles, it also appears to be very useful in general soil identification.

It should be remembered that all of the soil identification charts presented in this section are only guides. Local experience, local conditions, and knowledge of the geology and ground water conditions at a site all should be used with engineering judgement to develop an understanding of the subsurface conditions from the CPT/CPTU.

Table 4.2 Estimated Fines content from CPT Index I_C

I_c	% Fines
$I_c < 1.26$	0
$1.26 < I_c < 3.50$	$\%Fines = 1.75 I_c^{3.25} - 3.7$
$I_c > 3.50$	100

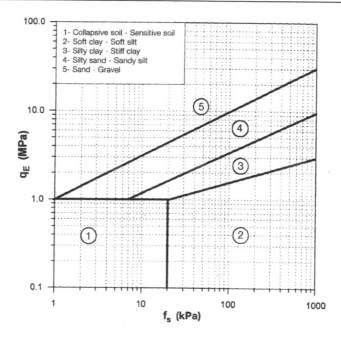

Figure 4.17 Simplified chart for identifying soil type from CPTU. (After Eslami & Fellenius 1997.)

4.8 INTERPRETATION OF TEST RESULTS IN COARSE-GRAINED SOILS

In freely draining coarse-grained soils, the advance of the CPT/CPTU represents drained penetration, and the results are not subject to pore water pressure effects that may be present in fine-grained soils.

4.8.1 Relative Density

Interpretation of the relative density in granular soils may be made directly using the cone tip resistance q_c and a number of published empirical correlations, as summarized by Robertson & Campanella (1983a). Most of the early work relating to the CPT was performed in large calibration chambers with different sands and showed that no single relationship exists between cone tip resistance and relative density for all sands. Soil variables such as stress history, mineralogy, grain-size distribution, angularity, compressibility, and aging influence the correlations.

Schmertmann (1978b) had presented a chart for estimating relative density from tip resistance obtained with an electric cone for normally consolidated, saturated, recent, uncemented, fine, SP sands. Relative density was related to the tip resistance, q_c, for different values of vertical effective stress, σ'_{vo}, and was based on K_o calibration chamber tests on six different sands. Schmertmann (1978b) noted that such a correlation would overestimate the relative density in the case of overconsolidated sands and suggested an approach to estimate the relative density taking into account the effects of effective lateral stresses higher than for the normally consolidated case.

Correlations presented by Baldi et al. (1981, 1982, 1986) and Jamiolkowski et al. (1988) appear to be commonly used in sands, Figures 4.18 and 4.19. Most correlations however are for relatively uniform, clean sands that often bear little resemblance to natural field sands that may originate from a wide range of geologies. Suggested correlations are given in Table 4.3.

Kulhawy & Mayne (1990) suggested correlations between a dimensionless normalized cone tip resistance parameter $\left(\text{QCD} = \left(q_c/pa\right)/\left(\sigma'_{vo}/pa\right)^{0.5}\right)$ and relative density taking into account stress history and compressibility, as shown in Figure 4.20; pa = atmospheric pressure.

More recently, Mayne (2014) has reevaluated available data and presented a global correlation between relative density and tip resistance, shown in Figure 4.21. Again, it should be noted that both stress history and compressibility influence these correlations, and estimates of D_r could be off by 15%–30%.

4.8.2 State Parameter

The concept of the state parameter to characterize the behavior of sand was suggested by Been & Jefferies (1985). The term "state" is used to describe the physical conditions under which the material exists since the material behavior will be controlled by these conditions. The most important physical conditions that define the current state of a soil and would therefore control its behavior are void ratio and stress level. The state parameter combines the influence of void ratio and stress level for each sand by reference to an ultimate or steady state.

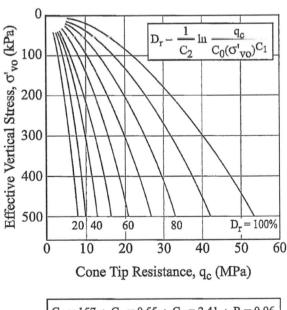

$$D_r - \frac{1}{C_2}\ln\frac{q_c}{C_0(\sigma'_{vo})^{C_1}}$$

$C_0 = 157 : C_1 = 0.55 : C_2 = 2.41 : R = 0.96$

q_c & σ'_{vo} (kPa)
(For $K_0 = 0.45$)

Figure 4.18 Correlation between tip resistance and relative density for NC sands. (After Baldi et al. 1986.)

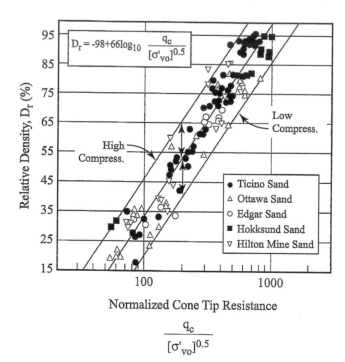

Figure 4.19 Correlation between relative density and normalized cone tip resistance in sands. (After Jamiolkowski et al. 1988.)

Table 4.3 Reported correlations for estimating relative density from CPT

Correlation	References
$D_r = (1/C_2)\ln\left[q_c/\left(C_o\,(\sigma'_{vo})C_1\right)\right]$ $C_o = 157$ $C_1 = 0.55$ $C_2 = 2.41$ (q_c and σ'_{vo} in kPa)	Baldi et al. (1986a)
$D_r = -98 + 66\log\left(q_c/\sigma'_{vo}\right)^{0.5}$ (q_c and σ'_{vo} in t/m²)	Jamiolkowski et al. (1988)
$D_R = 100\left[q_{t1}/\left(1/\left(305\ OCR^{0.2}\right)\right)\right]^{0.5}$ $q_{t1} = \left(q_t/pa\right)/\left(\sigma'_{vo}/pa\right)^{0.5}$	Mayne (2014)

The steady state line for a particular sand represents a condition of no dilation during shear and has been discussed in great detail in the literature (Castro 1969; Poulos 1981; Castro et al. 1985). The position and slope of a steady state line in e-σ′ space is a reference state that is different for different sands and depends on other material properties such as grain-size distribution, limiting void ratios, mineralogy, particle shape, compressibility, and friction angle at constant volume, φ'_{cv}. As illustrated in Figure 4.22, the state parameter is given by the difference between the current void ratio, and the void ratio at steady state at a given stress level and is denoted as ψ. Been et al. (1986, 1987) presented a comparison between the normalized net cone tip resistance (i.e., $(q_c - \sigma_{vo})/\sigma'_{vo}$) and State Parameter, ψ, for a number of sands, shown in Figure 4.23.

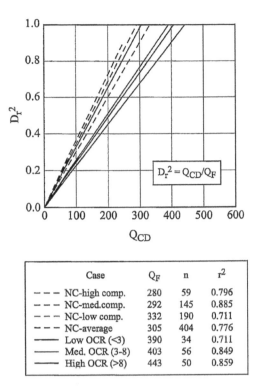

Case	Q_F	n	r^2
– – – NC-high comp.	280	59	0.796
– – – NC-med.comp.	292	145	0.885
– – – NC-low comp.	332	190	0.711
– – – NC-average	305	404	0.776
—— Low OCR (<3)	390	34	0.711
—— Med. OCR (3-8)	403	56	0.849
—— High OCR (>8)	443	50	0.859

Figure 4.20 Correlations between normalized tip resistance and relative density. (After Kulhawy & Mayne 1990.)

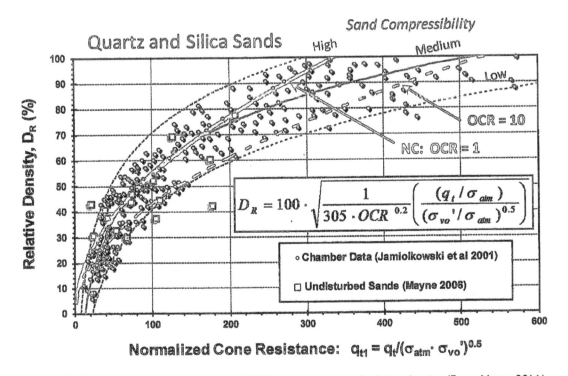

Figure 4.21 Correlation between normalized CPT top resistance and relative density. (From Mayne 2014.)

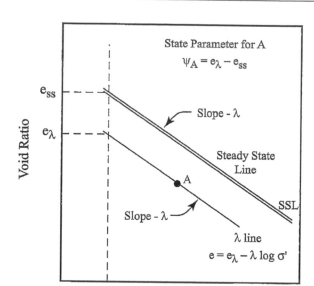

Figure 4.22 Definition of state parameter for sands. (After Been & Jeffries 1985.)

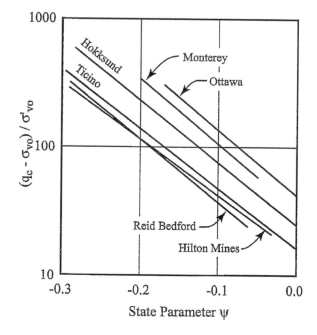

Figure 4.23 Steady state lines for a number of sands. (After Been et al. 1986)

4.8.3 Shear Strength (Drained Friction Angle)

Engineers may wish to make an estimate of the drained friction angle of granular soils for use in the design of foundations, for calculating earth pressures behind retaining structures or for other applications. Several approaches have been used to estimate φ' from CPT q_c including theoretical analysis by deep bearing capacity theory, theoretical analysis by cavity expansion theory, and empirical correlations based on other test results to give φ'.

4.8.3.1 φ′ from Deep Bearing Capacity Theory

Several studies (e.g., Muhs & Weiss 1971; Janbu & Senneset 1974; Chapman & Donald 1981) have noted a correlation between q_c and N_q, Terzaghi's bearing capacity factor for general shear. φ′ may be estimated using an appropriate chart for N_q found in most foundation engineering texts. The bearing capacity approach presented by Durgunoglu & Mitchell (1973) has been one of the most popular methods for determining friction angle from CPT tip resistance. The method assumes a cone roughness factor and requires knowledge of the normal stress around the cone. The method cannot account for soil compressibility and tends to underestimate the secant friction angle, φ'_s.

Robertson & Campanella (1983a) reviewed a number of calibration chamber studies and compared correlations between the bearing capacity number, $N_q = q_c/\sigma'_{vo}$, and drained friction angle, Figure 4.24. The correlation is somewhat sensitive to soil compressibility as well as the assumed shape of the failure surface. The correlation proposed by Robertson & Campanella (1983) is shown by the solid line of Figure 4.24 and can be expressed as

$$\varphi' = \tan^{-1}\left[0.1 + 0.38\log\left(q_c/\sigma'_{vo}\right)\right] \tag{4.22}$$

where q_c and σ'_{vo} are in the same units.

4.8.3.2 φ′ from State Parameter

The relationship between peak drained friction angle and state parameter for a number of sands is shown in Figure 4.25

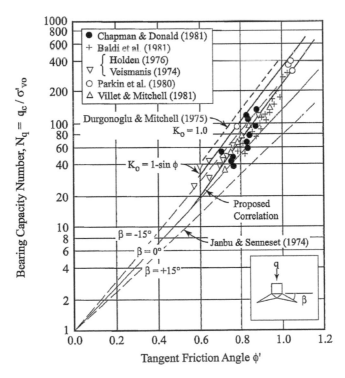

Figure 4.24 Correlation between tip resistance and friction angle for sands. (After Robertson & Campanella 1983.)

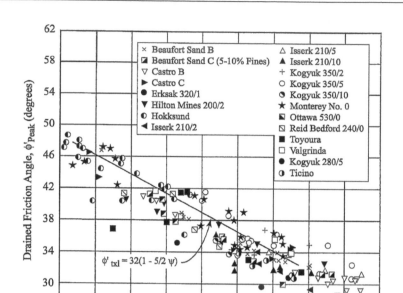

Figure 4.25 Relationship between peak drained friction angle and state parameter. (After Been & Jeffries 1985.)

Kulhawy and Mayne (1990) summarized the available friction angle data reported in the literature for sands compiled based on calibration chamber data that gave a general correlation between friction angle and normalized tip resistance as

$$\Phi' = 17.6 + 11.0 \log(q_{t1}) \tag{4.23}$$

An updated chart was presented by Mayne (2014) and included data from Uzielli et al. (2013) as shown in Figure 4.26. The results of Uzielli et al. (2013) use a power function as

$$\Phi' = 25.0(q_{t1})^{0.10}$$

4.8.4 Stress History and *In Situ* Stress

At the present time, there are insufficient reliable results available to accurately predict the *in situ* lateral stress or stress history of coarse-grained soils from CPT results. A number of calibration chamber tests results available in clean uncemented quartz sands show that there is a unique relationship between the measured tip resistance and the effective horizontal stress, σ_h' (e.g., Parkin 1988; Houlsby & Hitchman 1988). While it is well known that the prepenetration state of stresses largely controls the response obtained from the CPT in sands, it is not a simple matter to extract the magnitude of *in situ* lateral stresses from cone data.

4.8.5 Elastic Modulus

The elastic modulus, E_S, of soils has also been correlated to the results of tip resistance measurements (q_c) obtained from the CPT. Most early correlations between q_c and E_S were of the general form

$$E_S = \lambda q_c \tag{4.24}$$

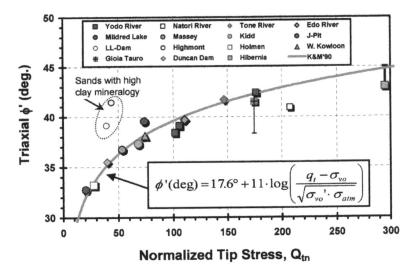

Figure 4.26 Updated correlation between normalized CPT tip resistance and friction angle. (From Mayne 2014.)

where

λ = a constant (empirical factor)

Mitchell & Gardner (1975) compiled a large number of reported correlations, which generally show very wide scatter. Table 4.4 summarizes a number of suggestions for estimating soil modulus from CPT.

E_s in units of q_c for CPT

The use of large calibration chamber tests on reconstituted samples of sands has helped to evaluate variables that can influence correlations between soil modulus and CPT results.

Table 4.4 Some reported correlations between E_s and q_c in sands

Correlation	Soil	References
$E_s = 2.9\ q_c$	Sand	Garga & Quin (1974)
$E_s = aq_c$ $1 < a < 4$	NC and OC sand	Dahlberg (1974)
$E_s = 11\ q_c$	Sand	Lambrechts & Leonards (1978)
$E_s = 2.5\ q_c$ $E_s = 3.5\ q_c$	Axisymmetric loading Plain strain loading	Schmertmann et al. (1978)
$E_s = 2.5\ q_c$	Sand	Roth et al. (1982)
$E_s = \alpha\ q_c$ $\alpha = 1.7\text{--}4.4$ Average = 2.5	Medium sand	Das Neves (1982)
$E_s = 8(q_c)^{0.5}$ q_c in MPa	Sand	Denver (1982)
$E_s = 2\text{--}4\ q_c$ $E_s = \left(1 + D_r^2\right)q_c$	Sand (normally consolidated) Sand (overconsolidated)	Bowles (1988)
$E_s = 6\text{--}30\ q_c$		
$E_s = 3\text{--}6\ q_c$	Clayey sand	
$E_s = 1\text{--}2\ q_c$	Silty sand	

E_s in units of q_c for CPT.

For a given sand, the ratio E/q_c is related to stress history and current stress level for sands at different relative densities. Robertson (1991) proposed that the ratio of E_S/q_c should be expressed as a function of the relative load intensity.

Because of the wide range in correlation constants that may exist between the results of *in situ* penetration tests and a singular value of soil modulus, it is doubtful that any method that relies on these techniques for the accuracy of settlement estimates will be of much value, other than those created by local correlations developed from full-scale field observations of performance.

4.8.6 Constrained Modulus

In one-dimensional compression, the tangent slope of the stress-strain curve is defined as the constrained modulus:

$$D = \Delta\sigma'_v / \Delta\varepsilon_v \tag{4.25}$$

From elastic theory, the constrained modulus is related to the Young's modulus and shear modulus as

$$M = E\left[(1-\upsilon)/(1+\upsilon)(1-2\upsilon)\right] = 2G\left[(1-\upsilon)/(1-2\upsilon)\right] \tag{4.26}$$

Based on a series of calibration chamber tests on a dry medium to fine quartz sand, Chapman & Donald (1981) found that the initial one-dimensional constrained modulus, M, could be related to q_c. For normally consolidated conditions: the expression $M = 3q_c$ provided a lower bound, with most data falling in the range $M = 3$–$4\ q_c$. For overconsolidated specimens (with OCR < 2), the results indicated $M = 8$–$15\ q_c$ with an average value of $M = 12\ q_c$.

For a given sand, the ratio M/q_c is related to stress history and current stress level for sands at different relative densities. For both normally consolidated and overconsolidated sands, the ratio M/q_c decreases with increasing relative density, all other factors being equal. Mayne (2006) presented a compilation of test results showing a correlation between D and net tip resistance as

$$D = \alpha'_c \left(q_t - \sigma_{vo}\right) \tag{4.27}$$

The value of $\alpha'_c = 5$ for soft to firm clays and normally consolidated sands. The global correlation between M and CPTU net tip resistance for a wide range of soils is shown in Figure 4.27.

4.8.7 Shear Wave Velocity and Small-Strain Shear Modulus

The small-strain shear modulus in soils may be measured using a number of field crosshole and downhole methods that are more or less considered standard techniques for dynamic testing. The shear wave velocity may also be measured directly using a seismic cone or piezocone (SCPT/SCPTU) as illustrated in Figure 4.28.

Estimates of either V_S or G_{max} may be obtained indirectly from CPT results through empirical correlations. G_{max} is obtained from V_s as

$$G_{max} = \rho V_S^2 \tag{4.28}$$

where
   ρ = soil density

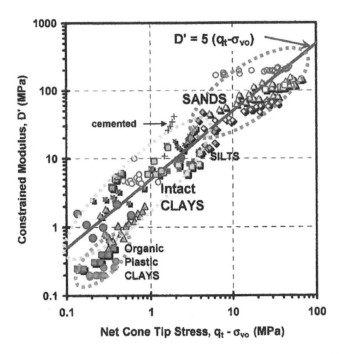

Figure 4.27 Correlation between constrained modulus and net CPTU tip resistance. (From Mayne 2006.)

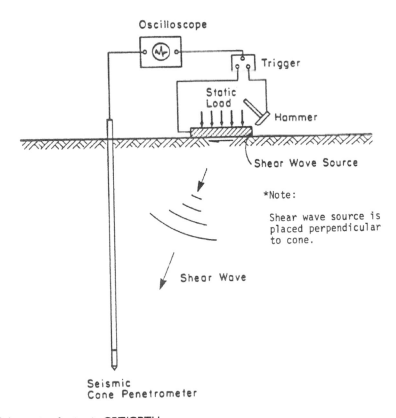

Figure 4.28 Schematic of seismic CPT/CPTU.

4.8.7.1 Shear Wave Velocity and Shear Modulus from q_c

Estimates of V_S and G_{max} in different coarse-grained soils have been suggested. Table 4.5 gives a number of reported empirical correlations. Several studies have also shown that the results of cone tests may be used to make direct estimates of G_{max} in coarse-grained soils. Robertson (1990) suggested a generalized approach relating G_{max} to normalized tip resistance that appears to be useful for estimating G_{max} for a wide range of soils, as shown in Figure 4.29. A similar correlation is given in Figure 4.30.

4.8.8 Liquefaction Potential

For over 40 years, the CPT/CPTU has been used to evaluate liquefaction potential of coarse-grained soils (Robertson & Campanella 1985; Robertson & Wride 1995; Stark & Olson 1995; Idriss & Boulanger 2004; Moss et al. 2006; Juang et al. 2008). The methods generally are based on cone tip resistance, q_c or q_t, or on direct or indirect measurement of shear wave velocity (Kayabali 1996; Andrus et al. 2004). More recently, methods have also been presented incorporating CPT/CPTU friction ratio (e.g., Mola-Abasi et al. 2018). The methods are similar to approaches that use SPT N-values discussed in Chapter 2.

In general, the available methods make use of databases of CPT/CPTU results obtained at sites that experienced liquefaction and separate them from sites that did not experience liquefaction for the same seismic event. Methods are typically for earthquake with magnitude 7.5 and are based on a plot of cyclic stress ratio (CSR) that causes liquefaction and normalized cone tip resistance, as shown in Figure 4.31 for clean sands and Figure 4.32 for silty sands. The normalized tip resistance is obtained as

$$q_{C1N} = \left((q_c - \sigma_{vo})/pa\right)\left(pa/\sigma'_{vo}\right)^n \tag{4.29}$$

Table 4.5 Some reported empirical correlations between V_S and q_c for coarse-grained soils

Correlation	References
$V_S = 134.1 + 0.0052q_c$	Sykora & Stokoe (1983)
$V_S = 277\ q_t^{0.13}\ \sigma'^{0.27}_{vo}$ (q_t in MPa)	Baldi et al. (1989)
$V_S = 13.18\ q_c^{0.192}\ \sigma'^{0.179}_{vo}$ (4.25a)	Hegazy & Mayne (1995)
$V_S = 12.02\ q_c^{0.319}\ f_s^{-0.0466}$	
$V_S = 25.3\ q_c^{0.163}\ f_s^{0.029} D^{0.155}$	Piratheepan (2002)
$V_S = 118.8\ \log{(f_s)} + 18.5$	Mayne (2006)
$G_{max} = 1634\ q_c^{0.250}\ \sigma'^{0.375}_{vo}$	Rix & Stokoe (1991)
$G_{max} = 2.26\ q_c + 59.2$ (G_{max} & q_c in MPa)	Fiorovante et al. (1991)
$G_{max} = 50\ q_c^{1.05}$	Anagnostopoulos et al. (2003)
$G_{max} = 800\ q_c^{0.250}\ \sigma'^{0.375}_{vo}$ (upper bound: cemented)	Schnaid et al. (200)
$G_{max} = 280\ q_c^{0.250}\ \sigma'^{0.375}_{vo}$ (lower bound: cemented) (upper bound: uncemented)	
$G_{max} = 110\ q_c^{0.250}\ \sigma'^{0.375}_{vo}$ (lower bound: uncemented)	
$G_{max} = 50\,pa\left[\left(q_t - \sigma_{vo}/pa\right)\right]^{0.6}$	Mayne (2006)

V_S in m/s; q_c, f_s, and σ'_{vo} in kPa; D = depth in m; G_{max} in MPa.

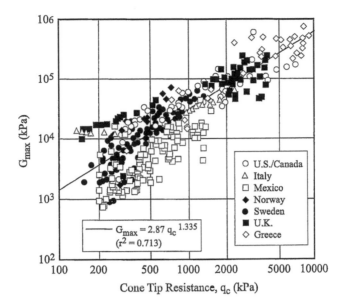

Figure 4.29 Correlation between G_{max} and CPT q_c. (After Robertson 1990.)

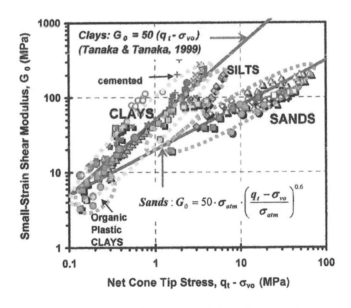

Figure 4.30 Small strain shear modulus related to net CPTU tip resistance. (From Mayne 2006.)

where
  pa = atmospheric pressure
  n = function of soil behavioral type (I_C)

  if $I_C < 1.64\, n = 0.5$

  if $I_C > 3.30\, n = 1.0$

  if $I_C\, 1.64 < I_C < 3.30\, n = (I_C - 1.64)\, 0.3 + 0.5$

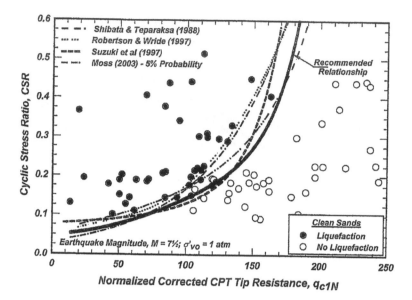

Figure 4.31 CPT-based liquefaction potential for clean sands.

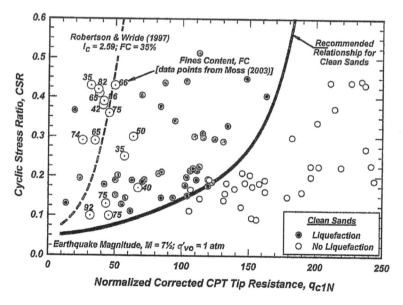

Figure 4.32 CPT-based liquefaction potential for silty sands.

4.9 INTERPRETATION OF CPT RESULTS IN FINE-GRAINED SOILS

While the mechanics of the CPT advance in granular soils is generally considered to represent drained penetration, the CPT in fine-grained soils is largely interpreted as undrained penetration. This is especially the case for soft saturated clays which generally occur in conditions of relatively high water table. In stiff, highly overconsolidated clays, such as a surficial clay crust, there may be some questions as to the exact drainage conditions surrounding the cone; however, as will be shown, the results generally can be assessed using similar procedures along with other fine-grained soils.

4.9.1 Undrained Shear Strength

Results of the CPT have been used extensively to provide an estimate of the undrained shear strength of clays using both the tip resistance and sleeve friction. Table 4.6 summarizes different approaches for estimating undrained shear strength in clays.

4.9.1.1 s_u from q_c

Most current correlations for determining s_u from CPT tip resistance q_c have the form

$$s_u = (q_c - \sigma_{vo})/N_k \tag{4.30}$$

where

σ_{vo} = *in situ* total vertical stress
N_k = empirical factor

The value of N_k depends on the reference value of s_u and can vary from about 10 to 40 but typically averages about 14 in a large number of clays. In soft clays, the reference has often been the filed vane test, whereas in stiff clays, the reference may be other field tests, e.g., plate load or laboratory tests, such as triaxial compression or unconfined compression. N_k is related to the rigidity index, $I_r = G/s_u$, of the clay. Baligh (1975) showed theoretically that for a 60° cone, the value of N_k only varied from about 14 to 18 over a wide range of I_r values.

An average value of $N_k = 17$ was suggested by Lunne et al. (1976) to provide a good comparison with corrected field vane results for a number of marine clays in Scandinavia. Stark & Delashaw (1990) obtained a value of N_k of 12 from unconsolidated-undrained (UU) triaxial tests. Jamiolkowski et al. (1982) summarized a large number of test sites correlating q_c to corrected field vane results and showed that N_k decreased with increasing plasticity with a range from about 5 to 25 for soft to medium clays. If soils with low plasticity (P.I. < 10) are eliminated, the mean value of N_k is approximately 15, which corresponds closely with that presented by Baligh, as previously mentioned.

Table 4.6 Summary of methods for estimating s_u from CPT/CPTU

General correlation	References
$s_u = (q_c - \sigma_{vo})/N_k$	Lunne et al. (1976) Jamiolkowski et al. (1982) Rad & Lunne (1988) Stark & Deleshaw (1990) Anagnostopoulos et al. (2003)
$s_u = (q_t - \sigma_{vo})/N_{kt}$	Jamiolkowski et al. (1988) Larsson & Mulabdic (1991)
$s_u = (u_2 - u_o)/N_u$	Robertson & Campanella (1983b) Campanella et al. (1985) Larsson & Mulabdic (1991)
$s_u = (q_t - u_2)/N_{kc}$	Senneset et al. (1982) Campanella et al. (1982) Mayne & Chen (1993, 1994)
$(s_u/\sigma'_{vo}) = Q_t/N_{KT}$	Robertson (2009)

4.9.1.2 s_u from q_T

Undrained shear strength may also be estimated from the corrected tip resistance, q_t, from the CPTU, which may be used to provide an estimate of

$$s_u = (q_t - \sigma_{vo})/N_{KT} \tag{4.31}$$

The value of the cone factor N_{KT} is a function of the rigidity index of the soil, I_r, ($I_r = G/s_u$). For typical values of I_r in clays (i.e., 50–500), theoretical values of N_{KT} only range from about 9.1 to 12.2. Values of N_{KT} reported in the literature range from about 7 to 30 with reference values of undrained shear strength coming from a variety of laboratory and *in situ* tests. The value of N_{KT} is also somewhat dependent on sensitivity and therefore may need to be reduced to $N_{KT} = 10$ (approximately) for sensitive clays. Wroth (1988) showed that for a given soil, N_{KT} is a constant and generally independent of OCR (with a typical value of about 12 over the range of 1 < OCR < 8).

Results obtained in soft clays indicated that s_u from laboratory direct simple shear tests, the value of N_{KT} can be roughly related to the liquid limit. For the range of LL between about 30 and 200, their data suggested that

$$N_{KT(DDS)} = 13.4 + 6.65 \, LL \tag{4.32}$$

where
LL = liquid limit (in decimal)

While there was considerable scatter, values of N_{KT} ranged between about 14 and 20 with an average of about 16. Using results from triaxial compression tests, the test data showed an average value of N_{KT} of about 11 which was also related to the liquid limit as follows:

$$N_{KT(TC)} = 3.6 + 13.2 \, (LL) \tag{4.33}$$

A typical value of $N_{KT} = 15$ is often used for preliminary evaluation of s_u unless local experience or other local correlations are available. In soft clays, a value of $N_{KT} = 15$ has been shown to provide good agreement with field vane tests (e.g., Cai et al. 2010; Chung et al. 2010).

4.9.1.3 s_u from u

Robertson & Campanella (1983b) suggested that based on cavity expansion theory, the undrained shear strength could be estimated from

$$4 < (\Delta u_1/s_u) < 7 \tag{4.34a}$$

(spherical cavity expansion with u measured at cone tip; u_1 position)
    or

$$3 < (\Delta u_2/s_u) < 5 \tag{4.34b}$$

(cylindrical cavity expansion with u measured behind cone tip; u_2 position)

where
Δu = excess pore water pressure = $u_T - u_o$

In general, then, s_u may be estimated from the empirical expression:

$$s_u = \Delta u / N_u \tag{4.35}$$

where
N_u = an empirical factor

Observed values of the factor N_u vary over a much narrower range than either N_K or N_{KT} and typically range from about 4 to 10 but are a function of the location of the porous element. The value of N_u also depends to some degree on the sensitivity, the pore pressure parameter at failure, A_f, and the rigidity index.

In stiff, highly overconsolidated clays, (e.g., OCR > 8) some problems may be encountered using Equation 4.65 if pore water pressures are measured behind the cone tip (Type 2). Because of high pore pressure gradients established between the cone tip and cone base, the pore water pressures measured at the cone base during the cone sounding may be low in highly overconsolidated clays. This is usually indicated by a significant rise in pore water pressure when the sounding is stopped, and a dissipation test is performed. This means that pore water pressures measured during penetration will be low, leading to estimates of s_u that are too low.

4.9.1.4 s_u from q_T and u

Senneset et al. (1982) and Campanella et al. (1982) suggested that the undrained shear strength could be estimated from the "effective" corrected cone tip resistance, i.e., $(q_t - u_2)$. The general expression for s_u may be given as

$$s_u = (q_t - u_2)/N_{ke} \tag{4.36}$$

Reported values of N_{ke} typically range from about 3 to 12 depending on the reference value of s_u.

Mayne & Chen (1993, 1994) presented an effective stress model for estimating s_u from CPTU using this approach. In this model, the value of N_{qu} is related to the effective stress friction angle, φ', the critical state failure, M, the volumetric strain ratio, X, and a modified Cam-clay equation factor. For most typical clays, the value of N_{qu} would vary from about 6.1 to 7.1.

4.9.1.5 s_u from Q

Robertson (2009) suggested estimating the normalized undrained shear strength from the normalized cone tip resistance assuming a value of $N_{KT} = 14$ as

$$(s_u/\sigma'_{vo}) = Q_t/14 \tag{4.37}$$

where

$$(Q_t) = (q_t - \sigma_{vo})/\sigma'_{vo}$$

4.9.1.6 s_u from f_s

Estimates of undrained shear strength using CPT friction sleeve measurements have also been made, as first suggested by Begemann (1965); however, it appears that this practice is currently less common than using the tip resistance. A number of reported correlations between f_s and s_u are summarized in Table 4.7. It has been shown that in some cases,

Table 4.7 Some reported correlations between CPT sleeve friction and undrained shear strength

Soil	Correlation	Cone type	Reference strength	References
	$s_u = f_s$	M	N/A	Sanglerat (1972)
CL-ML	$s_u = f_s/1.24$	M	CIUC	Gorman et al. (1973)
Bangkok clay	$s_u = f_s/0.47$	M	Vane	Brand et al. (1974)
CL-ML	$s_u = f_s$	M	CIUC & UU	Gorman et al. (1975)
CH-CL	$s_u = (k)f_s$ (k = 0.67–4 depending on L.I. & s_u)	E	Vane	Marr & Endley (1982)
CH-CL	$s_u = 1.72f_s$	E	N/A	Tumay et al. (1982)
ML-CL	$s_u = f_s/1.28$	M	N/A	Cancelli et al. (1982)
Bangkok clay	$s_u = f_s/0.61$ (soft marine clay) $s_u = f_s/0.53$ (stiff clay)	M (?)	PMT	Bergado & Khaleque (1986)
CL	$s_u = f_s/1.23$	E (?)	UU	Zervogiannis & Kalteziotis (1988)
CH-CL stiff	$s_{urem} = f_s$	E (?)	Remolded UU	Quiros & Young (1988)
ML	$s_u = 4.2f_s$ $s_u = 2.8f_s$	E	UU UC	Takesue et al. (1995)
ML-CL stiff	$s_u = f_s/1.26$ $s_u = f_s$	M E	UU	Anagnostopoulos et al. (2003)

the sleeve friction may actually more closely represent remolded strength. For example, Robertson et al. (1986) presented data for clays in the Vancouver area, which generally showed that the sleeve friction values from an electric cone were very close to remolded strength. This difference may be related to sensitivity.

4.9.1.7 s_u from σ'_p

It is also possible to estimate the undrained shear strength from the pre-consolidation stress, σ'_p, discussed in Section 4.8.3. For many clays, there is a unique relationship between undrained shear strength and stress history, which may be given approximately as

$$s_u = 0.23 \, \sigma'_p \qquad (4.38)$$

Therefore, using the CPT/CPTU to first estimate σ'_p, one can use Eq. 4.38 to estimate s_u.

4.9.2 Sensitivity

Schmertmann (1978b) suggested that sensitivity, S_t, could be estimated from

$$S_{t \, (\text{field vane})} = N_s/R_f \, (\%) \qquad (4.39)$$

where
  $S_{t \, (\text{field vane})}$ = sensitivity
  N_s = an empirical factor
  R_f = friction ratio

Schmertmann (1978b) suggested a value of $N_s = 15$ for R_f from a mechanical CPT and S_t obtained from Nilcon or Geonor type field vane. Robertson & Campanella (1983b) suggested

$N_s = 10$ for electric cone data. Greig et al. (1986) and Robertson et al. (1986) suggested $N_s = 6$ for data collected from clays in the Vancouver area. Robertson (2009) suggested that sensitivity could be estimated from CPTU results using the normalized friction ratio as

$$S_t = 7.1/F_r \tag{4.40}$$

where

$$F_r = \left[f_s/(q_t - \sigma_{vo}) \right] 100\%$$

If the sleeve friction can be taken as an estimate of the remolded strength, then sensitivity may also be obtained by combining sleeve resistance with an estimate of undisturbed strength from tip resistance or pore pressure.

4.9.3 Stress history – Preconsolidation Stress, σ'_p

The results of the CPT/CPTU may be used to make an estimate of the stress history of clays, by either estimating the preconsolidation stress, σ'_p, or the overconsolidation ratio, OCR. Both empirical and analytical models have been suggested predicting σ'_p and OCR using the corrected tip resistance or pore water pressure. More rigorous numerical models have also been used. Chen & Mayne (1994) presented a review of the various approaches suggested. In some cases, piezocones with multiple pore water pressure sensor locations are required. Table 4.8 summarizes the different approaches for estimating stress history for CPT/CPTU results.

4.9.3.1 σ'_p from q_c

A direct relationship between q_c and the preconsolidation stress, σ'_p, has been noted for different clays. Tavenas & Leroueil (1987) showed that for sensitive Canadian clays

$$\sigma'_p = q_c/3 \tag{4.41}$$

Mayne & Kemper (1988) suggested a general correlation between q_c and σ'_p as

$$\sigma'_p = q_c/\beta_c \tag{4.42}$$

Table 4.8 Summary of empirical approaches for estimating stress history from CPT/CPTU

Parameter	General correlation	References
σ'_p	$\sigma'_p = q_c/k_c$	Tavenas & Leroueil (1987) Mayne & Kemper (1988)
	$\sigma'_p = (q_c - \sigma_{vo})/k_k$	Mayne (1986) Tavenas & Leroueil (1987)
	$\sigma'_p = \alpha_t (q_t - \sigma_{vo})$	Powell et al. (1989)
	$\sigma'_p = \alpha_u (u_T - u_o)$	Roy et al. (1981) Powell et al. (1989) Mayne (1995)
	$\sigma'_p = \alpha_p (q_t - u_1)$	$\sigma'_p = 0.8(q_t - u_1)$
OCR	$OCR = k\left[(q_c - \sigma_{vo})/\sigma'_{vo} \right]$	Mayne & Kemper (1988)
	$OCR = 0.49 + 1.50 \left[(u_1 - u_2)/u_0 \right]$	Sully et al. (1988)
	$OCR = \beta\left[(q_t - u_1)/\sigma'_{vo} \right]$	Chen & Mayne (1994)

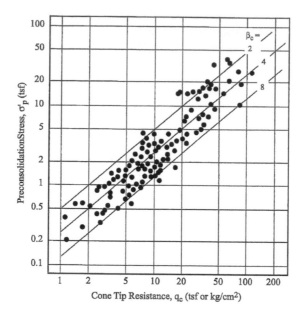

Figure 4.33 Correlation between σ'_p and CPT tip resistance. (After Mayne & Kemper 1988.)

As shown in Figure 4.33, values of k_c for both electric cones and for mechanical cones vary between about 2 and 8. A general trend to these data which may be useful for preliminary work may be taken as

$$\sigma'_p = q_c/4 \tag{4.43}$$

It has also been suggested (e.g., Mayne 1986; Tavenas & Leroueil 1987) that an estimate of σ'_p may be made using the net tip resistance, $q_c - \sigma_{vo}$.

4.9.3.2 σ'_p *from* q_t

Empirical correlations between the preconsolidation stress, σ'_p, and the corrected cone tip resistance, q_t, have been suggested for a number of different clay deposits. These correlations have the general form

$$\sigma'_p = \alpha_t \left(q_t - \sigma_{vo} \right) \tag{4.44}$$

The value of α_t typically is in the range of 0.20 to 0.40. Table 4.9 presents a summary of different reported values of α_t. The use of Equation 4.73 and α_t requires that pore pressure be measured behind the cone base, i.e., in the u_2 position so that q_c may be corrected to q_t. Data collected by Mayne (1995) are shown in Figures 4.34 and 4.35.

4.9.3.3 σ'_p *from* Δu

Based on the results of pile tests, Roy et al. (1981) suggested that σ'_p could be related directly to the pore pressure difference $\Delta u = u_T - u_o$ for sensitive clays in Canada. For Δu measured at the pile tip, they found that

$$\sigma'_p = 0.58 \, \Delta u \tag{4.45a}$$

Table 4.9 Reported values of α_t

α_t	Soil	References
0.20	Clay till, UK	Powell et al. (1989)
0.30	Soft clay	Powell et al. (1989)
0.29	Swedish clays	Larsson & Mulabdic (1991)
0.34	Norwegian clays	Lunne et al. (1992)
0.28	Eastern Canada clays	Leroueil et al. (1995)
0.33	U.S. clays	Mayne (1995a)
0.30–0.45	Polish clays	Borowczyk & Szymanski (1995)

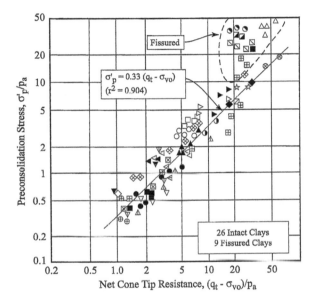

Figure 4.34 Estimation of σ'_p from CPTU net tip resistance. (After Mayne 1995.)

for Δu measured on the shaft,

$$\sigma'_p = 1.08 \, \Delta u \tag{4.45b}$$

Since the pore water pressure obtained from the CPTU is dependent on the location of the piezoelement, specific correlations for predicting σ'_p from Δu would be needed for different cone designs. Mayne (1995) presented a summary of the available data shown in Figures 4.36 and 4.37, which shows that

$$\sigma'_p = 0.47 \, \Delta u \; (\text{tip}) \tag{4.46a}$$

and

$$\sigma'_p = 0.54 \, \Delta u \; (\text{behind tip}) \tag{4.46b}$$

Note that fissured clays do not follow the trend lines of these data since the measured pore water pressures are often dependent on other soil behavior. In highly overconsolidated clays, Equation 4.46b may not provide a realistic estimate of σ'_p since Δu may be zero or negative during penetration (Figure 4.37).

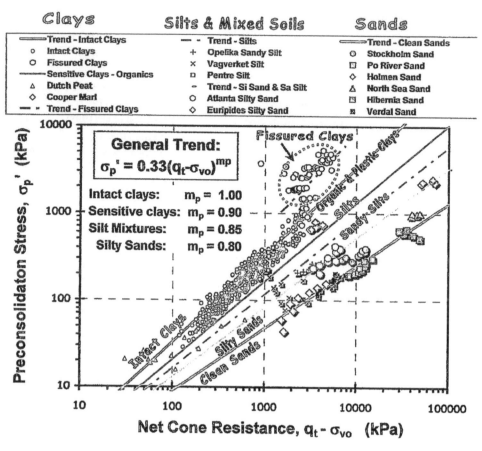

Figure 4.35 Global estimation of σ_p' from CPTU net tip resistance. (From Mayne 1995.)

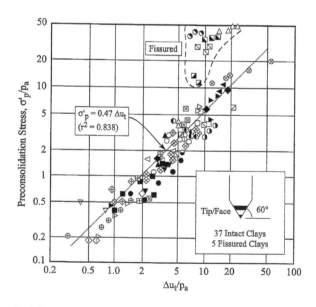

Figure 4.36 Estimation of σ_p' from Δu_1 (Type 1). (After Mayne 2006.)

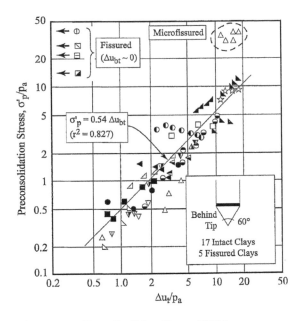

Figure 4.37 Estimation of σ'_p from Δu_2 (Type 2). (After Mayne 2006.)

4.9.3.4 σ'_p from q_t and u

Results presented by Mayne & Chen (1994) suggested that OCR may be determined from using both q_t and u. This approach may be used to make a direct estimate of σ'_p as

$$\sigma'_p - 0.8\left(q_t - u_1\right) \tag{4.47a}$$

$$\sigma'_p = 0.5\left(q_t - u_2\right) \tag{4.47b}$$

4.9.4 Stress History – OCR

4.9.4.1 OCR from q_c

Mayne & Kemper (1988) suggested that the OCR could be estimated directly from the tip resistance as

$$OCR = k\left[\left(q_c - \sigma_{vo}\right)/\sigma'_{vo}\right] \tag{4.48}$$

Values of k ranged from 0.3 to 0.8 for electric cones and 0.12 to 0.5 for mechanical cones.

4.9.4.2 OCR from q_t and u

A relatively recent development has been presented suggesting that OCR may be estimated by combined use of the net tip resistance and pore water pressure. Chen & Mayne (1994) suggested

$$OCR = 0.8\left(q_t - u_1\right)/\sigma'_{vo} \tag{4.49a}$$

$$OCR = 0.5\left(q_t - u_2\right)/\sigma'_{vo} \tag{4.49b}$$

4.9.4.3 OCR from Pore Pressure Difference

Sully et al. (1988) showed that for a dual element piezocone, the pore pressure difference, PPD, $(PPD = (u_1 - u_2)/u_o)$ provided an excellent correlation with OCR up to OCR = 10 as

$$OCR = 0.49 + 1.50 \text{ PPD} \tag{4.50}$$

However, for more heavily overconsolidated clays (10 < OCR < 40), the PPD did not work very well, and considerable scatter was observed.

4.9.5 In Situ Lateral Stress

4.9.5.1 K_0 from OCR

For routine work, an estimation of *in situ* lateral stresses may be made using an appropriate empirical correlation between K_o and another parameter, such as OCR, after first making an estimate of OCR. Mayne & Kulhawy (1982) had suggested that in an overconsolidated soil undergoing simple unloading, the value of K_o could be obtained from

$$\left(K_o\right)_{OC} = \left(K_o\right)_{NC}\left(OCR\right)^{\sin\varphi'} \tag{4.51}$$

In normally consolidated soils, $(K_o)_{NC}$ may be estimated from the effective stress friction angle as

$$\left(K_o\right)_{NC} = 1 - \sin\varphi' \tag{4.52}$$

For typical values of φ' for clay, ranging from 20° to 30°, $(K_o)_{NC}$ ranges from about 0.50 to 0.66. Therefore, an estimate of OCR could first be estimated and then used to estimate K_o from Equation 4.51.

4.9.5.2 Empirical Correlations to q_t and Δu

Empirical results between K_o obtained from self-boring pressuremeter tests (SBPMT) and q_T or Δu_1 have been presented by Kulhawy & Mayne (1990) and are shown in Figures 4.38 and 4.39. These observations suggest that

$$K_o = 0.10\left(q_t - \sigma_{vo}\right)/\sigma'_{vo} \tag{4.53}$$

$$K_o = 0.24\left(\Delta u_1/\sigma'_{vo}\right) \tag{4.54}$$

4.9.6 Shear Wave Velocity and Small-Strain Shear Modulus

4.9.6.1 Shear Wave Velocity from q_c and q_t

As noted in Chapter 2, small strain shear modulus may be obtained from the shear wave velocity as

$$G_{max} = \rho V_s^2 \tag{4.55}$$

Even though G_{max} is a small-strain property and q_c or q_t is a large-strain property, a reasonable relationship between these two parameters should be expected since both parameters

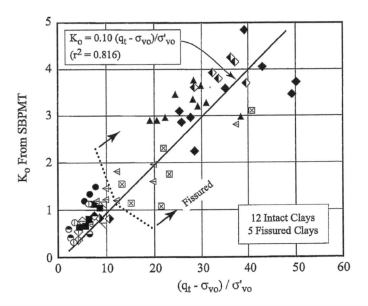

Figure 4.38 Correlation between K_o and CPTU tip Resistance in clays. (After Kulhawy & Mayne 1990.)

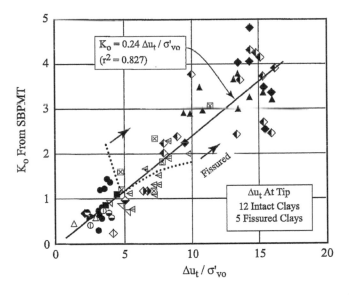

Figure 4.39 Correlation between K_o and CPTU pore pressure (Type 1) in clays. (From Idriss & Boulanger 2006.)

are a function of the existing stress state and the stress history. However, it has also been well established that G_{max} is dependent on the initial void ratio, e_o, of soil. In some cases, empirical correlations for either V_S or G_{max} may be improved if the initial voids ration, e_o, is included.

Several empirical correlations have been presented between uncorrected and corrected cone tip resistance and shear wave velocity for fine-grained soils. A number of these correlations are given in Table 4.10.

Table 4.10 Reported correlations between q_c or q_t and V_s

Correlation	Clay	References
$V_s = 1.75\, q_c^{0.627}$	Misc. Clays	Mayne & Rix (1995)
$V_s = 2.94\, q_t^{0.613}$	Norwegian soft clay	Long & Donohue (2010)
$V_s = 7.95\, q_t^{0.403}$	Jiangsu clays	Cai et al. (2014)
$V_s = 90\, q_t^{0.101} e_0^{-0.663}$		
$V_s = 4.54\, q_t^{0.487} \left(1+B_q\right)^{0.337}$		

V_s in m/s, q_c in kPa.

4.9.6.2 Shear Wave Velocity from f_s

For a wide range in soils, Mayne (2006) suggested that V_s could be estimated from the local sleeve resistance as

$$V_s = 118.8 \log(f_s) + 18.5 \tag{4.56}$$

with V_s in m/s and f_s in kPa.

4.9.6.3 Shear Modulus from q_c and q_t

A direct estimate of G_{max} may also be made from q_c or q_t for fine-grained soils. Table 4.11 gives some reported correlations.

Figure 4.40 shows a global correlation between tip resistance and V_s and CPTU sleeve resistance for a number of soils, including both clays and sands.

4.9.7 Constrained Modulus

As previously discussed, the one-dimensional constrained modulus, M, is often correlated to CPT/CPTU tip resistance as

$$M = 1/m_v = \alpha'_c q_c \tag{4.57}$$

Table 4.11 Some reported correlations between q_c or q_t and G_{max}

Correlation	Clay	References
$G_{max} = 2.8\, q_c^{1.4}$	Greek clays	Bouckovalas et al. (1989)
$G_{max} = 2.78\, q_c^{1.335}$	Norwegian clay	Mayne & Rix (1993)
$G_{max} = 406\, \left(q_c\right)^{0.695} / e_0^{1.130}$		
$G_{max} = 50\, (q_t - \sigma_{vo})$	Japanese clays	Tanaka et al. (1994)
$G_{max} = 65\, (q_t - \sigma_{vo})$	Bangkok clays	Shibuya et al. (1998)
$G_{max} = 1.96\, q_c^{0.579} \left(1+B_q\right)^{1.202}$	Norwegian soft clays	Long & Donohue (2010)
$G_{max} = 30.1\, q_t^{0.31} \left(1+B_q\right)^{3.14}$	Jiangsu clays	Cai et al. (2014)

V_s in m/s, G_{max}, q_c and q_t in kPa.

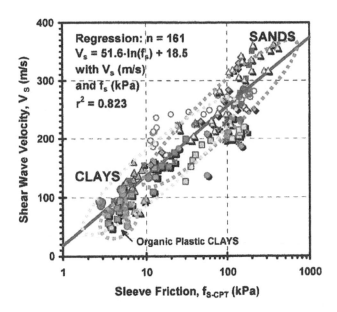

Figure 4.40 Correlation between V_S and cone sleeve resistance. (From Mayne 2007)

Kulhawy & Mayne (1990) compiled a piezocone database from a number of different clays and developed a correlation between constrained modulus and corrected CPTU tip resistance, q_t, as

$$M = 8.25(q_t - \sigma_{vo})$$
(4.58)

The test results are shown in Figure 4.41.

4.9.8 Coefficient of Consolidation

One of the unique features of the CPTU is that the pore water pressure measurement may be used to estimate the fluid flow characteristics of the soil. If the sounding is interpreted by stopping the advance of the cone at some desired test depth, then the pore water pressure may

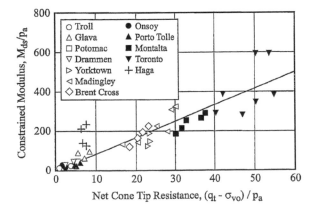

Figure 4.41 Correlation between constrained modulus and tip resistance in clays. (After Parez & Fauriel 1988.)

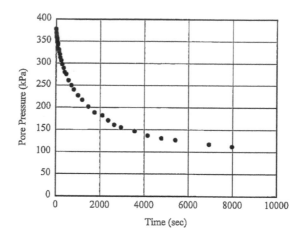

Figure 4.42 Typical CPTU pore pressure dissipation curve. (After Kulhawy & Mayne 1990.)

be monitored over time to provide a direct evaluation of the time-rate of pore pressure decay. An example of this decay is shown in Figure 4.42. This procedure is commonly referred to as a dissipation test. The results may be used to estimate the coefficient of consolidation and hydraulic conductivity. In practice, the advance is stopped, and the cone rods are locked off with time zero corresponding to the time when advance was halted.

Analytical and numerical solutions have been presented (e.g., Torstenson 1975, 1977; Baligh & Levadoux 1986; Gupta & Davidson 1986; Elsworth 1993; etc.) to give the coefficient of consolidation in terms of the cone radius, R, and the time factor T for different levels of consolidation, obtained from the normalized pore water pressure:

$$c_h = \left(TR^2\right)/t \tag{4.59}$$

The horizontal coefficient of consolidation is used in Equation 4.59 since the horizontal flow characteristics largely control the dissipation rate for all piezoelement locations. Table 4.12 gives a comparison of T values from different theories for 50% consolidation. The value of T is dependent on the cone apex angle, the rigidity index of the soil (E/s_u) and the pore pressure parameter A_f and is different for different piezoelement locations and the degree of dissipation.

The time factor presented by Gupta & Davidson (1986) for a 60° cone, with pore pressures measured at the cone base (Type 2) $E/s_u = 200$, and $A_f = 0.9$ ($t_{50} = 1.2$) corresponding to 50% dissipation provides values of c_h that compare well with average normally consolidated c_h values from oedometer tests. For a standard 10 cm² cone, Equation 4.59 becomes

$$c_h = \left(1.53 \times 10^{-3}\right)/t_{50} \ (m^2/s) \tag{4.60}$$

Table 4.12 Approximate CPTU time factors for 50% consolidation

Element location	t_{50}	References
Tip	1.2–3.2	Torstenson (1975, 1977)
Mid face	12.0	Chan (1982)
	3.6	Baligh & Levadoux (1986)
Cone base	5.2	Baligh & Levadoux (1986)
	1.2	Gupta & Davidson (1986)
	0.25	Houlsby & Teh (1988)

The time to reach 50% dissipation, t_{50}, is obtained from a plot of normalized excess pore water pressure, where U_n is defined as

$$U_n = (u_t - u_o)/(u_i - u_o) \tag{4.61}$$

where
$\quad$ u_t = measured pore water pressure at any time t after stopping the cone advance
$\quad$ u_o = *in situ* (initial) pore water pressure at the test depth
$\quad$ u_i = pore water pressure at time zero

An example of such a plot is shown in Figure 4.43 using the test data of Figure 4.42.

One problem that can occur with CPTU dissipation tests is that the results may show an initial rise in pore water pressure before the dissipation (decrease) actually starts. That is, there is a time lag to reach the maximum measured pore water pressure. This typically occurs in very stiff overconsolidated clays with Type 2 cones. For these cases, the author has taken the time where the observed maximum pore water pressure occurs as time zero and then adjusted all subsequent times to this value.

Even though different time factors are available for different consolidation levels, the author routinely carries all CPTU dissipation tests to a minimum of 50% dissipation in order to use t_{50} in Equation 4.60. A reasonably reliable estimate of u_o is needed to determine the normalized excess pore water pressure and can usually be obtained from piezometer data or knowledge of ground water conditions at the site.

Teh & Houlsby (1991) suggested that the theoretical time factor could be normalized by the rigidity index ($I_r = G/s_u$) to give a modified time factor:

$$T^* = (c_h t)/(R^2 I_r^{0.5}) \tag{4.62}$$

Using 50% dissipation, Robertson et al. (1992) prepared a chart relating c_h to t_{50} for different values of the rigidity index for both a 10 and 15 cm^2 piezocone given in Figure 4.44.

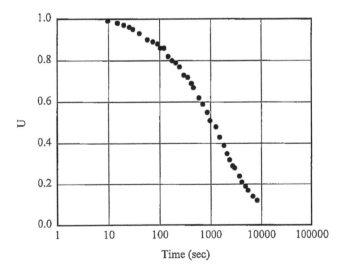

Figure 4.43 Normalized CPTU pore pressure dissipation curve.

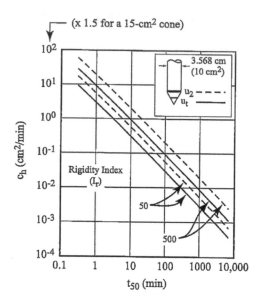

Figure 4.44 *Chart for estimating c_h from t_{50}. (After Robertson et al. 1992.)*

4.9.9 Hydraulic Conductivity

Using the results above to estimate c_h, the hydraulic conductivity may be estimated from dissipation tests as follows:

$$k_h = c_h \rho_w m_h \tag{4.63}$$

where
   k_h = horizontal hydraulic conductivity, (m/s)
   ρ_w = unit weight of water (kN/m³)
   m_h = coefficient of volume change, (m²/kN)

The value of m_h may be estimated using the relationship with cone tip resistance, q_c or q_t, as previously described.

   Baligh & Levadoux (1980, 1986) proposed that the horizontal hydraulic conductivity be evaluated from

$$k_h = (\rho_w \, RR \, c_h)/(2.3 \sigma'_{vo}) \tag{4.64}$$

where
   ρ_w = unit weight of water
   RR = recompression ratio
   c_h = coefficient of consolidation
   σ'_{vo} = initial *in situ* vertical effective stress

The recompression ratio must be evaluated in the laboratory using oedometer tests or by some other means. This presents a practical drawback to this approach.

   It has been suggested that the value of k_h may be estimated quickly based on the measured value of t_{50}, provided that all other variables remain constant. That is, the cone size, pore pressure element location, etc. are all the same at different sites. Even though soil properties may vary from site to site (e.g., I_r, A, OCR, etc.), it appears that from a practical standpoint, these are minor, especially considering the potential variability in hydraulic conductivity. Ventura (1983) suggested a simple expression relating hydraulic conductivity k and t_{50} as

$$k = \left(10^{-1}\right)\big/(z)(t_{50}) \text{ (cm/s)} \tag{4.65}$$

where
  z = depth of test in meters.

Parez & Fauriel (1988) suggested the chart shown in Figure 4.45 for estimating k_h based on t_{50} measurements obtained with a porous element located behind the cone tip. Results obtained by the author at a number of sites are shown in Figure 4.46, where the reference values of k_h were obtained from laboratory flexible wall tests on undisturbed samples with flow in the horizontal direction.

4.10 ADVANTAGES AND LIMITATIONS OF CPT/CPTU

The use of the CPT/CPTU offers some distinct advantages over a more conventional approach to geotechnical site investigations. Advantages of using the CPT/CPTU are listed in Table 4.13.
  There may also be both real and perceived limitations to using either the CPT or CPTU for site investigations. A number of limitations are given in Table 4.14.

4.11 CPT-SPT CORRELATIONS

Early comparisons between the CPT and SPT suggested that the ratio q_c/N tended to increase as the soil became coarser. Robertson et al. (1983) presented a chart correlating the ratio q_c/N with mean grain size, D_{50}, in mm. As shown in Figure 4.47, the ratio increases with mean grain size, consistent with previous observations. The ratio q_c/N was originally proposed as an attempt to convert q_c to N so that correlations previously developed

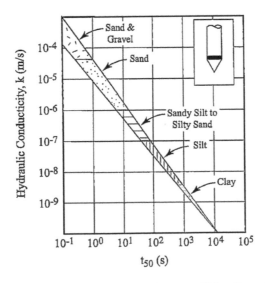

Figure 4.45 General trend between t_{50} and hydraulic conductivity. (After Parez & Fauriel 1988.)

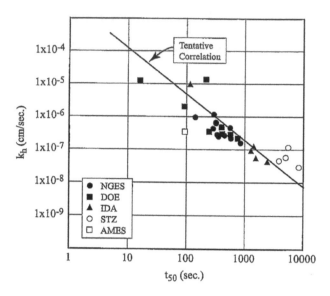

Figure 4.46 Observed results between t_{50} (Type 2) and laboratory hydraulic conductivity.

Table 4.13 Advantages of CPT/CPTU

Advantage	Notes
Speed	Deployment is usually faster than conventional drilling and sampling using boreholes; more locations can be investigated in the same total time
Continuous profiling	A near-continuous record of the subsurface conditions is obtained from changes in tip, sleeve, and pore pressure measurements; clays, sands, and gravelly soils provide uniquely distinct results; thin soil zones may be identified, which might otherwise be missed using regular drilling and sampling methods
Increased productivity	Because of the reduced time involved in the completion of a profile, a much larger number of areal points may be investigated at a lower cost compared to the more conventional drilling and sampling. This allows for adjustments in the site investigation to be made without lost time to the overall project
Minimal surface disturbance	Often, no test boring is required, and therefore, there is minimal cleanup and minimal surface restoration required at the end of testing; no drill cuttings are generated for removal or disposal
Economics	Overall cost of the site investigation program may be reduced by reducing the number of samples and laboratory tests required
Reduced time of investigation	Because the test is rapid, the data may provide a large amount of information about the subsurface conditions quickly and provide the opportunity to review results in a timely fashion
Large volume of high-quality data	The amount of information obtained from profiling far exceeds that obtained by traditional drilling, sampling, and laboratory testing
Testing soils that are difficult to sample	Characteristics of soils that are difficult or impossible to sample may be evaluated

between N and soil properties or foundation design could be used. Many comparisons have been presented between q_c and N showing a general trend but with considerable scatter. Zervogiannis & Kalteziotis (1988) suggested that results be expressed as

$$q_c = 4.66\,N\,D_{50}^{0.25}$$

(4.66)

Table 4.14 Potential disadvantages of CPT/CPTU

No sample	No sample is obtained for using in a visual-manual classification or for other laboratory testing. This is a disadvantage that many engineers are uncomfortable with. The test results are used to infer soil type
Complexity	Cone tests are more complex than either the SPT or DCPT. That is, they generally require more set-up and more initial attention to small mechanical or electrical details. The equipment is more complicated and often requires a portable computer or data acquisition system, which may also require an external power source, etc. Piezocones may require equipment for deairing and calibration
Availability	In many parts of the world, reliable CPT/CPTU equipment is not readily available
Initial cost	The initial investment in purchasing CPT/CPTU equipment represents a significant capital expense

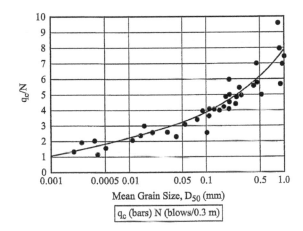

Figure 4.47 Early correlation between CPT q_c and SPT N. (After Robertson et al. 1983.)

It appears that the relationship between q_c/N and D_{50} by Robertson et al. (1983) may represent an approximate average for relatively clean coarse-grained soils, generally with content of fines less than about 10%. The ratio q_c/N also decreases with increasing content of fines as shown in Figure 4.48.

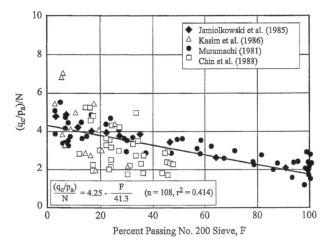

Figure 4.48 Influence of fines content on q_c/N. (After Kulhawy & Mayne 1990.)

The observed scatter in q_c/N relationships may result, in part, from differences in test procedures and equipment used to perform both the SPT and CPT. Differences in q_c values obtained from either electric or mechanical cones, as well as differences in reported N values obtained using various SPT equipment may create substantial variations in the q_c/N ratio for a given sand. In order to reduce some of this scatter, it may be more appropriate to evaluate the ratio q_c/N_{60}. At a given site, the ratio may not be constant for the same soil but may also be influenced by the mean stress or the overburden stress since q_c and N may be affected differently by the stress field.

There is strong evidence that suggests that there may be a better correlation between f_s and N (e.g., Kruizinga 1982; Takesue et al. 1996). This may be more intuitive, especially if we consider that for most soils, the SPT N value is dominated by sampler side friction.

4.12 CPT/CPTU IN FOUNDATION DESIGN

As with the SPT discussed in Chapter 2, there are many applications of the CPT/CPTU in geotechnical design.

4.12.1 Shallow Foundations

Several methods have been suggested for determining the ultimate bearing capacity and settlement of shallow foundations through the use of the CPT by direct and indirect or empirical methods (Meyerhof 1965; Schmertmann 1970, 1978a; Schmertmann et al. 1978; Goel 1982).

Tand et al. (1995) reviewed load tests from nine footings on lightly cemented medium dense sand and proposed a relationship between the ultimate bearing capacity and cone tip resistance as

$$q_{ult} = R_k q_c + \sigma_{vo} \tag{4.67}$$

where

q_{ult} = stress producing a relative settlement (s/B) equal to 0.05B
R_k = is a factor that varies from 0.14 to 0.19 depending on depth and width of the footing
q_c = average CPT tip resistance
σ_{vo} = total overburden stress at the base of the footing

Equation 4.67 is a similar approach to the method used to estimate the ultimate capacity of shallow foundations using the pressuremeter, discussed in Chapter 7. A similar method was presented by Tand et al. (1986) for shallow foundations on clay.

Eslaamizaad & Robertson (1996) presented a simplified approach for estimating the ultimate bearing capacity of shallow foundations on sands as

$$q_{ult} = K q_c \tag{4.68}$$

where

K = empirical correlation factor related to footing shape and embedment

A recommended chart for different shaped footings is shown in Figure 4.49.

A direct design approach for estimating both bearing capacity and settlement of shallow foundations on sands, similar to that discussed in Chapter 2, has also been presented

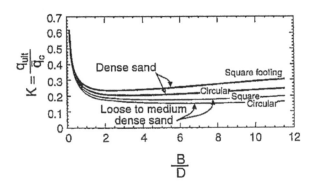

Figure 4.49 Correlation between ultimate bearing capacity for footings on sand. (After Eslaamizaad & Robertson 1996; from Lunne et al. 1997.)

(Mayne & Illingworth 2010; Mayne et al. 2012) using relative footing settlement. Results collected from a large number of footing load tests on a wide range of sands with different footing shapes showed that

$$q/q_c = 0.585(s/B)^{0.5} \tag{4.69}$$

where
  q = applied stress
  q_c = average cone tip resistance beneath the footing for a depth of 1.5B
  s = settlement
  B = footing width

Defining the ultimate bearing capacity as the stress producing a relative settlement of 10% of the footing with gives

$$q_{ult} = 0.18\,q_c \tag{4.70}$$

4.12.2 Deep Foundations

The CPT/CPTU has been used extensively to design deep foundations. There are several reasons for considering the CPT for direct design of deep foundations, especially driven piles: (1) the test procedure and equipment are standardized, (2) the cone resembles a model pile, (3) a continuous profile of the soil supporting the pile is usually obtained, and (4) the test and design procedures are relatively fast. However, because of differences in installation methods, scale effects, loading rates, soil disturbance, and other effects, it should be expected that there is a need for some adjustment to the measured cone response to the design of a pile.

At least 25 different methods have been suggested for estimating the capacity of deep foundations from CPT/CPTU results. The methods vary in their use of tip resistance and sleeve friction; however, most rely predominantly on cone tip resistance. Some methods have been developed using a database of full-scale pile load tests, exclusive to either soil type or the method of pile installation (e.g., Almeida et al. 1996; Eslami & Fellenius 1997; VanDijk & Kolk 2011). Niazi & Mayne (2013) presented a review and summary of existing methods.

The method of predicting the axial capacity of driven and jacked steel piles in clay presented by Almeida et al. (1996) is based on the results of 43 pile load tests at eight sites. The clays at the sites ranged in stiffness from very soft to very stiff, and the piles ranged in diameter from about 0.1 to 0.8 m. The design method is a direct empirical method and is based on estimating the unit end bearing and side resistance from the net cone tip resistance from

$$q_p = (q_t - \sigma_{vo})/k_2 \tag{4.71}$$

$$f_p = (q_t - \sigma_o)/k_1 \tag{4.72}$$

where
  q_p = pile unit end bearing
  f_p = Pile unit side resistance
  k_1 and k_2 = empirical constants

Almeida et al. (1996) recommended that the value of k_2 be taken as

$$k_2 = N_{KT}/9 \tag{4.73}$$

where
  N_{KT} = cone factor for undrained shear strength

The value of k_1 varies with the normalized net corrected tip resistance and shows considerable scatter; however, Almeida et al. (1996) suggested that an average value of observed data would give

$$k_1 = 12.0 + 14.9 \log\left[(q_t - \sigma_{vo})/\sigma'_{vo}\right] \tag{4.74}$$

Eslami & Fellenius (1997) presented a design method that also uses results from the CPTU. As opposed to other methods that use the arithmetic average of the cone tip resistance, this method uses the geometric average. In addition, the corrected cone tip resistance, q_t, is converted to "effective" cone tip resistance, q_E, by subtracting the measured CPTU pore water pressure, u_2, obtained at the cone base:

$$q_E = q_t - u_2 \tag{4.75}$$

The unit end bearing and the unit side resistance are obtained from

$$q_p = C_t q_{Eg} \tag{4.76}$$

$$f_p = C_s q_{Eg} \tag{4.77}$$

where
  q_{Eg} = the geometric average "effective" tip resistance
  C_t = pile toe correlation coefficient
  C_s = pile shaft correlation coefficient

Based on a comparison with pile tests and other studies, the toe correlation coefficient, C_t, is taken equal to 1.0. The value of C_s is related to soil type and is obtained from the

Table 4.15 Shaft correlation coefficient C_s for use in Eslami & Fellenius (1997)

Soil zone	Soil type	C_s Range (%)	C_s Approximation (%)
1	Soft, sensitive soils	7.37–8.64	8.0
2	Clay	4.62–5.56	5.0
3	Stiff clay and mixtures of clay and silt	2.06–2.80	2.5
4	Mixtures of silt and sand	0.87–1.34	1.0
5	Sand	0.34–0.60	0.4

recommended "soil profiling chart" previously shown in Figure 4.17. In this procedure, the soil is "classified" according to one of five different soil types. The value of C_s is then obtained from Table 4.15.

4.13 SUMMARY OF CPT/CPTU

The CPT and CPTU have established a solid position in *in situ* testing. The tests are generally reliable and can be performed with simple equipment. The cost of both regular cones and piezocones has come down in recent years as the technology has become more simplified. Provided the tests can be reliably deployed, the results are superior to other penetration tests, especially the SPT and DCP. A large database now exists for both coarse-grained and fine-grained soils for estimating specific soil properties and/or behavior. Engineers should work to incorporate both tests, where appropriate, into routine practice. CPT and CPTUs are no longer only applicable to large projects but can be used on many projects to enhance the site investigation.

REFERENCES

Almeida, M., Danziger, F., and Lunne, T., 1996. Use of the Piezocone Test to Predict the Axial Capacity of Driven and Jacked Piles in Clay. *Canadian Geotechnical Journal*, Vol. 33, pp. 23–41.

Anagnostopoulos, A., Koukis, G., Sabatakakis, N., and Tsiambaos, G., 2003. Empirical Correlations of Soil Parameters Based on Cone Penetration Tests (CPT) for Greek Soils. *Geotechnical and Geological Engineering*, Vol. 21, pp. 377–387.

Andrus, R., Stokoe, K., and Juang, C., 2004. Guide for Shear-Wave-Based Liquefaction Potential Evaluation. *Earthquake Spectra*, Vol. 20, No. 2, pp. 285–308.

Arango, I., 1997. Historical and Continued Role of the Standard Penetration Test Method in Geotechnical Earthquake Engineering. *3rd Seismic Short Course on Evaluation and Mitigation of Earthquake Induced Liquefaction Hazards*, San Fransico, CA.

Baldi, G., Bellotti, R., Ghionna, V., Jamiolkowski, M., and Pasqualini, E., 1982. Design Parameters for Sands from CPT. *Proceedings of the 2nd European Symposium on Penetration Testing*, Vol. 2, pp. 425–438.

Baldi, G., Bellotti, R., Ghionna, V., Jamiolkowski, M., and Pasqualini, E., 1986. Interpretation of CPT's and CPTU's Part 2, Drained-Penetration of Sands. *4th International Geotechnical Seminar Field Instrumentations and In Situ Measuarements, Singapore*, pp. 143–156.

Baldi, G., Bellotti, R., Ghionna, V., Jamiolkowski, M., and Lo Presti, D.C.F., 1989. Modulus of Sands from CPT's and DMT's. *Proceedings of the 12th International Conference on Soil Mechanics and Foundation Engineering*, Vol. 1, pp. 165–170.

Baligh, M.M., 1975. Theory of Deep Site Static Cone Penetration Resistance. *Report No. R-75-76,* Massachusetts Institute of Technology.

Baligh, M.M. and Levadoux, J.N., 1980. Pore Pressure Dissipation after Cone penetratin. Research Report No. R80-11. Department of Civil Engineering, MIT, Cambridge, MA, 367 pp.

Baligh, M.M. and Levadoux, J.N., 1986. Consolidation after Undrained Piezocone Penetration II: Interpretation. *Journal of the Geotechnical Engineering, ASCE,* Vol. 112, No. 7, pp. 727–745.

Been, K., Crooks, J.H.A., Becker, D.A., and Jefferies, M.G., 1986. The Cone Penetration Test in Sands: Part I, State Parameter and Interpretation. *Geotechnique,* Vol. 36, No. 2, pp. 239–250.

Been, K. and Jefferies, M.G., 1985. A State Parameter for Sands. *Geotechnique,* Vol. 35, No. 2, pp. 99–112.

Been, K., Jefferies, M.G., Crooks, J.H.A., and Rothenburg, L., 1987. The Cone Penetration in Sands: Part II, General Inference of State. *Geotechnique,* Vol. 37, No. 3, pp. 285–299.

Begemann, H.K.S., 1953. Improved Method of Determining Resistance to Adhesion by Sounding Through a Loose Sleeve Placed Behind the Cone. *Proceedings of the 3rd International Conference on Soil Mechanics and Foundation Engineering,* Vol. 1, pp. 213–217.

Begemann, H.K.S., 1965. The Friction Jacket Cone as an Aid in Determining the Soil Profile. *Proceedings of the 6th International Conference on Soil Mechanics and Foundation Engineering,* Vol. 1, pp. 17–20.

Bergado, D. and Kahaleque, M., 1986. Correlations of LLT Pressuremeter, Vane and Dutch Cone Tests in Bangkok Marine Clay, Thailand. ASTM STP 950, pp. 339–353.

Borowcszyk, M. and Szymanski, A., 1995. The Use of In Situ Tests for Determination of Stress History. *Proceedings of the 11th European Conference on Soil Mechanics and Foundation Engineering,* Vol. 1, pp. 17–22.

Bowles, J.E., 1988. *Foundation Analysis and Design,* 4th Edition. McGraw-Hill, Inc., New York.

Bouckovalas, G., Kalteziotis, N., Sabatakakis, N., and Zervogiannis, C., 1989. Shear Wave Velocity in a Very Soft Clay. *Proceedings of the 12th International Conference on Soil Mechanics and Foundation Engineering,* Vol. 1, pp. 191–194.

Brand, E., Moh, Z., and Wirojanagud, P., 1974. Interpretation of Dutch Cone Tests in Soft Bangkok Clay. *Proceedings of the European Symposium on Penetration Testing,* Vol. 2.2, pp. 51–58.

Cai, G., Liu, L., Tong, L., and Du, G., 2010. Field Evaluation of Undrained Shear Strength from Piezocone Tests in Soft Marine Clay. *Marine Georesources and Geotechnology,* Vol. 28, pp. 143–153.

Cai, G., Puppala, A., and Liu, S., 2014. Characterization on the Correlation between Shear Wave Velocity and Piezocone Tip Resistance of Jiangsu Clays. *Engineering Geology,* Vol. 171, pp. 96–130.

Campanella, R.G. and Roberston, P.K., 1981. Applied Cone Research, *Cone Penetration Testing and Experience, ASCE,* pp. 343–362.

Campanella, R.G., Robertson, P.K., and Gilllespie, D., 1983. Cone Penetration Testing in Deltaic Soils. *Canadian Geotechnical Journal,* Vol. 20, No. 1, pp. 23–35.

Campanella R.G., Robertson, P.K., Gillespie, D, and Grieg, J., 1985. Recent Developments in In Situ Testing of Soils. *Proceedings of the 11th International Conference on Soil Mechanics and Foundation Engineering,* Vol. 2, pp. 849–854.

Cancelli, A., Guadagnini, R., and Pellegrini, M., 1982. Friction-Cone Penetration Testing in Alluvial Clays. *Proceedings of the 2nd European Symposium on Penetration Testing,* Vol. 1, pp. 513–518.

Castro, G., 1969. Liquefaction of Sand. *PhD thesis, Division of Engineering and Applied Physics,* Harvard University.

Castro, G., Poulos, S.J., and Leathers, F.D., 1985. Re-examination of Slide of Lower San Fernando Dam. *Journal of the Geotechnical Engineering Division, ASCE,* Vol. 111, No. 9, pp. 1093–1107.

Chapman, G.A. and Donald, I.B., 1981. Interpretation of Static Penetration Tests in Sand. *Proceedings of the 10th International Conference on Soil Mechanics and Foundation Engineering,* Vol. 2, pp. 455–458.

Chan, A., 1982. Analysis of Dissipation of Pore Pressures after cone penetration. M.S. Thesis, Department of Civil Engineering, Louisiana State University, Baton Rouge, 120 pp.

Chen, B.S. and Mayne, P.W., 1994. Profiling the Over consolidation Ratio of Clays by Piezocone Tests. *Georgia Institute of Technology Report No. GIT-CEEGEO-94-1.*

Chung, S., Randolph, M., and Schneider, J., 2006. Effect of Penetration Rate on Penetrometer Resistance in Clay. *Journal of Geotechnical and Geoenvironmental Engineering, ASCE,* Vol. 132, No. 9, pp. 1188–1196.

Chung, S., Chung, J., Jang, W., and Lee, J., 2010. Correlations between CPT and FVT Results for Busan Clay. *Marine Georesources and Geotechnology,* Vol. 28, pp. 49–63.

Dahlberg, R., 1974. Penetration Pressure and Screw Plate Tests in a Preloaded Natural Sand Deposit. *Proceedings of the 1st European Symposium on Penetration Testing,* Vol. 2.2, pp. 69–87.

Das Neves, E.M., 1982. Direct and Indirect Determination of Alluvial Sand Deformability. *Proceedings of the 2nd European Symposium on Penetration Testing,* Vol. 1, pp. 95–99.

DeLima, D.C. and Tumay, M.T., 1991. Scale Effects in Cone Penetration Tests. *Geotechnical Engineering Congress, ASCE,* Vol. 1, pp. 38–51.

Denver. H., 1982. Modulus of Elasticity for Sand Determined by SPT and CPT. *Proceedings of the 2nd European Symposium on Penetration Testing,* Vol. 1, pp. 35–40.

DeRuiter, J., 1982. The Static Cone Penetration Test: State-of-the-Art Report. *Proceedings of the 2nd European Symposium on Penetration Testing,* Vol. 2, pp. 389–405.

Douglas, B.J., 1984. The Electric Cone Penetrometer Test: A User's Guide to Contracting for Services, Quality Assurance, Data Analysis. *The Earth Technology Corporation,* Long Beach, CA.

Douglas, B.J. and Olsen, R.S., 1981. Soil Classification Using Electric Cone Penetrometer. *Cone Penetration Testing and Experience, ASCE,* pp. 209–227.

Durgunoglu, H.T. and Mitchell, J.K., 1973. Static Penetration Resistance of Soils. Research Report Prepared for NASA Headquarters, Department of Civil Engineering, University of California, Berkeley, CA.

Dayal, U. and Allen, J., 1975. The Effect of Penetration Rate on the Strength of Remolded Clay and Sand Samples. *Canadian Geotechnical Journal,* Vol. 22, pp. 336–348.

Elsworth, D., 1993. Piezometer Cone Penetration Tests in Marine Subsoil Profiles. *Journal of Geotechnical Engineering, ASCE,* Vol. 119, No. 10, pp. 1601–1623.

Eslaamizaad, S. and Robertson, P., 1996. Cone Penetration Test to Evaluate Baring Capacity of Foundations in Sands. *Proceedings of the 49th Canadian Geotechnical Conference.*

Eslami, A. and Fellenius, B., 1997. Pile Capacity by Direct CPT and CPTu Methods Applied to 102 Case Histories. *Canadian Geotechnical Journal,* Vol. 34, pp. 886–904.

Garga, V.K., and Quin, J.T., 1974. An Investigation on Settlements of Direct Foundations on Sand. *Settlement of Structures, BGS,* pp. 22–36.

Goel, M.C., 1982. Correlation of Static Cone Resistance with Bearing Capacity. *Proceedings of the 2nd European Symposium on Penetration Testing,* Vol. 1, pp. 575–580.

Greig, J.W., Campanella, R.G. and Robertson, P.K., 1986. Comparison of Field Vane Results with Other In Situ Tests Results. *Soil Mechanics Series No 106,* Department of Civil Engineering, University of British Columbia.

Gorman, C.T., Hopkins, T., and Drnevich, V., 1973. In Situ Shear Strength Parameters by Dutch Cone Penetration Tests. Research Report 374, Kentucky Department of Transportation, 12 pp.

Gupta, R.C. and Davidson, J.L., 1986. Piezoprobe Determined Coefficient of Consolidation. *Soils and Foundations,* Vol. 26, No. 1, pp. 12–22.

Hegazy, Y. and Mayne, P.W., 1995. Statistical Correlations between V_S and Cone Penetration Data for Different Soil Types. *Proceedings of the International Symposium on Cone Penetration Testing, CPT '95,* pp. 173–178.

Houlsby, G.T. and Hitchman, R., 1988. Calibration Chamber Tests of a Cone Penetrometer in Sand. *Geotechnique,* Vol. 38, No. 1, pp. 39–44.

Houlsby, G.T. and Teh, T.I., 1988. Analysis of the Piezocone in Clay. *Proceedings of the 1st International Symposium on Penetration Testing,* Vol. 2, pp. 777–783.

Idriss, I. and Boulanger, R., 2006. Semi Empirical Procedures for Evaluating Liquefaction Potential during Earthquakes. *Soil Dynamics and Earthquake Engineering,* Vol. 26, Nos. 2–4, pp. 115–130.

Jacobs, P.A. and Coutts, J.S., 1992. A Comparison of Electric Piezocone Tips at the Bothkennar Test Site. *Geotechnique*, Vol. 42, No. 2, pp. 369–375.

Jamiolkowski, M., Lancellotta, R., Tordella, L., and Ballaglio, M., 1982. Undrained Strength for CPT. *Proceedings of the 2nd European Symposium on Penetration Testing*, Vol. 2, pp. 599–606.

Jamiolkowski, M., Ghionna, V.N., Lancellotta, R., and Pasqualini, E., 1988. New Correlations of Penetration Tests for Design Practice. *Proceedings of the 1st International Symposium on Penetration Testing*, Vol. 1, pp. 263–296.

Janbu, N. and Senneset, K., 1974. Effective Stress Interpretation of In-Situ Static Penetration Tests. *Proceedings of the 1st European Symposium on Penetration Testing*, Vol. 2, No. 2, pp. 181–194.

Jeffries, M.G. and Been, K., 2006. Soil Liquefaction – A Critical State Approach. Taylor & Francis, London.

Jekel, J.W., 1988. Wear of the Friction Sleeve and its Effect on the Measured Local Friction. *Proceedings of the 1st International Symposium on Penetration Testing*, Vol. 2, pp. 805–808.

Jones, G.A. and Rust, E.A., 1982. Piezometer Penetration Testing, CUPT. *Proceedings of the 2nd European Symposium on Penetration Testing*, Vol. 2, pp. 607–613.

Juang, C., Chen, C., and Mayne, P.W., 2008. CPTU Simplified Stress-Based Model for Evaluating Soil Liquefaction Potential. *Soils and Foundations*, Vol. 48, No. 6, pp. 755–770.

Kayabali, K., 1996. Soil Liquefaction Evaluation Using Shear Wave Velocity. *Engineering Geology*, Vol. 44, Nos. 1–4, pp. 121–127.

Kim, K., Prezzi, M., Salgado, R., and Lee, W., 2008. Effect of Penetration Rate on Cone Penetration Resistance in Saturated Clayey Soils. *Journal of Geotechnical and Geoenvironmental Engineering*, ASCE, Vol. 134, No. 8, pp. 1142–1153.

Kok, L., 1974. The Effect of the Penetration Speed and the Cone Shape on the Dutch Static Cone Penetration Test Results. *Proceedings of the 1st European Symposium on Penetration Testing*, Vol. 2.2, pp. 215–220.

Konard, J.M., 1987. Piezo-Friction-Cone Penetrometer Testing in Soft Clays. *Canadian Geotechnical Journal*, Vol. 24, No. 4, pp. 645–652.

Kruizinga, J., 1982. SPT-CPT Correlations. *Proceedings of the 2nd European Symposium on Penetration Testing*, Vol. 1, pp. 91–96.

Kulhawy, F.H. and Mayne, P.W., 1990. Manual for Estimating Soil Properties for Foundation Design. *Electric Power Research Institute Report 1493-6*.

Lambrechts, J.R. and Leonards, G.A., 1978. Effects of Stress History on Deformation of Sand. *Journal of the Geotechnical Engineering Division*, ASCE, Vol. 104, No. GT11, pp. 1371–1387.

Larsson, R. and Mulabdic, M., 1991. Piezocone Tests in Clay. *Swedish Geotechnical Institute Report No. 42*.

Lehane, B., O'Loughlin, C., Gaudin, C., and Randolph, M., 2009. Rate Effects on Penetrometer Resistance in Kaolin. *Geotechnique*, Vol. 59, No. 1, pp. 41–52.

Long, M. and Donohue, S., 2010. Characterization of Norwegian Marine Clays with Combined Shear Wave Velocity and Piezocone Penetration Test (CPTU) Data. *Canadian Geotechnical Journal*, Vol. 47, No. 7, pp. 709–718.

Lunne, T., Eide, O., and de Ruiter, J., 1976. Correlations Between Cone Resistance and Vane Shear Strength in Some Scandinavian Soft Medium Stiff Clays. *Canadian Geotechnical Journal*, Vol. 13, No. 4, pp. 430–441.

Lunne, T., Lacasse, S., and Rad, N.S., 1992. General Report: SPT, CPT, Pressuremeter Testing and Recent Developments. *Proceedings of the 12th International Conference on Soil Mechanics and Foundation Engineering*, Vol. 4, pp. 2339–2403.

Lunne, T., Robertson, P., and Powell, J., 1997. *Cone Penetration Testing in Geotechnical Practice*. Blackie Academic and Professional Publishers, New York.

Marr, L.S. and Endley, S., 1982. Offshore Geotechnical Investigation Using Cone Penetrometer. Offshore Technology Conference, Houston, Paper OTC4298-MS.

Massarsch, K., 2014. Cone Penetration Testing – A Historic Perspective. *Proceedings of the 3rd International Symposium on Cone Penetration Testing*, Vol. 1, pp. 97–134.

Mayne, P.W., 1986. CPT Indexing of In Situ OCR in Clays. *Use of In Situ Tests in Geotechnical Engineering*, ASCE, pp. 780–793.

Mayne, P.W., 1995. Profiling Yield Stress in Clays by In Situ Tests. *Transportation Research Record* No. 1479, pp. 43–50.

Mayne, P.W., 1995a. CPT Determination of Overconsolidation Ratio and Lateral Stresses in Clean Quartz Sands. *Proceedings of the International Symposium on Cone Penetration Testing*, Vol. 2, pp. 215–220.

Mayne, P.W., 2006. In Situ Test Calibrations for Evaluating Soil Parameters. *Proceedings the Characterization and Engineering Properties of Natural Soils*, Vol. 3, pp. 1323–1379.

Mayne, P.W., 2014. Interpretation of Geotechnical Parameters from Seismic Piezocone Tests. *Proceedings of the 3rd International Symposium on Cone Penetration Testing*, pp. 47–73.

Mayne, P.W. and Chen, B.S., 1993. Effective Stress Method for Piezocone Evaluation of s_u. *Proceedings of the 3rd International Conference on Case Histories in Geotechnical Engineering*, Vol. 2, pp. 1305–1311.

Mayne, P.W. and Chen, B.S., 1994. Preliminary Calibration of PCPT - OCR Model for Clays. *Proceedings of the 13th International Conference on Soil Mechanics and Foundation Engineering*, Vol. 1, pp. 283–286.

Mayne, P.W., Cop, M.R., Springman, S., Huang, A.B. and Zornberg, J., 2009. Geomaterial Behavior and Testing, *Proceedings of the 17th International Conference on Soil Mechanics and Geotechnical Engineering*, pp. 2777–2872.

Mayne, P.W. and Kemper, J.B., 1988. Profiling OCR in Stiff Clays by CPT and SPT. *Geotechnical Testing Journal*, ASTM, Vol. 11, No. 2, pp. 139–147.

Mayne, P.W. and Rix, G.J., 1993. G_{max} – q_c Relationships for Clays. *Geotechnical Testing Journal, ASTM*, Vol. 16, No. 1, pp. 54–60.

Mayne, P.W. and Illingworth, F., 2010. Direct CPT Method for Footing Response in Sands Using a Database Approach. *Proceedings of the 2nd International Symposium on Cone Penetration Testing*, 8 pp.

Mayne, P.W., Uzielli, M., and Illingworth, F., 2012. Shallow Footing Response on Sands Using a Direct Method Based on Cone Penetration Tests. *Full-Scale Testing and Foundation Design*, ASCE, pp. 664–674.

Meigh, A.C., 1987. *Cone Penetration Testing Methods and Interpretation*. Butterworths, London, 141 p.

Meyerhof, G.G., 1965. Shallow Foundations. *Journal of the Soil Mechanics and Foundation Division*, ASCE, Vol. 91. No. SM2, pp. 21–31.

Mitchell, J.K. and Gardner, W.S., 1975. In Situ Measurement of Volume Change Characteristics. *In Situ Measurement of Soil Properties, ASCE*, Vol. 2, pp. 279–345.

Mola-Abasi, H., Kordtabar, B., and Kordnaeij, A., 2018. Liquefaction Potential Prediction Using CPT Data by Triangular Chart Identification. *International Journal of Geotechnical Engineering*, Vol. 12, No. 4, pp. 377–382.

Moss, R., Seed, R., Kayen, R., Stewaet, J., Der Kiureghian, A., and Cetin, K., 2006. CPT-Based Probabilistic and Deterministic Assessment of In Situ Seismic Sol Liquefaction Potential. *Journal of Geotechnical and Geoenvironmental Engineering, ASCE*, Vol. 132, No. 8, pp. 1032–1051.

Muhs, H. and Weiss, K., 1971. Investigation of Ultimate Bearing Capacity and Settlement Behavior of Individual Shallow Footings in Non-Uniform Cohesionless Soil (in German). *DEGEBO (Journal of the German Soil Mechanics Society)*, University of Berlin, Vol. 26, pp. 1–37.

Muromachi, T., 1981. Cone Penetration Testing in Japan. *Cone Penetration Testing and Experience*, ASCE, pp. 49–75.

Niazi, F. and Mayne, P.W., 2013. Cone Penetration Test Based Direct Methods for Evaluating Static Axial Capacity of Single Piles. *Geotechnical and Geological Engineering*, Vol. 31, pp. 979–1009.

Oliveira, J., Almeida, M.S., Motta, H., and Almeida, M.C., 2011. Influence of Penetration Rate on Penetrometer Resistance. *Journal of Geotechnical and Geoenvironmental Engineering, ASCE*, Vol. 1237, No. 7, pp. 695–703.

Parez, L. and Fauriel, R., 1988. Le Piezocone Amerliorations Apportes a la Reconnaissance des Sols. *Review Francais Geotechnology*, Vol. 44, pp. 13–27.

Parkin, A.K., 1988. The Calibration of Cone Penetrometers. *Proceedings of the 1st International Symposium on Penetration Testing*, Vol. 1, pp. 221–244.

Piratheepan, P., 2002. Estimating Shear Wave Velocity from SPT and CPT Data. MS Thesis, Clemson University.

Poulos, S.J., 1981. The Steady State of Deformation. *Journal of the Geotechnical Engineering Division, ASCE*, Vol. 107, No. GT 5, pp. 553–562.

Powell, J.J.M. and Quarterman, R.S.T., 1988. The Interpretation of Cone Penetration Tests in Clays with Particular Reference to Rate Effects. *Proceedings of the 1st International Symposium on Penetration Testing*, Vol. 2, pp. 903–909.

Powell, J.J.M., Quarterman, R.S.T., and Lunne, T., 1989. Interpretation and Use of the Piezocone Test in U.K. Clays. *Proceedings of the Symposium on Penetration Testing in the U.K.*, pp. 151–156.

Quiros, G. and Young, A., 1988. Comparison of Filed Vane, CPT and Laboratory Strength Santa at Santa Barbara Cannel Site, ASTM STP 1014, pp. 307–317.

Rad, N.S. and Lunne, T., 1988. Direct Correlations Between Piezocone Test Results and Undrained Shear Strength of Clay. *Proceedings of the 1st International Symposium on Penetration Testing*, Vol. 2, pp. 911–917.

Robertson, P., 1990. Soil Classification Using the Cone Penetration Test. *Canadian Geotechnical Journal*, Vol. 27, No. 1, pp. 151–158.

Robertson, P., 2009. Interpretation of Cone Penetration Tests – A Unified Approach. *Canadian Geotechnical Journal*, Vol. 46, pp. 1337–1355.

Robertson, P.K. and Campanella, R.G., 1983. Interpretation of Cone Penetration Tests - Part I: Sands. *Canadian Geotechnical Journal*, Vol. 20, No. 4, pp. 718–733.

Robertson, P., and Campanella, R.G., 1983a. Interpretation of Cone Penetration Tests - Part II: Clay. *Canadian Geotechnical Journal*, Vol. 20, pp. 734–745.

Robertson, P. and Campanella, R., 1985. Liquefaction Potential of Sand Using the CPT. *Journal of Geotechnical Engineering, ASCE*, Vol. 111, No. 3, pp. 384–403.

Robertson, P.K., Campanella, R.G. and Wightman, A., 1983. SPT-CPT Correlations. *Journal of the Geotechnical Engineering Division, ASCE*, Vol. 109, No. 11, pp. 1454–1459.

Robertson, P.K., Sully, J.P., Woeller, D.J., Lunne, T., Powell, J.J.M., and Gillespie, D.G., 1992c. Estimating Coefficient of Consolidation from Piezocone Tests. *Canadian Geotechnical Journal*, Vol. 29, No. 4, pp. 539–550.

Robertson, P. and Wride, C., 1998. Evaluating Cyclic Liquefaction Potential Using the Cone Penetration Test. *Canadian Geotechnical Journal*, Vol. 35, No. 3, pp. 442–459.

Robertson, P., Campanella, R., Gillespie, D., and Greig, J., 1986. Use of Piezometer Cone Data. *Use of In Situ Tests in Geotechnical Engineering, ASCE*, pp. 1263–1280.

Roth, W.H., Swantko, T.D., Patil, U.K., and Berry, S.W., 1982. Monorail Piers on Shallow Foundations, Settlement Analysis Based on Dutch Cone Data. *Proceedings of the 2nd European Symposium on Penetration Testing*, Vol. 2, pp. 821–826.

Roth, W.H., Swantko, T.D., Patil, U.K., and Berry, S.W., 1982. Monorail Piers on Shallow Foundations, Settlement Analysis Based on Dutch Cone Data. *Proceedings of the 2nd European Symposium on Penetration Testing*, Vol. 2, pp. 821–826.

Roy, M., Blanchet, R., Tavenas, F., and LaRochelle, P., 1981. Behavior of a Sensitive Clay During Pile Driving. *Canadian Geotechnical Journal*, Vol. 18, No. 1, pp. 67–85.

Sanglerat, G., 1972. *The Penetration and Soil Exploration*. Elsevier Publishing Co., Amsterdam.

Schmertmann, J.H., 1970. Static Cone to Compute Static Settlement Over Sand. *Journal of the Soil Mechanics and Foundation Division, ASCE*, Vol. 96, No. SM3, pp. 1011–1043.

Schmertmann, J.H., 1978a. Use of the SPT to Measure Dynamic Soil Properties? - Yes, But...! *ASTM Special Technical Publication 654*, pp. 341–355.

Schmertmann, J.H., 1978b. Guidelines for Cone Penetration Test: Performance and Design, *U.S. D.O.T., FHWA Report TS78-209*.

Schmertmann, J.H., Hartman, J.D. and Brown, P.R., 1978. Improved Strain Influence Factor Diagrams. *Journal of the Geotechnical Division, ASCE*, Vol. 104, No. GT8, pp. 1131–1135.

Schnaid, F., Lehane, B., and Fahey, M., 2004. In Situ Test Characterization of Unusual Geomaterials. *Proceedings of the 2nd International Symposium on Soil Characterization*, Vol. 1, pp. 49–74.

Senneset, K, and Janbu, N., 1985. Shear Strength Parameters Obtained from Static Cone Penetration Tests. *ASTM Special Technical Publication* Vol. 883, pp. 41–54.

Senneset, K., Janbu, N., and Svano, G., 1982. Strength and Deformation Parameters from Cone Penetration Tests. *Proceedings of the 2nd European Symposium on Penetration Testing*, Vol. 2, pp. 863–870.

Shibuya, S., Hanh, L., Wilailak, K., Lohani, T., Tanaka, H., and Hamouche, K., 1998. Characterizing Stiffness and Strength of Soft Bangkok Clay from In Situ and Laboratory Tests. *Proceeding of the 1st International Conference on Site Characterization*, Vol. 2, pp. 1361–1366.

Stark, T.D. and Delashaw, J.E., 1990. Correlations of Unconsolidated - Undrained Triaxial Tests and Cone Penetration Tests. *Transportation Research Record No. 1278*, pp. 96–102.

Stark, T. and Olson, S., 1995. Liquefaction Resistance Using CPT and Field Case Histories. *Journal of Geotechnical and Geoenvironmental Engineering, ASCE*, Vol. 121, No. GT12, pp. 856–869.

Sully, J.P., Campanella, R.G., and Robertson, P.K., 1988. Overconsolidation Ratio of Clays from Penetration Pore Water Pressures. *Journal of Geotechnical Engineering, ASCE*, Vol. 114, No. GT 2, pp. 209–215.

Suzuki, Y., Koyamada, K., Tokimatsu, K., Taya, Y., and Kubota, Y., 1995b. Empirical Correlation of Soil Liquefaction Based on Cone Penetration Test. *Proceedings of the International Conference on Earthquake Geotechnical Engineering*, Vol. 1, pp. 369–374.

Sykora, D. and Stokoe, K., 1983. Correlations of In Situ Measurements in Sands of Shear Wave Velocity. *Soil Dynamics and Earthquake Engineering*, Vol. 20, pp. 125–136.

Takesue, K., Sasao, H., and Makihara, Y., 1996. Cone Penetration Testing in Volcanic Soil Deposits. *Proceedings of the International Conference on Advances in Site Investigation Practice*, pp. 452–463.

Tanaka, H., Tanaka, M., Iguchi, H., and Nishida, K., 1994.Shear Modulus of Soft Clays Measured by Various Kinds of Tests. *Proceedings of the International Symposium on Pre-Failure Deformation of Geomaterials*, Vol. 1, pp. 235–240.

Tand, K.E., Funegard, E.G., and Briaud, J.-L., 1986. Bearing Capacity of Footings on Clay – CPT Method. *Use of In Situ Tests in Geotechnical Engineering, ASCE*, pp. 1017–1033.

Tand, K.E., Warden, P., and Funegard, E., 1995. Predicted-Measured Bearing Capacity of Footings on Sand. *Proceedings of the International Symposium on Cone Penetration Testing*, Vol. 2, pp. 589–594.

Tavenas, F. and Leroueil, S., 1987. Laboratory and In Situ Stress-Strain-Time Behavior of Soft Clays: A State-of-the-Art. *Proceedings of the International Symposium on Geotechnical Engineering of Soft Clays*, Mexico City, pp. 1–146.

Teh, C.I. and Houlsby, G.T., 1991. An Analytical Study of the Cone Penetration Test in Clay. *Geotechnique*, Vol. 41, No. 1, pp. 17–34.

Tekamp, W.G.B., 1982. The Influence of the Rate of Penetration on the Cone Resistance "q_c" in Sand. *Proceedings of the 2nd European Symposium on Penetration Testing*, Vol. 2, pp. 627–633.

Torstensson, B.A., 1975. Pore Pressure Sounding Instrument. *In Situ Measurement of Soil Properties, ASCE*, Vol. 2, pp. 48–54.

Torstensson, B.A., 1977. Time-Dependent Effects in the Field Vane Test. *Proceedings of the International Symposium on Soft Clays*, Bangkok, pp. 387–397.

Treen, C.R., Robertson, P.K., and Woeller, D.J., 1992. Cone Penetration Testing in Stiff Glacial Soils Using a Downhole Cone Penetrometer. *Canadian Geotechnical Journal*, Vol. 29, No. 3, pp. 448–455.

Tumay, M.T., Acar, Y., Deseze, E., and Yilmaz, R., 1982. Soil Exploration in Soft Clays with the Quasi-Static Electric Cone Penetrometer. *Proceedings of the 2nd European Symposium on Penetration Testing*, Vol. 2, pp. 915–921.

Uzielli, M., Mayne, P.W. and Cassidy, M.J., 2013. Probabilistic Assessment of Design Strengths for Sands from In-Situ Testing Data. *Modern Geotechnical Design Codes of Practice, Advances in Soil Mechanics and Geotechnical Engineering*, Vol. 1, pp. 214–227.

Van de Graaf, H.C. and Jekel, J.W.A., 1982. New Guidelines for the Use of the Inclinometer with the Cone Penetration Test. *Proceedings of the 2nd European Symposium on Penetration Testing*, Vol. 2, pp. 581–584.

VanDijk, B. and Kolk, H., 2011. CPT Based Method for Axial Capacity of Offshore Piles. *Frontiers in Offshore Geotechnics*, Vol. II, pp. 555–559.

Vlasblom, A., 1985. The Electrical Penetrometer; A Historical Account of Its Development. *Delft Soil Mechanics Laboratory Publication No. 92*, 51 pp.

Wissa, A.E.Z., Martin, R.T., and Garlanger, J.E., 1975. The Piezometer Probe. *In Situ Measurement of Soil Properties*, ASCE, Vol. 1, pp. 536–545.

Wroth, C.P., 1988. Penetration Testing - A More Rigorous Approach to Interpretation. *Proceedings of the 1st International Symposium of Penetration Testing*, Vol. 1, pp. 303–311.

Zervogiannis, C. and Kalteziotis, N., 1988. Experiences and Relationships from Penetration Testing in Greece. *Proceedings of the 1st International Symposium on Penetration Testing*, Vol. 2, pp. 1063–1071.

Chapter 5

Field Vane Test (FVT)

5.1 INTRODUCTION

The field Vane Test (FVT) was developed for determining the undrained shear strength of soft clays and was used in Sweden as early as 1919 by Olsson (Kallstenius 1956). The vane was developed in part to overcome problems with obtaining undisturbed samples, especially in soft clay, for determining laboratory undrained shear strength. In the early 1950s, observations often showed that the shear strength from laboratory tests were low compared to back-calculated strengths from field cases. In soft clays, early use of the FVT often showed higher strengths as compared to traditional laboratory tests.

The results of undrained strength measurements obtained with the FVT in soft clays are often used as the reference value of *in situ* undrained strength for many engineers. Correlations developed to estimate undrained strength from other *in situ* tests, for example, the CPTU or DMT, often rely on reference values obtained from the FVT.

The most common application of the FVT is in soft to medium stiff saturated clays; however, there are a number of reported uses of the field vane in other materials, including organic soils and peat (e.g., Helenelund 1967; Northwood & Sangrey 1971; Landva 1980; Faust et al. 1983; Muhuai et al. 1983). FVTs may be difficult to perform in very stiff overconsolidated clays or fissured clays. Test results in very stiff clays can be difficult to interpret as a result of questions related to drainage conditions, failure conditions, etc. FVTs conducted in these materials generally do not fit into the framework of normal FVT interpretation developed for soft clays.

5.2 MECHANICS

The FVT is conducted by inserting a thin four-blade-vane into the ground and then rotating the vane to create a shear failure in the soil, as shown in Figure 5.1. The usual geometry of most vanes is rectangular with a height to diameter ratio (H/D) of 2 although special vanes with different H/D ratios may be used to evaluate anisotropy. Thin blades are normally used in an attempt to reduce the amount of disturbance to the soil structure and state of stress prior to shearing. The undrained shear strength of the soil is not measured directly in the test, but the maximum torque needed to cause rotation and failure of the soil is measured. To estimate the undrained shear strength from the torque, a number of simplifying assumptions need to be made regarding the failure mechanism.

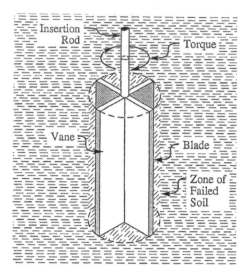

Figure 5.1 Principle of Vane Shear Test.

5.3 EQUIPMENT

There are generally four ways that the vane may be inserted to perform the test: (1) using an unprotected vane advanced at the base of an open or cased borehole; (2) using an unprotected vane with a double-rod system without a borehole; (3) using a protective housing in which both the rods and vane are protected without using a borehole; or (4) using both unprotected rods and an unprotected vane without a borehole with a rod slip coupling located just above the vane. A schematic of these methods is shown in Figure 5.2. These methods are primarily related to vane tests performed on land.

The pushing thrust to advance the vane is usually provided by hydraulic pressure from a drill rig or a small reaction frame operated by either a hand crank mechanism or a portable hydraulic cylinder. The torque is normally measured using some form of a calibrated torque head operated by a simple hand crank located at the ground surface. The different techniques for deploying the vane have developed as a result of efforts to make the testing more efficient while at the same time retaining high quality testing. Each of the methods may be used in conjunction with a borehole, a condition that may arise out of necessity to advance through surface materials, such as fill, sand layers, or very stiff crusts. Geise et al. (1988) described a remote offshore vane that uses the housing as torque reaction down hole. Other successful uses of field vane offshore have been reported (e.g., Young et al. 1988)

5.3.1 Unprotected Vane Through Casing

When pushed at the bottom of a borehole as shown in Figure 5.2a, the vane is normally inserted through a casing or hollow-stem augers since an open borehole may collapse or squeeze, especially in soft clay. The use of rod centralizers inside the casing helps keep the vane rods vertical and centered at the base of the hole and also helps prevent bending of the rods during pushing. The casing or hollow stem augers then become the reaction base for the torque head.

This method of testing can be slow since the vane must be withdrawn after each test and the casing and borehole must be advanced to a new test depth before performing the next test.

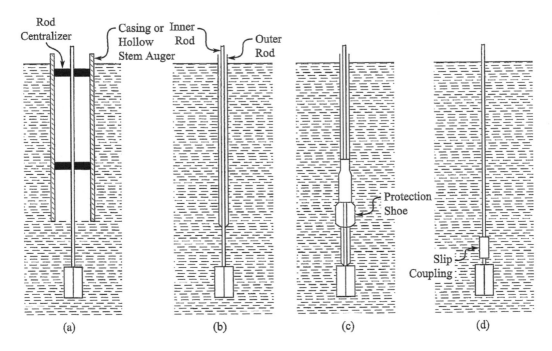

Figure 5.2 Methods of inserting field vane. (a) unprotected vane through a cased borehole, (b) unprotected vane with protected rods, (c) protected vane and rods, and (d) unprotected vane and rods with slip coupling.

Open hole drilling with just a surface casing may also be used. The measured torque will include a small unknown component of soil friction on the rods just above the vane, which is difficult to account for in the data reduction. Additionally, the drilling may produce unknown disturbance and stress relief effects at the bottom of the hole, both of which may lead to errors in interpretation. Because of these difficulties, this method of testing is considered to be the least desirable approach to FVT.

5.3.2 Protected Rods and Unprotected Vane

The use of a double-rod system to advance the rods and vane as shown in Figure 5.2b increases testing efficiency and eliminates the need for a borehole, provided that the system can be advanced by pushing from the surface. Some difficulties may be encountered pushing through fill or a very stiff crust near the surface, and usually an open or cased hole is needed. The outer protective rods act to prevent the inner vane rods from bending during pushing and help eliminate friction on the inner rods. Advance of the system is stopped above the test depth, and then the inner rod is pushed further to force the vane out ahead of the outer protective rods while the outer rods are held fixed.

5.3.3 Protected Rods and Protected Vane

The use of a protective vane housing was presented by Cadling & Odenstad (1950) as a means of eliminating rod friction and protecting the vane from bending but still allowing the vane to be advanced efficiently. The system is shown in Figure 5.2c and allows the vane and rods to advance while being fully protected. The advantage of using a protective vane housing is that the vane itself may be machined with very thin blades in an attempt to reduce soil disturbance (e.g., Geonor SGI Vane Borer H-10). To perform the test, the system is pushed to a depth just above the test depth, and then the vane is advanced just ahead of the protective housing while the inner rods remain free from contact with the soil.

5.3.4 Unprotected Rods and Unprotected Vane with Slip Coupling

In this case, both the vane rods and vane are unprotected as shown in Figure 5.2d and are pushed from near the surface without a borehole (e.g., Nilcon Vane Borer M-1000). A rotary slip coupling is located between the rods and vane. The slip coupling allows a rotation of the rods of about 15° before the vane is engaged. The torque measurement during this initial rotation represents the friction acting on the rods alone without any torque being applied to the vane. Once the vane is engaged, the measured torque represents both rod friction and soil resistance from rotation of the vane. The portion of the measured torque attributed to just the vane is obtained by subtraction. This method of testing is efficient, especially in soft and very soft clays where the rods and vane may be inserted with little pushing effort provided the rods do not bend. Figure 5.3 shows a photo of the torque head of the Nilcon Vane Borer. A sample recording of the measured torque showing the rod friction, peak strength, post-peak strength, and remolded strength scribed onto the paper disc is shown in Figure 5.4.

5.3.5 Vanes

Different manufacturers supply different size vanes to suit different soil conditions. Common sizes include vanes with nominal diameters of 50 mm (2 in.), 65 mm (2.5 in.), and 80 mm (3.1 in.). Some vanes have a tapered blade cross section and are tapered at the base and have rounded corners at the top as shown in Figure 5.5. Rectangular vanes with blades

Figure 5.3 Vane shear test in progress using Nilcon M-1000 Vane Borer Torque Head attached to casing.

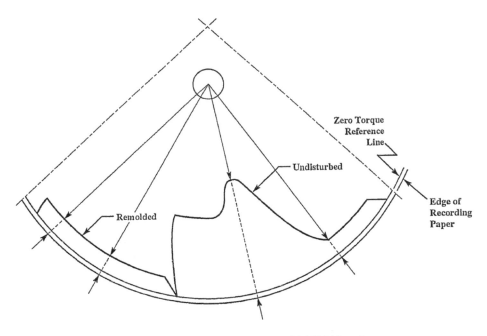

Figure 5.4 Sample wax chart recording of torque from Nilcon M-1000 Vane Borer.

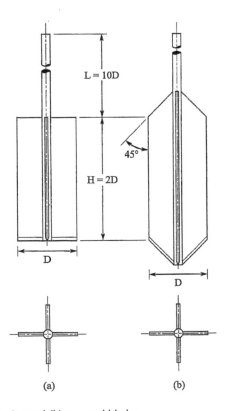

Figure 5.5 Vane with (a) rectangular and (b) tapered blades.

Table 5.1 ASTM D2573 recommended dimensions of field vanes

Casing size	Diameter in. (mm)	Height in. (mm)	Thickness of blade in. (mm)	Diameter of rod in. (mm)	Perimeter ratio (%)
AX	1.5 (38.1)	3 (76.2)	0.063 (1.6)	0.5 (12.7)	5.3
BX	2 (50.8)	4 (101.6)	0.063 (1.6)	0.5 (12.7)	4.0
NX	2.5 (63.5)	5 (127.0)	0.125 (3.2)	0.5 (12.7)	6.4
101.6	3.6 (92.1)	7.2 (184.1)	0.125 (3.2)	0.5 (12.7)	4.4

of constant cross section are preferred to reduce uncertainties with interpretation of the results. Table 5.1 gives different sizes of vanes suggested by ASTM D2573. A vane height to diameter ratio of 2 is specified in ASTM D2573.

It is noted by ASTM D2573 that "the selection of the vane size is directly related to the consistency of the soil being tested, that is, the softer the soil, the larger the vane diameter"; however, this may cause some problems with interpretation as will be discussed. Also, because of the dimensions given in Table 5.1, the level of disturbance created by all of the vanes is not consistent, since the perimeter ratio (defined in Section 5.6.1) for each vane is not constant.

5.4 TEST PROCEDURES

A recommended method for performing the FVT is described by ASTM in standard test method D2573-08 *Standard Test Method for Field Vane Shear Test in Cohesive Soil*. The International Organization for Standardization (ISO) also provides a recommended standard for FVTs ISO/DIS 22476-9. The test is performed by first inserting the vane to the desired test depth using one of the procedures described or other suitable technique. In the case where a borehole or a protective vane housing is used, it is recommended by the ASTM standard that the hole or housing should be no closer than five borehole diameters or five vane housing diameters from the test location. This provision is given to reduce disturbance effects from drilling. It is further suggested that the vane should be inserted in a single thrust without applying torque to the vane rods. No recommendation is given for rate of installation. It is recommended that tests be performed at intervals of not less than 0.61 m (2 ft) throughout the soil profile, except when a very small vane is used.

Once the vane is in position, torque is applied to the vane at a rate not to exceed 0.1°/s (6°/min). ASTM D2573 specifies that the waiting time between inserting the vane and beginning rotation should be no more than 5 min. Normally, the test should be performed within about 1–2 min after the installation. Using the recommended rate of vane rotation of 0.1°/s will give a time to reach peak torque ranging from 2 to 5 min except in very soft plastic clays where the time to failure may be more in the order of 10–15 min. In brittle soils that reach peak strength at very low strain levels, failure may occur very quickly.

After the maximum torque has been obtained, ASTM D2573 recommends that the rods be rotated a minimum of ten revolutions to remold the soil. Within about 1 min after this remolding process, the remolded strength is obtained by rotating the vane in the same manner as used to obtain the peak strength. Remolded strengths, s_{ur}, are used along with peak strength measurements, s_u, to provide an indication of soil sensitivity, S_t, as follows:

$$S_t = s_u/s_{ur} \qquad\qquad (5.1)$$

5.5 FACTORS AFFECTING TEST RESULTS

Several factors can affect the results obtained with the FVT and therefore can affect the interpretation of the undrained shear strength value obtained. These factors include both variations in the equipment used and variations in the test procedure.

5.5.1 Installation Effects

5.5.1.1 Disturbance

Insertion of the vane causes some degree of disturbance in nearly all soils. The degree of fabric disturbance may depend on the specific geometry of the vane and other components (i.e., blade thickness, diameter, rod size, etc.) and is also related to soil properties such as stress history, plasticity, and sensitivity. Disturbance occurs because a finite amount of soil must be displaced in order to allow the vane to occupy that space. Disturbance created by inserting the vane may also be expected to be greater for soils with high sensitivity.

Photographic observations of fabric disturbance by vane insertion and rotation (e.g., Arman et al. 1975; Chandler 1988; Roy & Leblanc 1988) indicate that the area of disturbance is comparable to that postulated by Cadling & Odenstad (1950). It may be that the most significant consequence of fabric disturbance is that the actual failure surface produced by the vane is somewhat larger than the vane itself. This is likely to be more significant in stiff overconsolidated clays and clays of low plasticity. The effective diameter of the failure surface appears to be about 5% larger than the vane in these soils. This would lead to a direct decrease in the estimated shear strength of about 16%.

Disturbance produced by inserting a vane should relate to the geometry of the vane for the same soil. The more soil that must be displaced to allow the vane to be inserted, the more the amount of disturbed soil. Thicker blades on the vane will produce more disturbance for a vane with a constant diameter. Similarly, vanes with the same blade thickness but different diameters should show less disturbance effects as the diameter increases. In order to rationally quantify the level of disturbance caused by the vane, Cadling & Odenstad (1950) suggested using the term "perimeter ratio", illustrated in Figure 5.6 and defined as follows:

$$\alpha = 4e/\pi D \tag{5.2}$$

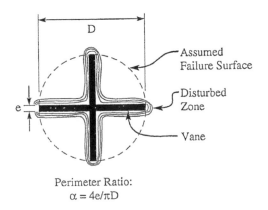

Perimeter Ratio:
$\alpha = 4e/\pi D$

Figure 5.6 Definition of perimeter ratio for vane.

where

α = perimeter ratio (usually expressed as %)
e = thickness of vane blade
D = diameter of vane.

If the zone of disturbed soil adjacent to the vane blades is related to the blade thickness, then for the same diameter vane, a larger amount of undisturbed soil for testing will result from using a vane with thinner blades. The undrained strength measurement from a thinner vane should be more representative of undisturbed conditions. Equation 5.2 suggests that in order to reduce disturbance effects, either the blade thickness should be reduced, or the vane diameter must be increased, or both. Typical commercial vanes have perimeter ratios in the range of 4%–8%. The perimeter ratios for the recommended vane dimensions given in Table 5.1 range from 4.0% to 6.4%

La Rochelle et al. (1973) illustrated the influence of vane perimeter ratio on the resulting strength profiles obtained in sensitive Champlain Sea clay (S_t = 12) by using vanes of the same diameter but different blade thicknesses. The resulting strength profiles are shown in Figure 5.7 and show that for most of the tests, the measured undrained shear strength increases as the vane blade thickness decreases, i.e., as α decreases.

By assuming that the true *in situ* undisturbed strength would be represented by a vane with a blade thickness of 0 (i.e., α = 0), the results presented in Figure 5.7 were used to extrapolate the measured undrained strength to a blade thickness of zero or "zero disturbance", as shown in Figure 5.8. An increase in the estimated undisturbed strength over the measured value using a vane with α = 5% of about 15% is indicated. Similar results have been presented by Roy & Leblanc (1988) for sensitive marine clay in Canada (S_t = 4–14) and Cerato & Lutenegger (2004) for varved clay in Massachusetts (S_t = 4–6) as shown in Figure 5.9. Results from these studies suggest that the extrapolation to zero disturbance can result in an increase in estimated shear strength in the order of 10%–15% over a vane with α = 5%. That is, a conventional vane gives a strength about 10%–15% lower than the true undisturbed strength simply because of disturbance from insertion.

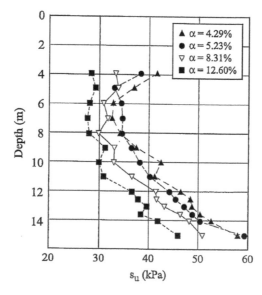

Figure 5.7 Influence of perimeter ratio (disturbance) on measured shear strength. (After La Rochelle et al. 1973.)

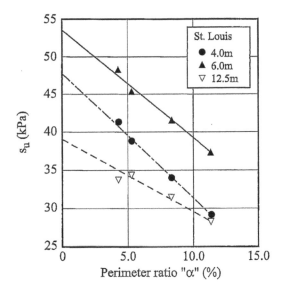

Figure 5.8 Extrapolation of vane data with different values of perimeter ratio to vane of zero blade thickness to estimate undisturbed strength. (After La Rochelle et al. 1973.)

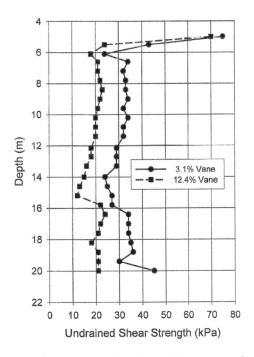

Figure 5.9 Influence of vane area ratio on measured undrained shear strength.

5.5.1.2 Insertion Pore Water Pressures

In addition to producing soil disturbance in the vicinity of the vane blades, inserting the vane in saturated clays is assumed to occur under undrained conditions. The insertion of the vane produces a change in the effective stress field surrounding the vane. Several laboratory studies using instrumented vanes have shown that significant excess pore water pressures can be generated during vane insertion in both natural and reconstituted soft clays (e.g., Matsui & Abe 1981; Kimura & Saitoh 1983; Roy & Mercier 1989). The magnitude of generated pore water pressure is in the order of $0.5–0.75\sigma_{vo}'$. These studies have also shown that there is very little dissipation of excess pore water pressure in the normal waiting time between insertion of the vane and the start of the vane rotation. It was observed that complete dissipation of the excess pore water pressure required as much as four to five hours, as discussed further in Section 5.6.2.

5.5.2 Delay (Consolidation) Time

As previously noted in Section 5.4, the ASTM procedure for conducting the FVT specifies that the waiting period between the end of insertion and the beginning of rotation should be no more than 5 min although this time typically is in the order of 1–2 min for routine work. In order to reduce the effects of installation on the measured strength, especially the effect of excess pore water pressures, it might be appropriate to allow a sufficient amount of time to lapse between vane insertion and rotation. During this time, excess pore water pressures created during installation will dissipate, and the soil will reconsolidate and possibly gain strength through thixotropic hardening taking place in the disturbed zone surrounding the vane blades.

Aas (1965) compared FVT results obtained at two sites in Norway between tests where the vane had been left in the ground for periods between 1 and 3 days before shearing and tests where the test was performed immediately. For two quick clays (S_t = 40–70) with low plasticity (P.I. = 6–8), the reported increase in undrained shear strengths from waiting ranged between 30% and 260% but typically were in the order of 40%. Flaate (1966a; 1966b) also found that the delay time between insertion of the vane and testing could have a significant effect on the measured undrained shear strength. He showed that in Norwegian marine clays, a delay time of up to 8 h between vane insertion and testing could produce an increase in undrained shear strength in the order of 20% above the strength that is normally obtained with a 5 min delay time (Flaate 1966a).

Results obtained in soft clays in Sweden were reported by Torstensson (1977), shown in Figure 5.10, and for soft marine clays in Canada by Roy & Leblanc (1988), shown in Figure 5.11. In both cases, the results show a rapid increase in the undrained shear strength during the first hour or so after vane insertion, and thereafter, a more gradual increase. Very little increase appears to have occurred after a 1-day time delay, even up to 7 days. The shape of the strength increase curves appears very similar to inverted time consolidation curves

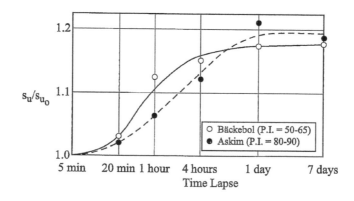

Figure 5.10 Influence of delay time (consolidation) on measured shear strength. (After Torstensson 1977.)

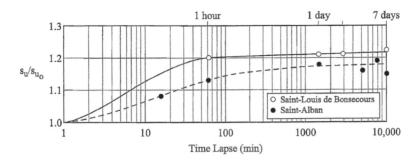

Figure 5.11 Influence of delay time (consolidation) on measured shear strength. (After Roy & Leblanc 1988.)

from oedometer tests and CPTU dissipation tests. This suggests that dissipation of excess pore water pressures around the vane accounts for the majority of the increase in measured strength as opposed to "aging", at least up to about a week.

The results presented in Figures 5.10 and 5.11 indicate that an increase in undrained shear strength of about 20% occurs above the reference undrained shear strength obtained without delay using the standard vane testing procedure. Results of two vane profiles obtained in soft varved clay by the author for routine tests and a delay of 1 day prior to shearing showed an average increase in strength of about 15% after 1 day.

5.5.3 Rate of Shearing

The rate of shearing may affect the measured peak torque in the field vane in a similar manner to undrained laboratory strength tests on clays. Cadling & Odenstad (1950) showed that vane strengths measured at a rotation rate of 1°/s (60°/min) were about 20% higher that strengths measured at a rate of 0.1°/s (6°/min) for soft clays in Sweden. A number of laboratory and field studies have been performed to investigate the influence of rate of shearing on the resulting undrained shear strength from the vane test (e.g., Aas 1965; Wiesel 1973; Torstensson 1977; Perlow & Richards 1977; Nathan 1978; Sharifounnasab & Ullrich 1985; Roy & Leblanc 1988; Hirabayashi et al. 2017). Many of these results have been discussed by Chandler (1988).

The rate of shearing of the cylinder of assumed failure is really determined in large part by the angular shear velocity at the outer edges of the vane blades. This angular velocity is influenced not only by the rotation rate of the vane but also the vane size and can be described as follows:

$$V = r\Omega \tag{5.3}$$

where
   V = angular shear velocity (mm/s)
   r = vane blade radius (mm)
   Ω = vane rotation rate (°/s).

Equation 5.3 states that if the same *rotation rate* is used with two vanes of different diameters, a different *shear rate* results. This may help explain differences observed sometimes in results obtained using different size vanes at the same site. Vanes of different sizes will give different results if the same rotation rate is used. Figure 5.12 shows the influence of angular velocity on measured strength for two clays with different P.I.

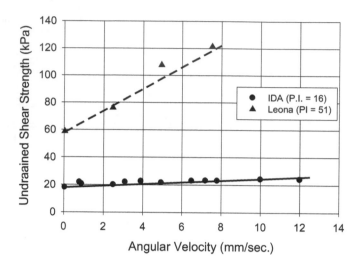

Figure 5.12 Influence of soil type on strain rate effect.

Since angular shear velocity increases with increasing vane diameter at the same rotation rate, significant differences in the shearing rate at failure exists between large and small field vanes and should be considered when comparing the results obtained with laboratory vanes and field vanes at similar rotation rates. The suggestion made by ASTM D2573 that the selection of the vane size is dependent on soil consistency can result in different values of measured undrained shear strength.

Peuchen & Mayne (2007) compiled results for a wide range of soils and suggested that the relationship between vane rotation rate and the normalized vane strength could be described by a power function as follows:

$$s_u/s_{uo} = (\theta / \theta_o)^\beta \tag{5.4}$$

where

s_u = undrained vane strength at any rotation rate
s_{uo} = undrained strength at a rate of 6°/min
θ = any rotation rate
θ_o = reference rotation rate of 6°/min
β = empirical exponent.

For most clays, the value of beta in Equation 5.4 varies from between 0.05 and 0.10, as shown in Figure 5.13.

Chandler (1988) noted that the shearing rate could affect soil behavior by allowing some drainage to occur during the testing. Suggestions for the rate of shearing for both high P.I. and low P.I. soils to maintain undrained conditions are shown in Figure 5.14.

5.5.4 Progressive Failure

Because of the geometry and the mode of failure in the test, the FVT may be subject to errors in measurement of undrained shear strength associated with progressive failure. When the maximum torque is recorded in the test, the soil on the top and bottom of the cylinder has already gone past peak strength and is somewhere in the post-peak region. It is also likely

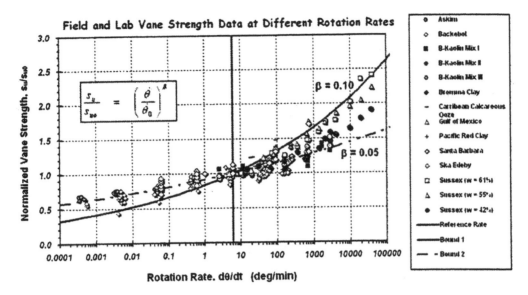

Figure 5.13 Influence of strain rate on shear strength. (From Peuchen & Mayne 2007.)

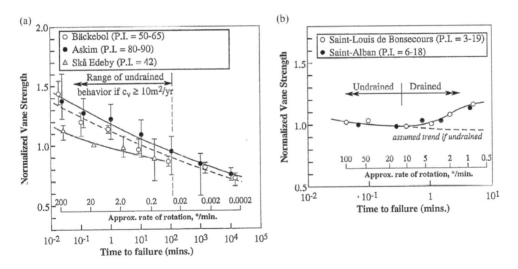

Figure 5.14 Time to failure to maintain undrained conditions: (a) high P.I. clays and (b) low P.I. clays. (After Chandler 1988.)

that in some soils, failure takes place progressively around the perimeter of the vane as the soil directly in front of the blades reaches failure first. The soil does not behave in a perfectly elastoplastic manner.

De Alencar et al. (1988) suggested that the progressive failure mechanism depends on the rate of post-peak softening and that the maximum rotation to failure and post-peak torque rotation are influenced by strain-softening behavior. At the present time, this effect on the resulting measured strength is not considered significant enough to create problems with interpretation in most soils; however, it may be partly responsible for the observations leading to the use of vane correction factors.

5.5.5 Vane Size

It has previously been shown that different diameter vanes, with the same H/D ratio, may give different results because a smaller vane produces a higher shearing velocity at the same rotation rate. However, for different vanes with the same blade thickness, a larger diameter vane will have a smaller perimeter ratio (Equation 5.2) and therefore will also tend to produce less disturbance.

5.5.6 Vane Shape

The shape of the vane may also affect the test results, especially if vanes of nonstandard geometry, i.e., different than H/D = 2 are used. Some early vane results reported in the literature were performed with vane having H/D = 1 and therefore may need to be reevaluated. Additionally, some tests have been conducted with vanes that have one end tapered.

Table 5.2 summarizes the important factors related to equipment and test procedure that may influence FVT results and provides references that the reader may wish to consult for more specific details.

Table 5.2 Factors that may affect the results obtained from the FVT

Factor	References
Vane shape	Cadling & Odenstad (1950)
	Osterberg (1956)
	Bazett et al. (1961)
Vane size	Cadling & Odenstad (1950)
	Kietkajornkul & Vasinvarthana (1989)
	Ahnberg et al. (2004)
Disturbance	Lo (1965)
	Lo & Milligan (1967)
	La Rochelle et al. (1973)
	Loh & Holt (1974)
	Andrawes et al. (1975)
	Roy & Leblanc (1988)
	Cerato & Lutenegger (2004)
Rate of shearing	Skempton (1948)
	Cadling & Odenstad (1950)
	Bazett et al. (1961)
	Aas (1965)
	Wiesel (1973)
	Monney (1974)
	Smith & Richards (1975)
	Perlow & Richards (1977)
	Torstensson (1977)
	Schapery & Dunlap (1978)
	Sharifounnasab & Ullrich (1985)
	Law (1985)
	Roy & Leblanc (1988)
	Biscotin & Pestana (2001)
	Peuchen & Mayne (2007)
	Schlue et al. (2010)
Delay time (consolidation)	Aas (1965)
	Torstensson (1977)
	Roy & Leblanc (1988)

5.6 INTERPRETATION OF UNDRAINED STRENGTH FROM FVT

The principal measurement obtained from the FVT is the maximum torque required to rotate the vane. In order to extract an estimate of the undrained shear strength, it is necessary to make a number of assumptions regarding the failure surface, the distribution of shear stresses on the ends and sides of the vane, etc. In addition to the factors listed in Table 5.2, a number of factors that may affect the interpretation of vane test results from an individual test are listed in Table 5.3.

Flaate (1966a) listed the following assumptions traditionally made for computing the undrained shear strength from the torque measured in the field vane:

1. The test is completely undrained, and no consolidation takes place during the installation and throughout completion of the test.
2. No disturbance is caused during advancement of the borehole or vane.
3. The remolded zone around the vane is very small.
4. The shear strength is fully mobilized and uniform over the entire cylindrical failure surface at peak torque (i.e., there is no progressive failure).
5. The maximum shear strength, s_u, is isotropic.

The undrained shear strength is calculated from the peak torque and based on the geometry of the assumed failure surface shown in Figure 5.2 and is obtained as follows:

Table 5.3 Factors that may influence interpretation of field vane results

Factor	References
Disturbance	Osterberg (1956)
	Eden & Hamilton (1956)
	Flaate (1966a)
Progressive failure	Burmister (1957)
	Wiesel (1967)
	Wiesel (1973)
	Donald et al. (1977)
	De Alencar et al. (1988)
Strength anisotropy	Aas (1965)
	Aas (1967)
	Wiesel (1967)
	Blight (1970)
	Donald et al. (1977)
	Silvestri & Aubertin (1988)
Pore pressure and drainage	Duncan (1967)
	Ladd et al. (1977)
	Matsui & Abe (1981)
	Chandler (1988)
Assumed stress distribution	Flaate (1966a)
	Donald et al. (1977)
	Menzies & Merrifield (1980)
	Wroth (1984)
Assumed geometry of failure surface	Skempton (1948)
	Arman et al. (1975)
	Chandler (1988)
	Roy & Leblanc (1988)
	Veneman & Edil (1988)
	Gylland et al. (2013)

$$T = s_u (\pi DH)(D/2) + 2s_u (\pi D^2/4)(D/a) \tag{5.5}$$

Solving for s_u gives

$$s_u = (2T)/(\pi D^2(H + D/a)) \tag{5.6}$$

where
   T = torque
   D = diameter of the vane
   H= height of the vane
   a= a shape factor to account for assumed distribution of shear stress.

Assuming a = 3 for uniformly distributed end shear gives

$$s_u = T/(\pi(D^2H/2 + D^3/6)). \tag{5.7}$$

For a standard vane with H/D = 2, Equation 5.7 reduces to

$$s_u = 0.86T/(\pi D^3). \tag{5.8}$$

One of the assumptions used in the calculation of undrained shear strength is that the strength is isotropic; i.e., the same unit strength is developed along the sides as well as the top and bottom of the vane. Additionally, it is assumed that the shear stress distribution on the cylindrical (vertical) surface and on the top and bottom (horizontal) surfaces is uniform, giving a rectangular distribution. Results of a three-dimensional finite element analysis presented by Donald et al. (1977) showed that while there was generally uniform distribution (rectangular) along the sides or vertical surface of the vanes, the distribution along the vane top and bottom was not rectangular. Experimental results obtained by Menzies & Merrifield (1980) using an instrumented vane showed similar results.

Wroth (1984) considered the effect of this more realistic stress distribution on vane results assuming that the shear stress distribution on the top and bottom surfaces could be expressed by a polynomial. As a result of this analysis, Wroth (1984) showed that for H/D = 2,

$$s_u = 0.94T/(\pi D^3) \tag{5.9}$$

Equation 5.9 suggests that most of the torque measured in the FVT is a result of the soil shearing resistance along the vertical surface, which means that the undrained strength obtained from the conventional interpretation (Equation 5.8) will be underestimated by about 9%.

5.7 ANISOTROPIC ANALYSIS

As described in Section 5.7, the normal assumption regarding the interpretation of FVT results is that the undrained strength is isotropic. A number of investigations have been performed using the FVT to evaluate the directional anisotropy of undrained shear strength (Table 5.4). In many cases, vanes of different geometries, as shown in Figure 5.15, have been used to determine the shear strength in different directions.

Table 5.4 Investigations of undrained shear strength anisotropy
using field vane

Soil Type	References
Norwegian clays	Aas (1965; 1967)
Fill	Wiesel (1967; 1973)
	Blight (1970; 1982)
Bangkok clay	Eide & Holmberg (1972)
	Richardson et al. (1975)
Louisiana clays	Ahmed (1975)
French clays	LeMasson (1976)
Marine clay	Menzies & Mailey (1976)
Brazil clays	Costa Filho et al. (1977)
Residual soil	Josseaume et al. (1977)
Marine clay	Hanzawa (1979)
Alluvial clay	Toh & Donald (1979)
Alluvial Clay-Thailand	Memon (1980)
Champlain clay	Silvestri & Aubertin (1983; 1988)
Finland clay	Slunga (1983)
Rio de Janeiro clay	Garga (1988)
Rio de Janeiro clay	Garga & Khan (1992)
Champlain clay	Silvestri et al. (1993)

The deficiency in the conventional approach to interpretation of FVT results suggests that vanes with different H/D give different s_u. This means that test data from early use of the field vane or with laboratory vane devices with H/D not equal to 2 (e.g., Skempton 1948; Bazett et al. 1961) may need to be reevaluated in relation to modern test results and interpretation. Analyses by the approaches suggested by both Aas (1967) and Wiesel (1973) using vanes of different H/D may be used to determine the undrained strength on the horizontal plane (s_{uh}) and vertical plane (s_{uv}). Figure 5.16 shows results obtained in marine clay using vanes of different geometries.

Several investigations (e.g., Bjerrum 1973; Richardson et al. 1975) have related the strength anisotropy ratio $K_s = s_{uh}/s_{uv}$ to soil plasticity. However, other test results (e.g., Silvestri & Aubertin 1983) show that the anisotropy ratio does not always follow the trend suggested by Bjerrum (1973). The anisotropy ratio may be related to the P.I. in soft near normally consolidated nonstructured or brittle soils (e.g., Ladd et al. 1977) where K_o is in the range of 0.5–0.7; however, there is also considerable evidence demonstrating that K_s is also related to K_o in overconsolidated soils (e.g., Garga & Khan 1992).

5.8 MEASURING POSTPEAK STRENGTH

In most cases, the torque measurement obtained during the test will show a drop off and stabilize after reaching the peak torque with continued vane rotation, as shown in Figure 5.4. This postpeak strength generally levels off to a relatively constant value with vane rotation in the order of 60°–90°. Following complete remolding by rotating the vane through 10–15 complete revolutions, the remolded strength shows considerably lower strength. The postpeak strength in the vane test has been referred to by others (Tammirinne 1981; Pyles 1984;

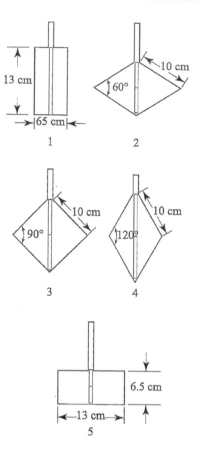

Figure 5.15 Different shaped vanes to determine strength anisotropy.

Chaney & Richardson 1988; Johnson et al. 1988) as the "residual" strength or "ultimate" strength (e.g., Hight et al. 1992); however, it may be preferable to use the term postpeak. Similar reduction from peak to postpeak strength from FVTs has been noted by Kolk et al. (1988).

The difference between the peak strength and postpeak strength may be related to the loss of soil structure and is analogous to the Undrained Brittleness Index (Bishop 1971). Johnson et al. (1988) noted that for deep water clays in the Gulf of Mexico, the ratio of peak to "residual" strength decreased with increasing Liquidity Index, L.I. The postpeak strength may be appropriate for large strain design problems, such as the stiff clay crusts under embankment or other shallow loadings, or for undrained analysis of driven displacement piles, where significant structure is lost during pile installation (Miller & Lutenegger 1993). Figure 5.17 shows peak, postpeak, and remolded strengths measured in a lacustrine clay.

5.9 FIELD VANE CORRECTION FACTORS

In 1972, Bjerrum suggested that for design of embankments on soft clays, the results obtained from FVTs in clays should be corrected for stability analysis. Bjerrum (1972) compiled several existing case histories of embankment failures. Using average vane strength

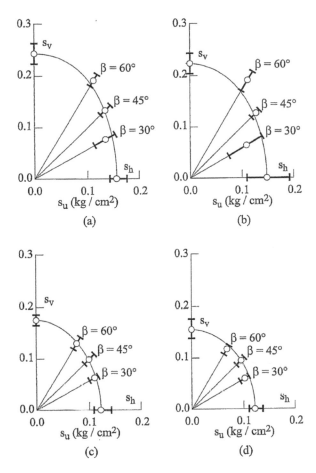

Figure 5.16 Determination of undrained shear strength on different planes: (a) depth = 1 m, depth = 2 m, depth = 3 m, and depth = 4 m. (after Richardson et al. 1975).

data and assuming that the factor of safety would be equal to 1.0 at failure, Bjerrum plotted the calculated factor of safety obtained using the FVT results as a function of the soil P.I. as shown in Figure 5.18. A field vane correction factor was then suggested by Bjerrum (1972) to force the field vane results to produce a factor of safety equal to 1.0. Although other vane correction factors have been suggested (e.g., Pilot 1972; Dascal & Tournier 1975; Helenelund 1977; Larsson 1980; Larsson et al. 1987), the figure presenting the proposed correction factor as originally given by Bjerrum (1972) is shown in Figure 5.19. According to Bjerrum (1972), the corrected undrained strength is obtained as follows:

$$\left(s_u\right)_{corr} = \mu\left(s_u\right)_{FV} \tag{5.10}$$

Bjerrum (1973) attributed much of the difference between the back-calculated shear strength and the measured field vane strength to effects resulting from strain rate and anisotropy as well as from progressive failure. Relative to normal full-scale loading, the testing rate used in the FVT is too fast. These rate effects would be more pronounced in clays of high plasticity. The measured strength would be too high, leading to a correction factor less than 1.0 to reduce the measured strength to operational field strength. Bjerrum (1973) attempted to

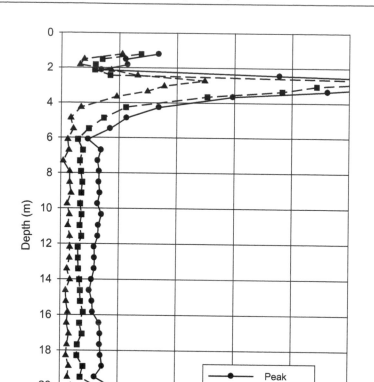

Figure 5.17 Peak, postpeak, and remolded strength profile from FVT in lacustrine clay.

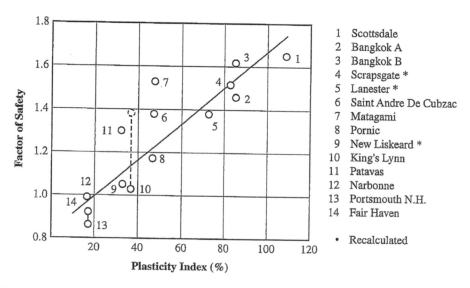

Figure 5.18 Back-calculated factors of safety from embankment cases (After Bjerrum 1972.)

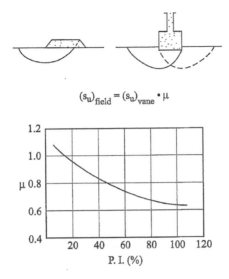

$$(s_u)_{field} = (s_u)_{vane} \cdot \mu$$

Figure 5.19 Bjerrum (1972) vane correction factor.

separate the effects of strain rate and anisotropy suggesting that Equation 5.10 be modified as follows:

$$(s_u)_{corr} = (s_u)_{FV} \, \mu_A \, \mu_R \qquad (5.11)$$

where

μ_A = correction for anisotropy
μ_R = correction for strain rate.

Sageseta & Arroyo (1982) and Arroyo & Sagaseta (1988) illustrated that the factors of safety of a number of individual embankment case histories could be brought nearer in line with F.S. = 1.0 if a reduction were applied to field vane strength obtained in the upper weathered crust. Azzouz et al. (1983) recommended a new correction factor for use in embankment design as shown in Figure 5.20. The results of analyses included end effects and increased the plane strain factor of safety by about 10%. Therefore, the proposed new correction factor is seen to give corrected field vane strength about 10% lower than recommended by Bjerrum (1972). As pointed out by Surendra & Mundell (1984), the recommendation of Azzouz et al. (1983) only attempts to account for the method of the stability analysis used and not to any other factors associated with potential inherent drawbacks of the FVT.

Chandler (1988) suggested that the strain rate correction factor could be obtained as follows:

$$\mu_R = 1.05 - b(P.I.)^{0.5} \, (\text{for P.I.} < 5\%) \qquad (5.12)$$

where
$b = 0.015 - 0.0075 \log t_f$

where
t_f = time to failure (between 10 and 10,000 min).

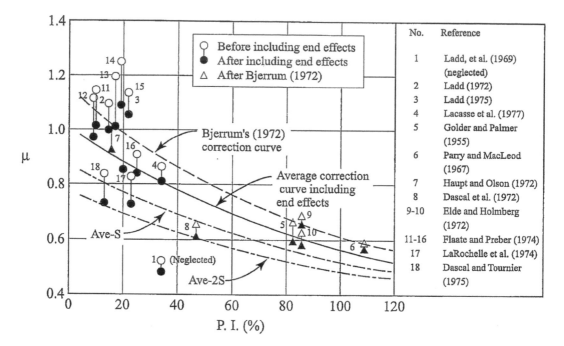

Figure 5.20 Vane correction factor suggested by Azzouz et al. (1983).

The scatter in data points originally presented by Bjerrum (1972) and expanded upon by others (e.g., Azzouz et al. 1983; Aas et al. 1986; Tang et al. 1992; Tanaka & Tanaka 1993; Tanaka 1994) is significant, suggesting that a more detailed explanation is needed before a correction be applied to field vane results. Aas et al. (1986) suggested that much of the scatter and uncertainty may be explained by the fact that Bjerrum did not distinguish between clays of different types of stress history. No mention was made by Bjerrum (1972; 1973) of the differences in behavior of either "young" vs. "aged" clays, or normally consolidated vs. overconsolidated clays. Correction factors recommended by Bjerrum (1972) for the design of embankments on clay should be viewed with caution and should not be used without great care for other design problems, e.g., undrained behavior of piles.

5.10 INTERPRETATION OF STRESS HISTORY FROM FVT

Jamiolkowski et al. (1985) presented the results of normalized field vane undrained shear strength as a function of laboratory OCR and suggested a reasonable fit to the data as follows:

$$s_{uo(FVT)}/\sigma'_{vo} = s_1(OCR)^m \tag{5.13}$$

where
   s_1 = undrained vane strength ratio for OCR = 1.

Data from nine different sites gave mean values of $s_1 = 0.22$ and m = 1 (disregarding one extreme value of $s_1 = 0.74$; m = 1.51). The value of s_1 may be related to P.I. as suggested by Bjerrum (1973).

Mayne & Mitchell (1988) used a database consisting of 262 data points of both field vane results and laboratory measured overconsolidation ratio, OCR, and suggested that the preconsolidation stress σ'_p, could be estimated as follows:

$$\sigma'_p = \alpha_{FV}\, s_{u(FV)} \tag{5.14}$$

The parameter α_{FV} was found to be related to the soil P.I. as follows:

$$\alpha_{FV} = 22(P.I.)^{-0.48} \tag{5.15}$$

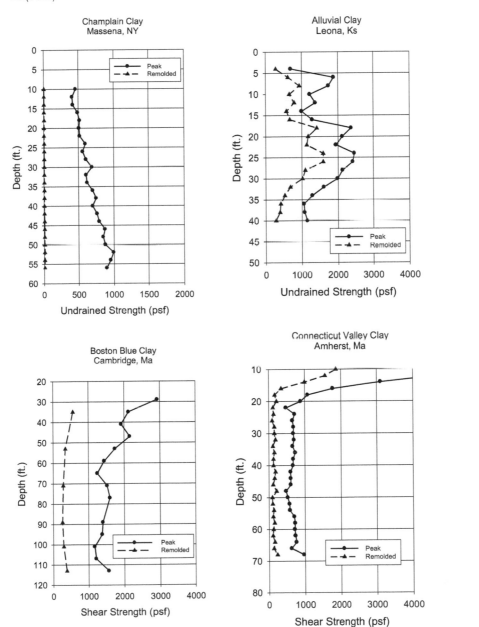

Figure 5.21 Vane shear test results from four clay sites.

5.11 SUMMARY OF FVT

The FVT is still a valuable and an important *in situ* test. In the author's opinion, it is the preferred field method for estimating the undrained shear strength of soft to medium stiff clays. The results obtained from the vane are influenced by delay time, vane geometry (especially blade thickness), and rate of shearing. It is recommended that a rectangular vane with H/D = 2 and constant blade thickness be used for routine work. A vane with a perimeter ratio generally less than about 6% should be used, and in all cases, the exact dimensions of the vane should always be reported to allow the engineer to adjust for disturbance. The vane should be rotated a sufficient amount in order to obtain the postpeak strength, and a remolded test should be conducted at each test depth to obtain the sensitivity. Figure 5.21 shows typical FVT results obtained at four different sites.

REFERENCES

Aas, G., 1965. A Study of the Effect of Vane Shape and Rate of Strain in the Measured Values of In-Situ Shear Strength of Clays. *Proceedings of the 6th International Conference on Soil Mechanics and Foundation Engineering*, Vol. 1, pp. 141–145.

Aas, G., 1967. Vane Tests for Investigation of Anisotropy of Undrained Shear Strength of Clays. *Proceedings of the Geotechnical Conference Oslo*, Vol. 1, pp. 3–8.

Aas, G., Lacasse, S., Lunne, T., and Hoeg, K., 1986. Use of In Situ Tests for Foundation Design on Clay. *Use of In Situ Tests in Geotechnical Engineering*, ASCE, pp. 1–30.

Ahmed, N., 1975. Evaluation of In Situ Testing Methods in Soils. *Dissertation*, Department of Civil Engineering, Louisiana State University.

Ahnberg, R., Larsson, R., and Berglund, C., 2004. Influence of Vane Size and Equipment on the Results of Field Vane Tests. *Proceedings of the International Conference on Geotechnical and Geophysical Site Characterization*, ISC-2, pp. 271–277.

Andrawes, K.Z., Krishnamurthy, D.N., and Barden, L., 1975. Anisotropy of Strength in Clays Containing Plates of Increasing Size. *Proceedings of the 4th Southeast Asian Conference on Soil Engineering*, Kuala Lumpur, Malaysia, Vol. 1, pp. 6–12.

Arman, A., Poplin, J.K., and Ahmad, N., 1975. Study of the Vane Shear. *Proceedings of the Conference of In Situ Measurement of Soil Properties*, ASCE, Vol. 1, pp. 93–120.

Arroyo, R. and Sagaseta, C., 1988. Discussion of Undrained Shear Strength in the Surficial Weathered Crust. *Canadian Geotechnical Journal*, Vol. 25, No. 1, pp. 173–174.

Azzouz, A., Baligh, M.M., and Ladd, C.C., 1983. Corrected Field Vane Strength for Embankment Design. *Journal of Geotechnical Engineering*, ASCE, Vol. 109, No. 5, pp. 730–734.

Bazett, D.J., Adams, J.I., and Matyas, E.L., 1961. An Investigation of Studies in a Test Trench Excavated in Fissured Sensitive Marine Clay. *Proceedings of the 5th International Conference on Soil Mechanics and Foundation Engineering*, Paris, Vol. 1, pp. 431–436.

Biscotin, G. and Pestana, J.M., 2001. Influence of Peripheral Velocity on Vane Shear Strength of an Artificial Clay. *Geotechnical Testing Journal*, ASTM, Vol. 24, No. 4, pp. 423–429.

Bishop, A.W., 1971. Shear Strength for Undisturbed and Remolded Soil Specimens. *Proceedings of the Roscoe Memorial Symposium*, pp. 3–58.

Bjerrum, L., 1972. Embankments of Soft Clay. *Performance of Earth and Earth-Supported Structures*, ASCE, Vol. 2, pp. 1–54.

Bjerrum, L., 1973. Problems of Soil Mechanics and Construction on Soft Clays. *Proceedings of the 8th International Conference on Soil Mechanics and Foundation Engineering*, Vol. 3, pp. 111–159.

Blight, G.E., 1970. In Situ Strength of Rolled Hydraulic Fill. *Journal of the Soil Mechanics and Foundation Division*, ASCE, Vol. 96, No. SM3, pp. 881–899.

Blight, G.E., 1982. Residual Soils in South Africa. *Engineering and Construction in Tropical and Residual Soils*, ASCE, pp. 147–171.

Burmister, D.M., 1957. General Discussion on Vane Shear Testing. ASTM Special Technical Publication 193, pp. 65–68.

Cadling, L. and Odenstad, S., 1950. The Vane Borer. *Proceedings of the Royal Swedish Geotechnical Institute*, No. 2, pp. 1–87.

Cerato, A. and Lutenegger, A.J., 2004. Disturbance Effects of Field Vane Tests in a Varved Clay. *Proceedings of the International Conference on Geotechnical and Geophysical Site Characterization, ISC-2*, pp. 861–867.

Chandler, R.J., 1988. The In-Situ Measurements of the Undrained Shear Strength of Clays Using the Field Vane. ASTM Special Technical Publication 1014, pp. 13–44.

Chaney, R.C. and Richardson, G.N., 1988. Measurement of Residual/Remolded Vane Shear Strength of Marine Sediments. ASTM Special Technical Publication 1014, pp. 166–181.

Costa Filho, L.M., Werneck, M.L.G., and Collet, H.B., 1977. The Undrained Strength of a Very Soft Clay. *Proceedings of the 9th International Conference on Soil Mechanics and Foundation Engineering*, Vol. 1, pp 69–72.

Dascal, O. and Tournier, J.P., 1975. Embankments on Soft and Sensitive Clay Foundations. *Journal of the Geotechnical Engineering Division, ASCE*, Vol. 101, No. GT3, pp. 297–314.

De Alencar, J., Chan, D., and Morgenstern, N.R., 1988. Progressive Failure in the Vane Test. ASTM STP 1014, pp. 150–165.

Donald, I.B. Jordan, D.O., Parker, R.J., and Toh, C.T., 1977. The Vane Test - A Critical Appraisal. *Proceedings of the 9th International Conference on Soil Mechanics and Foundation Engineering*, Vol. 1, pp. 81–88.

Duncan, J.M., 1967. Undrained Strength and Pore-water Pressures in Anisotropic Clays. *Proceedings of the 5th Australia-New Zealand Conference on Soil Mechanics and Foundation Engineering*, pp. 68–71.

Eden, W.J. and Hamilton, J.J., 1956. The Use of Field Vane Apparatus in Sensitive Clays. ASTM Special Technical Publication 193, pp. 41–53.

Eide, O. and Holmberg, S., 1972. Test Fills to Failure on the Soft Bangkok Clay. *Proceedings of the Specialty Conference on Performance of Earth and Earth-Supported Structures, ASCE*, Vol. 1 Part 1, pp. 159–180.

Faust, J., Moritz, K., and Stiefken, H., 1983. Vane Shear Test in Peat. *Proceedings of the International Symposium on In Situ Testing of Soil and Rock*, Vol. 2, pp. 283–286.

Flaate, K., 1966a. Factors Influencing the Results of Vane Tests. *Canadian Geotechnical Journal*, Vol. 3, No. 1, pp. 18–31.

Flaate, K., 1966b. Field Vane Tests with Delayed Shear. Norwegian Road Research Lab, Med. 29, 9pp.

Garga, V., 1988. Experience with Field Vane Testing at Sepetiba Test Fills. ASTM Special Technical Publication 1014, pp. 267–276.

Garga, V.K. and Khan, M.A., 1992. Interpretation of Field Vane Strength of an Anisotropic Soil. *Canadian Geotechnical Journal*, Vol. 29, No. 4, pp. 627–637.

Geise, J.M., Hoope, J., and May, R.E., 1988. Design and Offshore Experience with an In Situ Vane. ASTM STP 1014, pp. 318–338.

Gylland, A.S., Jostad, H.P., Nordal, S., and Emdal, A., 2013. Micro-level Investigation of the In Situ Shear Vane Failure Geometry in Sensitive Clay. *Geotechnique*, Vol. 63, No. 14, pp. 1264–1270.

Hanzawa, H., 1979. Undrained Strength Characteristics of an Alluvial Marine Clay in Tokyo Bay. *Soils and Foundations*, Vol. 19, No. 4, pp. 69–84.

Helenelund, K.V., 1967. Vane Tests and Tension Tests on Fibrous Peat. *Proceedings of the Geotechnical Conference Oslo*, Vol. 1, pp. 199–203.

Helenelund, K.V., 1977. Methods for Reducing Undrained Shear Strength of Soft Clay. Swedish Geotechnical Institute Report 3.

Hight, D.W., Bond, A.J., and Legge, J.D., 1992. Characterization of the Bothkennar Clay: An Overview. *Geotechnique*, Vol. 42, No. 2, pp. 303–347.

Hirabayashi, H., Tanaka, M., and Tomita, R., 2017. Effect of Rotation Rate on Field Vane Shear Strength. *Proceedings of the 27th International Ocean and Polar Engineering Conference*, pp. 742–747.

Jamiolkowski, M., Ladd, C.C., Germaine, J., and Lancellotta, R., 1985. New Developments in Field and Lab Testing of Soils. *Proceedings of the 11th International Conference on Soil Mechanics and Foundation Engineering*, Vol. 1, pp. 57–154.

Johnson, G.W., Hamilton, T.K., Ebelhar, R.J., Mueller, J.L., and Pelletier, J.H., 1988. Comparison of In Situ Vane, Cone Penetrometer, and Laboratory Test Results for Gulf of Mexico Deep Water Clays. ASTM Special Technical Publication 1014, pp. 293–305.

Josseaume, H., Blondeau, F., and Pilot, G., 1977. Study of Undrained Behaviour of Three Soft Clays. *Application to Embankment Design. Proceedings of the International Symposium on Soft Clays*, pp. 122–130.

Kallstenius, T., 1956. Swedish Vane Borer Design, ASTM STP 193, pp. 60–62.

Kietkajornkul, C. and Vasinvarthana, V., 1989. Influence of Different Vane Types on Undrained Strength of Soft Bangkok Clay. *Soils and Foundations*, Vol. 29, No. 2, pp. 146–152.

Kimura, T. and Saitoh, K., 1983. Effects of Disturbance Due to Insertion Vane Shear Strength of Normally Consolidated Cohesive Soils. *Soils and Foundations*, Vol. 23, No. 2, pp. 113–124.

Kolk, H.J., ten Hoope, J., and Ims, B.W., 1988. Evaluation of Offshore In-Situ Vane Test Results. ASTM Special Technical Publication 1014, pp. 339–353.

Ladd, C.C., Foott, R., Ishihara, K., Schlosser, F., and Poulos, H.G., 1977. Stress-Deformation and Strength Characteristics. *Proceedings of the 9th International Conference on Soil Mechanics and Foundation Engineering*, Vol. 2, pp. 421–494.

Landva, A.O., 1980. Vane Testing in Peat. *Canadian Geotechnical Journal*, Vol. 17, No. 1, pp. 1–19.

La Rochelle, P., Roy, M., and Tavenas, F., 1973. Field Measurements of Cohesion in Champlain Clays. *Proceedings of the 8th International Conference on Soil Mechanics and Foundation Engineering*, Vol. 1.1, pp. 229–236.

Larsson, R., 1980. Undrained Shear Strength in Stability Calculation of Embankments and Foundations on Soft Clays. *Canadian Geotechnical Journal*, Vol. 17, No. 4, pp. 591–602.

Larsson, R., Bergdahl, U., and Erikson, L., 1987. Evaluation of Shear Strength in Cohesive Soils with Special Reference to Swedish Practice and Experience. *Geotechnical Testing Journal, ASTM*, Vol. 10, No. 3, pp. 105–112.

LeMasson, H., 1976. A New Method for In Situ Measurement of Anisotropy of Clays. *Bulletin de Liaison des Laboratoires des Ponts et Chaussées*, Vol. 2, pp. 107–116.

Law, K.T., 1985. Use of Field Vane Tests Under Earth Structures. *Proceedings of the 11th International Conference on Soil Mechanics and Foundation Engineering*, Vol. 2, pp. 893–898.

Lo, K.Y., 1965. Stability of Slopes in Anisotropic Soils. *Journal of the Soil Mechanics and Foundation Division, ASCE*, Vol. 91, No. SM1, pp. 85–106.

Lo, K.Y. and Milligan, V., 1967. Shear Strength Properties of Two Stratified Clays. *Journal of the Soil Mechanics and Foundation Division, ASCE*, Vol. 93, No. SM1, pp. 1–15.

Loh, A.K. and Holt, R.T., 1974. Directional Variation and Fabric of Winnipeg Upper Brown Clay. *Canadian Geotechnical Journal*, Vol. 11, No. 3, pp. 430–437.

Matsui, T. and Abe, N., 1981. Shear Mechanisms of Vane Test in Soft Clays. *Soils and Foundations*, Vol. 21, No. 4, pp. 69–80.

Mayne, P. and Mitchell, J.K., 1988. Profiling of Overconsolidation Ratio in Clays by Field Vane. *Canadian Geotechnical Journal*, Vol. 25, pp. 150–157.

Memon, A., 1980. Undrained Shear Strength Anisotropy of Clays. *Proceedings of the International Symposium on Landslides*, New Delhi, pp.109–112.

Menzies, B.K. and Mailey, L.K., 1976. Some Measurements of Strength Anisotropy in Soft Clays Using Diamond Shaped Shear Vanes. *Geotechnique*, Vol. 26, No. 3, pp. 535–538.

Menzies, B.K. and Merrifield, C.M., 1980. Measurements of Shear Stress Distribution on the Edges of a Shear Vane Blade. *Geotechnique*, Vol. 22, No. 3, pp. 451–457.

Miller, G.A. and Lutenegger, A.J., 1993. Analysis of Small Diameter Pipe Piles Using the Field Vane. *Proceedings of the 3rd International Conference on Case Histories in Geotechnical Engineering*, Vol. 1, pp. 154–160.

Monney, N.T., 1974. An analysis of the Vane Shear Test at Varying Rates of Shear. *Deep Sea Sediments: Marine Science*, Vol. 2, pp. 151–167.

Muhuai, Y., Wulin, W., Changwei, X., and Zhao, L., 1983. Field Shear Rheologic Test of Peaty Soils. *Proceedings of the International Symposium on In Situ Testing on Soil and Rock*, Vol. 2, pp. 431–434.

Nathan, S.V., 1978. Discussion of Influence of Shear Velocity on Vane Shear Strength. *Journal of the Geotechnical Engineering Division, ASCE*, Vol. 104, No. GT1, pp. 151–153.

Northwood, R.P. and Sangrey, D.A., 1971. The Vane Test in Organic Soils. *Canadian Geotechnical Journal*, Vol. 8, pp. 68–76.

Osterberg, J.O., 1956. Introduction to Vane Testing of Soil. ASTM Special Technical Publication 193, pp. 1–7.

Peuchen, J. and Mayne, P., 2007. Rate Effects in Vane Shear Testing. *Proceedings of the 6th International Conference on Offshore Site Investigation and Geotechnics*, pp. 259–266.

Perlow, M. and Richards, A.F., 1977. Influences of Shear Velocity on Vane Shear Strength. *Journal of the Geotechnical Engineering Division, ASCE*, Vol. 103, No. GT1, pp. 19–32.

Pilot, G., 1972. Study of Five Embankment Failures on Soft Soils. *Performance of Earth and Earth Supported Structures, ASCE*, Vol. 1, Part 1, pp. 81–99.

Pyles, M.R., 1984. Vane Shear Data on Undrained Residual Strength. *Journal of Geotechnical Engineering, ASCE*, Vol. 110, No. 4, pp. 543–547.

Richardson, A.M., Brand, E.W., and Memon, A., 1975. In Situ Determination of Anisotropy of a Soft Clay. *Proceedings of the Conference on In Situ Measurement of Soil Properties, ASCE*, Vol. 1, pp. 337–349.

Roy, M. and Leblanc, A., 1988. Factors Affecting the Measurement and Interpretation of the Vane Strength in Soft Sensitive Clays. ASTM STP 1014, pp. 117–128.

Roy, M. and Mercier, M., 1989. The Pore Pressures in Vane Test. *Proceedings of the 12th International Conference on Soil Mechanics and Foundation Engineering*, Vol. 1, pp. 305–308.

Sageseta, C. and Arroyo, R., 1982. Limit Analysis of Embankments on Soft Clay. *Proceedings of the International Symposium on Numerical Methods in Geomechanics*, pp. 618–625.

Schapery, R.A. and Dunlap, W.A., 1978. Prediction of Storm Induced Sea Bottom Movement and Platform Forces. *Proceedings of the 10th Offshore Technology Conference*, OTC Paper No. 3259.

Schlue, B., Moerz, T., and Kreiter, S., 2010. Influence of Shear Rate on Undrained Vane Shear Strength of Organic Harbor Mud. *Journal of Geotechnical and Geoenvironmental Engineering, ASCE*, Vol. 136, No. 10, pp. 1437–1447.

Sharifounnasab, M. and Ullrich, R.C., 1985. Rate of Shear Effect on Vane Shear Strength. *Journal of Geotechnical Engineering, ASCE*, Vol. 111, No. 1, pp. 135–139.

Silvestri, V. and Aubertin, M., 1983. The In Situ Anisotropy of a Sensitive Canadian Clay. *Proceedings of the International Symposium on In Situ Testing*, Paris, France, Vol. 2, pp. 391–395.

Silvestri, V. and Aubertin, M., 1988. Anisotropy and In Situ Vane Tests. ASTM STP 1014, pp. 88–103.

Silvestri, V., Aubertin, M., and Chapuis, R.P., 1993. A Study of Undrained Shear Strength Using Various Vanes. *Geotechnical Testing Journal, ASTM*, Vol. 16, No. 2, pp. 228–237.

Skempton, A.W., 1948. Vane Test in the Alluvial Plains of River Froth Near Grange Mouth. *Geotechnique*, Vol. 1, No. 2, pp. 111–124.

Slunga, E., 1983. On the Increase in Shear Strength in Soft Clay Under an Old Embankment. *Proceedings of the 8th European Conference on Soil Mechanics and Foundation Engineering*, Vol. 1, pp. 83–89.

Smith, A.D. and Richards, A.F., 1975. Vane Shear Strength at Two High Rotation Rates. *Civil Engineering in the Oceans III, ASCE*, Vol. 1, pp. 421–433.

Surendra, M. and Mundell, J.A., 1984. Discussion of Corrected Field Vane Strength for Embankment Design. *Journal of Geotechnical Engineering, ASCE*, Vol. 110, No. 8, pp. 1150–1152.

Tammirinne, M., 1981. Some Aspects Concerning the Use of Vane Test. *Proceedings of the 10th International Conference on Soil Mechanics and Foundation Engineering*, Vol. 4, pp. 764–766.

Tanaka, H., 1994. Vane Shear Strength of a Japanese Marine Clay and Applicability of Bjerrum's Correction Factor. *Soils and Foundations*, Vol. 34, No. 3, pp. 39–48.

Tanaka, H. and Tanaka, M., 1993. Vane Shear Strength of Japanese Marine Clay. *Proceedings of the 11th Southeast Asian Geotechnical Conference*, pp. 231–238.

Tang, W., Mesri, G., and Halim, I., 1992. Uncertainty of Mobilized Undrained Shear Strength. *Soils and Foundations*, Vol. 32, No. 4, pp. 107–116.

Toh, C.T. and Donald, I.B., 1979. Anisotropy of an Alluvial Clay. *Proceedings of the 6th Asian Regional Conference on Soil Mechanics and Foundation Engineering*, Vol. 1, pp. 99–106.

Torstensson, B.A., 1977. Time-Dependent Effects in the Field Vane Test. *Proceedings of the International Symposium on Soft Clays*, Bangkok, pp. 387–397.

Veneman, P. and Edil, T., 1988. Micromophological Aspects of the Vane Shear Test. ASTM STP 1014, pp. 182–190.

Wiesel, C.E., 1967. Discussion on the Shear Strength of Soft Clays. *Proceedings of the Geotechnical Conference at Oslo*, Vol. 2, pp. 130–131.

Wiesel, C.E., 1973. Some Factors Influencing In Situ Vane Test Results. *Proceedings of the 8th International Conference on Soil Mechanics and Foundation Engineering*, Vol. 1.2, pp. 475–479.

Wroth, C.P., 1984. Interpretation of In Situ Tests. *Geotechnique*, Vol. 34, No. 4, pp. 449–489.

Young, A.G., McClelland, B., and Quiros, G.W., 1988. In-Situ Vane Shear Testing at Sea. ASTM Special Technical Publication 1014, pp. 46–67.

Chapter 6

Dilatometer Test (DMT)

6.1 INTRODUCTION

The Dilatometer Test (DMT) was introduced by Marchetti (1975; 1980) and is sometimes described as a flat plate penetrometer. The test is simple, robust, and easy to operate, and the test is applicable in a wide range of soils. The data may be reduced quickly, and the results are relatively simple to interpret. The test uses an active test phase to expand a circular membrane laterally against the soil after direct push penetration. The test can be used both as a stratigraphic logging tool and a specific property measurement tool, and the test can also be used as a prototype tool for direct design of axial and laterally loaded driven displacement piles. It has also been found reliable for estimating bearing capacity and settlement of shallow foundations.

6.2 MECHANICS

The DMT consists of a rectangular blade with a width of 94 mm (3.7 in.) and a thickness of 14 mm (0.55 in.) that has a sharpened (20°) leading point as shown in Figure 6.1. The blade is made of hardened stainless steel, and therefore, it can be used to penetrate a wide range of soils, from very soft clay to dense sand and till. The blade is pushed into the ground much in the same way as advancing the CPT or CPTU. There is a thin circular stainless steel membrane or diaphragm mounted on one face of the plate as shown in Figure 6.1. This membrane is the only moving part of the instrument in contact with the soil. Once penetration of the blade has reached the test depth, the membrane is expanded outward against the soil using controlled gas pressure from a console at the ground surface.

6.3 EQUIPMENT

In addition to the blade, the test equipment includes a simple control console that is used to perform the test and a gas supply, usually nitrogen. The console consists of a pressure gage and a needle valve to control the gas flow to the blade. The blade and the control console are connected by a small diameter coaxial electrical/pneumatic line that provides communication between the two. Figure 6.2 shows a complete arrangement of the test set-up showing all of the main components.

Tests are normally performed at depth intervals of 12 in. (0.30 m), and therefore, the test provides a semi-continuous profile of soil response. Since its introduction, the test has been used around the world in a wide range of soils, from very soft fine-grained deposits to very stiff granular deposits. Table 6.1 gives a representative collection of the reported use

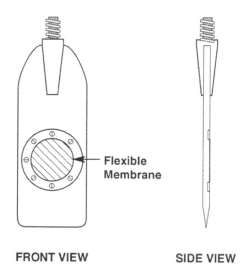

FRONT VIEW   SIDE VIEW

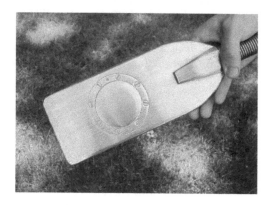

Figure 6.1 Schematic and photo of a dilatometer blade.

of the DMT in different soils and shows that the DMT has applications in a broad range of materials, including cohesive and cohesionless, saturated and partially saturated, normally consolidated and overconsolidated, "quick" and very stiff, and natural and compacted.

There appear to be only minor limitations in using the DMT in natural geologic materials. In boulderly glacial sediments or gravelly deposits, it may be difficult to advance the blade, and there may be damage to the blade and/or membrane. Offshore use of the DMT has been reported by Marchetti (1980), Burgess (1983), Sonnenfeld et al. (1985), and Lunne et al. (1987). Akbar & Clarke (2001) and Akbar et al. (2005) described a modified DMT for use in very stiff glacial deposits, substituting the flexible stainless steel membrane with a rigid steel piston.

6.4 TEST PROCEDURE

The blade is typically advanced to the test depth using quasi-static pushing thrust in the same way that the CPT/CPTU tests are advanced. A penetration rate of 2 cm/s is recommended for the DMT to be consistent with the CPT/CPTU. The blade can be advanced

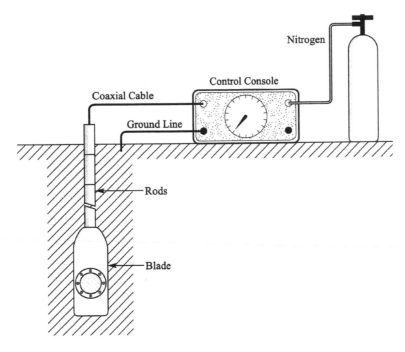

Figure 6.2 Test arrangement for DMT.

with a CPT rig or using the hydraulic feed off the back of a drill rig. Similar techniques previously described in Chapter 4 for advancing the CPT are used. Adapters can be made to allow the use of either CPT rods or conventional drill rods. The coaxial cable is run through the center of the push rods. Normally, an enlarged rod section is placed behind the blade to act as a "friction reducer" to allow the blade to be advanced as far as possible. As a general rule of thumb, it takes about twice as much pushing thrust to advance the DMT blade as it takes to advance a 10 cm² CPT. Even though the blade is extremely robust and made of high strength steel, it is not considered appropriate to advance the blade by driving, as with an SPT hammer.

The active or expansion phase of the test occurs after penetration and usually starts immediately after stopping penetration and releasing the downward thrust. The expansion phase of the test normally consists of two parts: (1) lift-off and (2) 1 mm expansion. A detailed description of test procedures has been developed and is given in ASTM D6635 *Standard Test Method for Performing the Flat Dilatometer.*

6.4.1 Lift-off Pressure

Under atmospheric pressure, i.e., out of the ground, the center of the flexible membrane is actually not in contact with the blade and is slightly concaved outward (Figure 6.3a). After penetration into the soil, the soil pressure pushes the membrane flat against the blade as shown in Figure 6.3b. When this happens, an electrical contact is made inside the blade between the membrane and the blade, completing a simple electrical circuit. This condition is identified by an audio signal produced by a buzzer at the control console. The first part of the expansion phase of the test is performed by slowly increasing the gas pressure inside the body of the blade (on the back side of the diaphragm) using a flow control needle valve on the control console. This forces the membrane outward against the soil, until the electrical

Table 6.1 Some reported use of DMT in different materials

Material	References
Sensitive marine clay	Lacasse & Lunne (1983) Fabius (1985) Bechai et al. (1986) Hayes (1986) Lutenegger & Timian (1986) Benoit et al. (1990) Masood & Kibria (1991) Lutenegger (2015)
Soft nonsensitive clays	Minkov et al. (1984) Chang (1986) Saye & Lutenegger (1988b) Chang (1991) Mayne and Frost (1991) Ortigao et al. (1996) Kim et al. (1997) Chu et al. (2002) Arulrajah et al. (2004)
Lacustrine clay	Chan & Morgenstern (1986) Martin & Mayne (1997) Lutenegger (2000) Ozer et al. (2006)
Glacial tills and/or very stiff overconsolidated clays	Davidson & Boghrat (1983) Schmertmann & Crapps (1983) Boghrat (1987) Powell & Uglow (1986; 1988) Lutenegger (1990)
Sand	Schmertmann (1982) Baldi et al. (1986) Clough & Goeke (1986) Lacasse & Lunne (1986) Schmertmann et al. (1986) Konrad (1991) Monaco et al. (2005) Tsai et al. (2009) Lee et al. (2011) Amoroso et al. (2015a)
Deltaic silt	Campanella & Robertson (1983) Konrad et al. (1985)
Loess	Lutenegger & Donchev (1983) Hammandshiev & Lutenegger (1985) Lutenegger (1986) Mlynarek et al. (2015)
Peat	Hayes (1983) Kaderabek et al. (1986) Nichols et al. (1989)
Compacted fill	Borden et al. (1985) Borden et al. (1986)
Soft/medium rock	Sonnenfeld et al. (1985)
Residual soils	Chang (1988) Borden et al. (1988) Wang & Borden (1996) Martin & Mayne (1998) Brown & Vinson (1998) Giacheti et al. (2006) Anderson et al. (2006b) Cruz & Viana de Fonseca (2006)

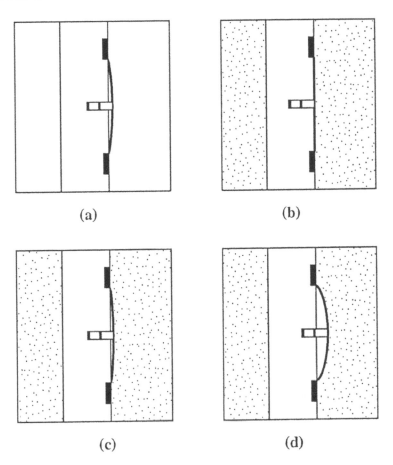

Figure 6.3 DMT membrane position during phases of the test. (a) atmospheric, (b) after penetration, (c) lift-off, and (d) I mm expansion.

contact is broken, and the audio buzzer stops, Figure 6.3c. The pressure where this occurs is noted and designated at the lift-off pressure or "A-Reading."

In order to actually obtain the A-Reading, the operator watches the pressure gage on the console and listens for the audio buzzer to go off. The operator can either mentally store the reading or repeat it aloud to an assistant who records the reading. The pressure gages should be of sufficient precision for the anticipated range of pressures to be encountered for a particular soil.

At this point, the center of the membrane has just lifted off the face of the blade. Under atmospheric pressure, the center of the membrane is concaved outward and must be artificially pulled in by a vacuum to make contact with the blade. This represents the intrinsic membrane resistance and must be accounted for in determining the actual pressure exerted on the soil. Therefore, the A-Reading must be corrected for membrane stiffness, denoted as ΔA, such that

$$P_0 = A - \Delta A. \tag{6.1}$$

Since ΔA is really negative (i.e., a vacuum behind the membrane is required to bring it against the blade), ΔA must be numerically *added* to the A-Reading to give the corrected pressure, P_0. The value of P_0 is referred to as the "lift-off" pressure.

6.4.2 1 mm Expansion Pressure

After obtaining the A-Reading and without stopping the gas flow, gas pressure is increased until the center of the membrane has moved 1-mm into the soil away from the plane of the blade, as noted in Figure 6.3d. At this point, an additional electrical contact is made inside the blade and the audio buzzer signal of the control console again sounds, signifying that the electrical circuit had once again been completed. The pressure where this occurs is noted as the "B-Reading." As with the A-Reading, since the membrane has some intrinsic stiffness, a correction must be applied to the B-Reading to give a corrected soil pressure. The correction for the 1 mm expansion (ΔB) is obtained by expanding the membrane under atmospheric pressure until 1 mm expansion is obtained. This represents the inertial membrane stiffness. The 1 mm expansion pressure, corrected for membrane stiffness, is designated as P_1 and is obtained as follows:

$$P_1 = B - \Delta B. \tag{6.2}$$

As with the A-Reading, in order to determine the B-Reading, the operator listens for the audio buzzer to sound while watching the pressure gage. At this point, usually both the A-Reading and B-Reading are recorded. Figure 6.4 shows a cross section through the blade showing an internal moving plunger that follows the membrane during expansion and is used to make the electrical contact at lift-off and the 1 mm expansion.

Both pressures, P_0 and P_1, were used by Marchetti (1980) to establish correlations for pertinent engineering properties, which will be discussed later in this chapter. After the B-Reading is obtained, the gas pressure is immediately released by a vent valve on the control console so that no further expansion of the membrane occurs. Continued expansion of the membrane can cause yielding of the metal and a change in the calibration constants, ΔA and ΔB. The gas flow is regulated with the needle valve on the console so that it takes about 15–30 s to obtain the A-Reading and another 15–30 s to obtain the B-Reading. After venting, the blade is ready to be advanced to the next test location with the vent open or is removed.

The corrections to the A- and B-Readings to obtain the corrected pressures are slightly more complex than indicated by Equations 6.1 and 6.2. To account for the specific internal mechanics of the blade, the corrected readings P_0 and P_1 are actually obtained from

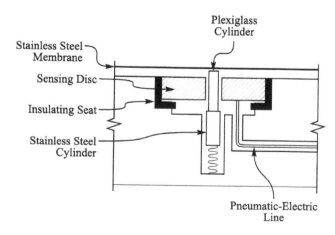

Figure 6.4 Internal mechanics of a dilatometer blade.

$$P_0 = 1.05(A - Z_m + \Delta A) - 0.05(B - Z_m - \Delta B) \tag{6.3}$$

$$P_1 = B - Z_m - \Delta B \tag{6.4}$$

where
Z_m = the initial pressure gauge reading

6.4.3 Recontact Pressure

An additional development in conducting the DMT was suggested in the mid-1980s to obtain a third pressure reading, designated as the "C-Reading." Following the B-Reading, instead of rapidly venting the gas pressure inside the blade, the pressure is slowly decreased by controlled deflation using another in-line flow control needle valve on the control console. If there is still external pressure pushing on the outside of the membrane, this pressure will be identified when the membrane recontacts the plane of the blade and takes the same position as in the beginning of the test, i.e., Figure 6.3b, and again the audio buzzer signal will sound. The C-Reading is corrected for membrane stiffness to give the "recontact" pressure, P_2, determined as

$$P_2 = C - \Delta A - Z_m \tag{6.5}$$

Even though the C-Reading was not a part of the initial work presented by Marchetti (1980), the author is a strong proponent of obtaining the C-Reading and recommends that this measurement be included as a routine part of every DMT test.

The operation of the DMT consists of a simple sequence of pressure measurements. The control console is relatively simple to operate, and most technicians master the controls quickly. This means that the test results at a given site are more likely to represent natural variations in soil conditions rather than variations in test procedures. The test results are generally very reproducible and essentially operator-independent.

6.5 DATA REDUCTION

As described in the previous section, the DMT allows the measurements of three separate pressure values. In the original form of the test, Marchetti (1980) had only the first two pressure readings, P_0 and P_1, which he used to develop empirical correlations for various soil properties. Based on these two readings, Marchetti established three dilatometer "indices" that were defined as follows:

I_D – *Material Index*

$$I_D = (P_1 - P_0)/(P_0 - u_o) \tag{6.6}$$

K_D – *Lateral Stress Index*

$$K_D = (P_0 - u_o)/(\sigma_{vo} - u_o) \tag{6.7}$$

E_D – *Dilatometer Modulus*

$$E_D = (P_1 - P_0)D \tag{6.8}$$

where

σ_{vo} = *in situ* total vertical stress
u_o = *in situ* pore water pressure
D = a constant.

These three parameters were used by Marchetti (1980) to establish correlations to pertinent soil properties using other field and laboratory data.

U_D – Dilatometer Pore Pressure Index

Lutenegger & Kabir (1988) suggested that the recontact pressure, P_2, could be used to identify site stratigraphy by making a comparison between P_2 and P_0; the rationale being that in a very soft saturated fine-grained soil, the two values would be nearly identical. They suggested an additional DMT index, referred to as the Pore Pressure Index defined as follows:

$$U_D = (P_2 - u_o)/(P_0 - u_o).$$ (6.9)

Theoretically, values of U_D could range from 0 to 1, but because of the difference in time required to obtain the two readings P_0 and P_2, the actual values range from 0 to about 0.8. Variations in U_D for a given soil reflect the tendency for generating positive pore water pressures and the rate of pore water pressure dissipation, both of which can be expected to vary with both stress history and soil type.

K_i – Initial Lateral Stress Index

Lutenegger (2006b; 2015) suggested that another useful parameter could be defined by considering the values of P_0 and P_2 together. The initial lateral stress ratio, K_i, is defined as:

$$K_i = (P_0 - P_2)/\sigma'_{vo}$$ (6.10)

Correlations between the various DMT indexes and soil properties and results of more recent work with the DMT will be discussed in more detail in subsequent sections. Initially, however, it would be useful to examine the fundamental mechanics of the penetration phase of the test and the influence of ground stress and soil behavior on the measured values of P_0, P_1, and P_2.

6.5.1 Lift-off and Penetration Pore Pressures

In a uniform soil deposit, the *in situ* vertical and horizontal effective stresses increase linearly with depth. As a result, the *in situ* mean or octahedral stress increases with depth. Accordingly, the DMT lift-off pressure should likewise increase with depth. The components of P_0 in any soil include the initial total stress and the change in stress as a result of the penetration of the blade such that

$$P_0 = \sigma'_{hO} + u_o + \Delta\sigma'_h + \Delta u$$ (6.11)

where

σ'_{hO} = initial *in situ* horizontal effective stress
u_o = initial *in situ* pore water pressure
$\Delta\sigma'_h$ = change in horizontal effective stress
Δu = change in pore pressure.

The magnitude and sign of each of the components of Equation 6.11 will be a function of the existing stress conditions of the soil and the soil behavior, including stress history (OCR), stiffness, void ratio, etc. Roque et al. (1988) included an additional term in Equation 6.11 to account for "attraction".

Two methods have been used to study the influence of pore pressures on the DMT: (1) by modifying the blade to accept a porous element and pore pressure system along with the expandable diaphragm (Campanella et al. 1985; Robertson et al. 1988) and (2) by using a separate identical blade (piezoblade) that has only a porous element and pore pressure system as a replacement for the expandable diaphragm (Davidson & Boghrat 1983; Lutenegger & Kabir 1988) as shown in Figure 6.5. Both methods use a pressure transducer to measure the pore pressure during and after penetration.

In order to determine what portion of the measured value of P_0 is pore water pressure in cohesive soils $(u_o + \Delta u)$, we can examine the results of piezoblade and instrumented DMT results. Davidson & Boghrat (1983) demonstrated that the amount of pore water pressure dissipated on the face of the blade within the time the test is performed was a function of soil type as related to the DMT Material Index, I_D. The amount of pore pressure dissipated in the first minute following penetration decreases with increasing fineness of the soil. That is, in granular soils with high coefficients of hydraulic conductivity, nearly all of the excess pore pressure was gone.

By contrast, in soft fine-grained soils, in which the coefficient of hydraulic conductivity can be several orders of magnitude lower than sands, there was only a very small amount of excess pore pressure dissipation after the first minute. As shown in Figure 6.6, this behavior can be related to the DMT Material Index, I_D. Note that these tests were performed in stiff overconsolidated soils, and the pore pressures measured were predominantly negative.

The results of Figure 6.6 show that for practical interpretation, the DMT should be considered an undrained test in clays, and P_0 may contain a large component of Δu. In sands, the test is almost completely drained, and P_0 will generally reflect a condition where $\Delta u = 0$.

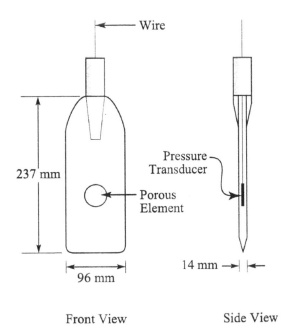

Front View   Side View

Figure 6.5 Piezoblade.

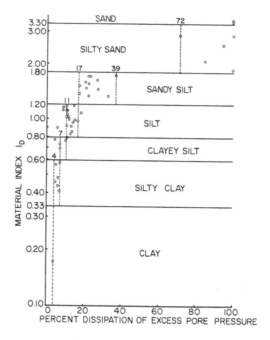

Figure 6.6 Per cent dissipation of pore water pressure on piezoblade I min after insertion. (After Davidson & Boghrat 1983.)

For soils in between, or of mixed composition, the amount of pore pressure present during the test will depend on the stress history and drainage characteristics of the soil. The magnitude of penetration (excess) pore water pressure (Δu) on the face of the DMT blade should be related to soil properties such as stress history, strength, and stiffness, much in the same way as was described for pore pressure measurements obtained with the CPTU.

The magnitude of the lift-off pressure is clearly affected by the amount of excess pore water pressure at the time P_0 is obtained. Mayne (1987) demonstrated that the value of P_0 was nearly identical to the pore water pressure measured with a CPTU for clays with over-consolidation ratio (OCR)<3. In normally consolidated and lightly overconsolidated clays, the dominant component of P_0 is $\Delta u + u_0$. In overconsolidated soils, a significant component of P_0 is composed of an increase in horizontal effective stress, $\Delta \sigma'_h$ and less of an influence of excess pore pressure. As the absolute value of P_0 increases, the effect of excess pore water pressure decreases. That is, as the soil becomes stronger and stiffer, the influence of excess pore water pressure on the DMT results becomes less. The influence of excess pore water pressures is roughly related to the DMT Material Index, I_D, as previously noted, with the magnitude of excess pore water pressure decreasing as I_D increases.

6.5.2 I mm Expansion Pressure

The expansion of the membrane out to 1 mm provides for a loading phase of the test after obtaining the lift-off pressure. In this way, an attempt is made in the test to obtain soil behavior under loading after the penetration phase of the test. Since the half thickness of the blade is about 7 mm, the 1 mm expansion actually constitutes a fairly large deformation condition. In a uniform deposit, the difference between P_1 and P_0 should increase with increasing depth, as the confining stress increases, which reflects the increase in soil stiffness with depth.

Campanella & Robertson (1991) presented results of tests obtained using a carefully instrumented DMT blade that could measure the full pressure-displacement behavior of the membrane. In soft clays, the shape of the diaphragm expansion curve is relatively flat as indicated in Figure 6.7. The major component of expansion is indicated to be additional pore water pressure. This suggests that the penetration phase of the test has already created a failure condition in the soil and that P_1 will be close to P_0. This also means that I_D, defined by Equation 6.6, will be very small since the difference between P_0 and P_1 is small. Also, E_D will be small, again because the pressure difference, $P_1 - P_0$, is small.

Therefore, in soft clays, one should be able to predict the undrained strength from either P_0 or P_1 or from the combination of the two. An example of the penetration phase of the test in clays was modeled by Yu et al. (1993), which illustrates the shape of the curve during penetration and 1 mm expansion, generally following the measured shape as shown in Figure 6.7. Any further expansion from the initial 7 mm (½ the blade thickness) penetration, e.g., an additional 1mm to obtain P_1, produces only a minor increase in pressure.

By contrast, typical results obtained in a stiff overconsolidated clay, shown in Figure 6.8, indicate that negative pore water pressures can be generated during penetration, and that there can be a substantial difference in P_0 and P_1. This suggests that the penetration phase of the test may not create a limit pressure condition, i.e., the soil is not at failure during penetration, but may approach a limiting or failure condition during the 1 mm expansion to P_1. The difference between P_1 and P_0 will be larger than in soft clays, and therefore, both I_D and E_D will be larger, indicating a stiffer material.

In sands, although expansion from P_0 to P_1 still shows an essentially linear behavior, the test results show a quite different behavior, as shown in Figure 6.9. P_1 is much larger than P_0; no limiting condition is achieved in either loose or dense sand, and essentially no excess pore water pressures are generated, either during penetration or expansion. This means that both I_D and E_D will be large, and the test may be considered drained. Additional expansion curves obtained using an instrumented DMT diaphragm on sand in a calibration chamber have been presented by Bellotti et al. (1997). Expansion of the diaphragm from P_0 to P_1 is essentially linear, the slope is related to OCR (for constant relative density), and the unload-reload slope is much stiffer than the initial loading slope. Similar observations of the full

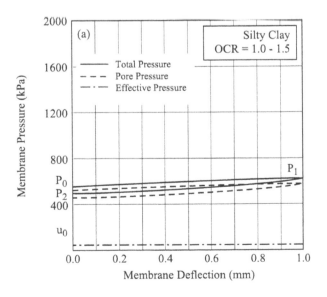

Figure 6.7 DMT membrane expansion in soft clay.

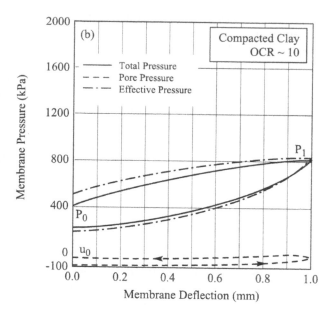

Figure 6.8 DMT membrane expansion in stiff clay.

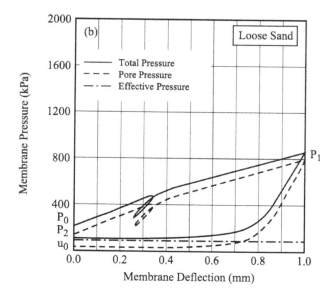

Figure 6.9 DMT membrane expansion in loose sand.

pressure expansion curve have been made using instrumented DMT blades (Kay & Chiu 1993; Kaggwa et al. 1995; Stetson et al. 2003).

Lutenegger (1988) had shown that for a limited number of tests in clays, the lift-off pressure from the DMT was close to the limit pressure, P_L, from the PMT. Tests for a wider range of clays, summarized by Lutenegger & Blanchard (1993), indicated that in soft clays, P_0 is close to P_L, but in stiff clays, the 1 mm expansion pressure P_1 is close to the PMT limit pressure P_L. In soft clays, the correlation between either P_0 or P_1 and P_L should be about the same, since P_0 and P_L are close (Figure 6.10).

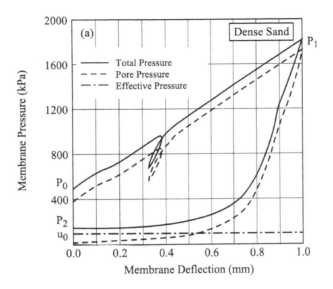

Figure 6.10 DMT membrane expansion in dense sand.

Similar observations have been made using results obtained from prebored and self-boring PMTs (e.g., Powell & Uglow 1986; Chang 1988; Kalteziotis et al. 1991; Wong et al. 1993; Hamouche et al. 1995; Ortigao et al. 1996), as shown in Figure 6.11. Figure 6.12 gives a compilation of data from tests performed by the author at a number of test sites, which shows similar results.

6.5.3 Recontact Pressure

The recontact pressure, P_2, is obtained *after* first obtaining the A and B readings. Therefore, on deflation since the soil had already been deformed 1 mm away from the face of the blade, the pressure pushing in on the membrane will be composed predominantly of pore pressure and a small amount of soil effective relaxation stress. This behavior was suggested by Campanella et al. (1985). In a uniform deposit, it would be expected that the recontact

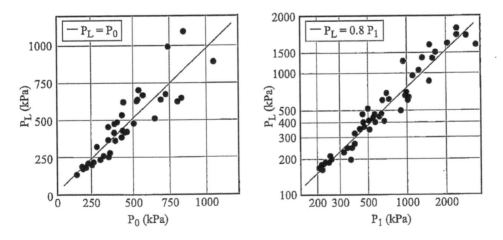

Figure 6.11 Comparison between DMT P_0 and P_1 and PMT P_L in soft and stiff clay.

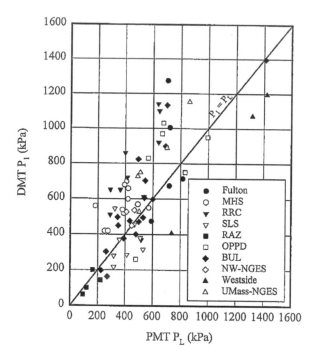

Figure 6.12 Comparison between DMT P_1 and PMT P_L in medium and stiff clay. (After Kalteziotis et al. 1991.)

pressure would increase linearly with depth, following the hydrostatic pore water pressure line in sands but reflecting excess pore water pressure in clays.

Results obtained at several research sites (e.g., Robertson et al. 1988; Lutenegger & Kabir 1988) show that the recontact pressure was closely related to the excess pore pressure in clays and close to the *in situ* equilibrium pressure in sands. Based on these and other observations, it can be assumed that the close-up pressure, P_2, is a reasonable measure of the pore water pressure acting on the face of the DMT blade during the test.

6.6 PRESENTATION OF TEST RESULTS

Presentation of DMT results should always show plots of the corrected pressure readings, P_0, P_1, and P_2 versus depth. Plots of P_0 and P_1 vs. depth should also include a plot of the *in situ* vertical effective stress as a reference. Similarly, the plot of P_2 vs. depth should include a plot on *in situ* pore water pressure as a reference. Examples are shown in Figures 6.13–6.16.

Figure 6.13 shows DMT results obtained at a site in Keene, NH, which consists of alluvial sands overlying a soft lake clay deposit. A casing was set through some random fill before the DMT was started. The transition between the two layers can easily be seen at a depth of about 14 m. Comparison of the P_2 reading with the *in situ* pore water pressure shows hydrostatic conditions in the sand but elevated (excess) pore water pressures in the clay. The DMT Indices, shown in Figure 6.14, indicate the transition from the coarse-grained to fine-grained materials by the Material Index, I_D, and the DMT Modulus, E_D.

Figures 6.15 and 6.16 show results obtained at a site in Massena, NY, which consists of very soft sensitive marine clay. The soils are uniform throughout the profile.

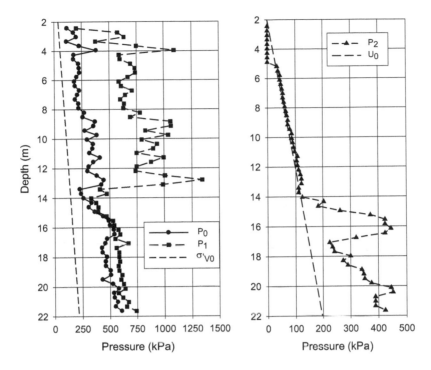

Figure 6.13 Presentation of DMT pressures P_0, P_1, and P_2 – Keene, N.H.

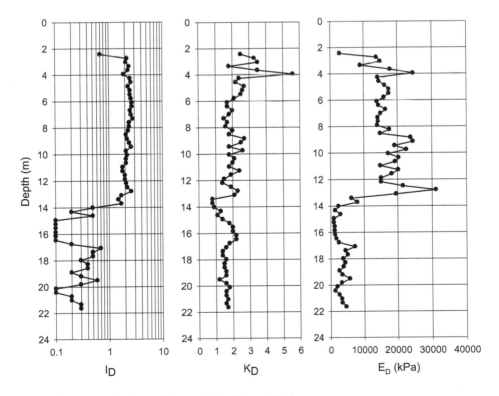

Figure 6.14 DMT indices for data of Figure 6.13 – Keene, N.H.

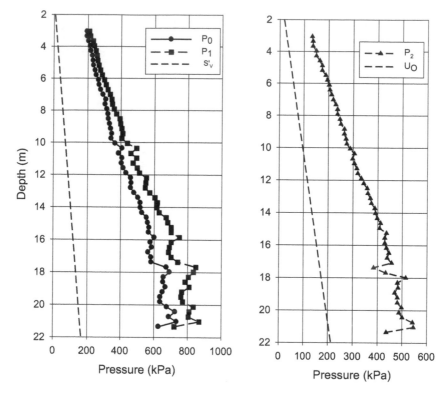

Figure 6.15 Presentation of DMT pressures P_0, P_1, and P_2 – Massena, N.Y.

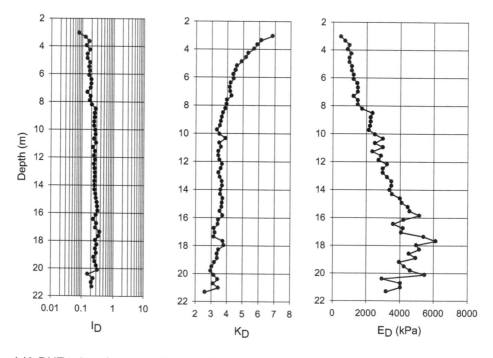

Figure 6.16 DMT indices for data of Figure 6.15 – Massena, N.Y.

6.7 INTERPRETATION OF TEST RESULTS

6.7.1 Evaluating Stratigraphy

A specific application of the results of a DMT profile is in determining changes in subsurface stratigraphy. This can be done primarily from the soil's point of view since the test results will be dependent on the behavior of the soil during and after penetration by the blade. This part of the test should not be considered as a detailed primary use of the instrument in contrast to a CPT and CPTU profile in which the results are nearly continuous and thus more detailed. As such, the DMT is not meant to be competitive with the CPT or CPTU but more complimentary in terms of the overall scope of an *in situ* testing program, since the DMT can provide additional soil data to help in the delineation of the subsurface profile and specific property evaluation.

Soil identification or "classification" using the DMT should be thought of in terms of soil behavior as opposed to specific composition, such as is done with laboratory or visual-based soil classification systems. This is essentially the same as the systems used to provide soil identification from CPT/CPTU results, wherein the results of the test are used in a rational way to help determine the nature of the material which is being tested. As described in Chapter 4, soil identification systems most commonly used with the CPT/CPTU make use of tip resistance, sleeve resistance, and pore water pressure to identify the soil type. Similarly, with the DMT, the lift-off pressure, 1 mm expansion, and recontact pressure can be used to identify the soil.

One way to evaluate changes in site stratigraphy is to examine the plot of the pressure readings P_0, P_1, and P_2 vs. depth as shown in Figure 6.13 and look for changes in test results. Changes in stratigraphy are indicated by abrupt changes in the trends of all three values. The variation in the recontact pressure, P_2, vs. depth appears to be one of the easiest ways of noting changes in stratigraphy much in the same way that pore pressure from a CPTU is used, especially while the test is being performed. A comparison with the *in situ* pore water pressure profile gives a rapid indication of fine-grained vs. granular soil layers, since the P_2 profile in a saturated granular soil will essentially follow the *in situ* pore water pressure profile.

Marchetti (1975) had proposed a simple system based on the DMT Material Index, I_D, that could be used to establish the soil type as shown in Table 6.2. As I_D increases, the soil type changes from very soft cohesive to very stiff granular, with several intergrades. From Equation 6.6, I_D defines the relative change in pressure from P_0 to P_1 normalized with respect to P_0 and accounting for the *in situ* pore water pressure. The variation in K_D and E_D can also be used to show changes in soil behavior.

The Pore Pressure Index, U_D, defined by Equation 6.9 has also been shown to be a useful parameter in determining site stratigraphy (Lutenegger & Kabir 1988). Variations in U_D indicate changes in soil drainage conditions and the tendency for the soil to generate excess positive pore water pressures during blade penetration. This means that a plot of U_D vs. depth helps indicate changes in soil stratigraphy. U_D can be used to distinguish "permeable" layers ($U_D < 0.2$) from "impermeable" layers ($U_D > 0.6$). Intermediate materials will typically fall in between, (i.e., $0.2 < U_D < 0.6$). Figure 6.17 shows U_D profiles from the data of Figures 6.13 and 6.15.

Table 6.2 Soil identification using material index, I_D

Soil type	Peat or sensitive clay	Clay		Silt			Sand	
			Silty	Clayey		Sandy	Silty	
I_D	< 0.10	0.10	0.35	0.60	0.90	1.2	1.8	3.3

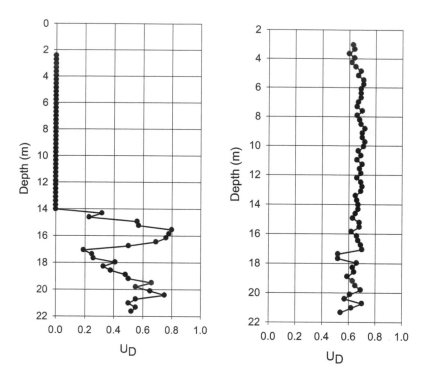

Figure 6.17 Variation in U_D from data of Figures 6.13 (left) and 6.15 (right).

Robertson (2009) identified an approximate relationship between the CPT Soil Behavioral Type Index, I_C, previously described in Chapter 4, and the DMT Material Index, I_D, which is shown in Figure 6.18 and which can be expressed as follows:

$$I_C = 2.5 - 1.5 \log(I_D) \tag{6.12}$$

There is also some evidence that the DMT Material Index, I_D, is related to the friction ratio, obtained from the CPT.

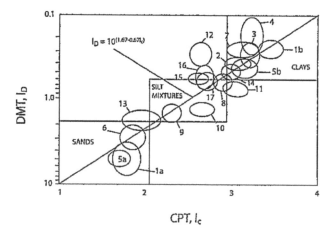

Figure 6.18 Approximate relationship between CPT I_C and DMT I_D. (From Robertson 1990.)

6.7.2 Interpretation of DMT Results in Fine-Grained Soils

The DMT has also been used extensively in fine-grained soils as previously summarized in Table 6.1. In some respects, using a DMT has some advantages over using a CPTU in that deairing of a porous element is not needed. Table 6.3 presents a summary of a number of reported correlations between soil properties and DMT results for fine-grained soils. The following sections discuss interpretation of DMT results in fine-grained soils.

6.7.2.1 Undrained Shear Strength

A number of methods have been suggested for estimating the undrained shear strength of saturated fine-grained soils from the results of the DMT.

Table 6.3 Reported correlations between soil properties and DMT in fine-grained soils

Soil parameter	DMT index	References
s_u or s_u/σ'_{vo}	$I_D, K_D, P_0, P_1, P_2, E_D$	Marchetti (1980) Lacasse & Lunne (1988) Roque et al. (1988) Su et al. (1993) Kamei & Iwasaki (1995) Lutenegger (2006a) Lechowicz et al. (2017)
K_o	I_D, K_D	Marchetti (1980) Powell & Uglow (1988) Lacasse & Lunne (1988) Mayne & Kulhawy (1990)
σ'_p	P_0	Mayne (1987; 1995) Ozer et al. (2006)
OCR	I_D, K_D, K_i	Marchetti (1980) Davidson & Boghrat (1983) Powell & Uglow (1988) Lacasse & Lunne (1988) Su et al. (1993) Lutenegger (2015)
M	I_D, E_D, K_D	Marchetti (1980) Ozer et al. (2006) Lechowicz et al. (2017)
E	E_D	Marchetti (1980) Davidson & Boghrat (1983) Borden et al. (1985) Lutenegger (1988) Su et al. (1993)
G_{max}	E_D, P_0, P_1	Kalteziotis et al. (1991) Tanaka et al. (1994) Shibuya et al. (1998)
c_h	P_0, P_2	Robertson et al. (1988) Schmertmann (1989) Marchetti & Totani (1989)
k_h (subgrade reaction modulus)	P_0, K_D	Robertson et al. (1988) Gabr & Borden (1988) Schmertmann (1989)
CBR	E_D	Borden et al. (1986)

6.7.2.1.1 Marchetti Method (1980)

Marchetti (1980) had suggested that a simple empirical relationship could be used to predict the normalized undrained strength of cohesive soils from the DMT lift-off pressure, P_0, according to

$$s_u/\sigma'_{vo} = 0.22(0.5\ K_D)^{1.25} \qquad (6.13)$$

where
  σ'_{vo} = vertical effective stress
  $K_D = (P_0 - u_o)/\sigma'_{vo}$
  u_o = *in situ* pore water pressure

A comparison between Equation 6.13 and the results of laboratory unconfined compression and triaxial compression and *in situ* field vane tests obtained by Marchetti (1980) provided reasonable accuracy as shown in Figure 6.19. This initial technique has been used by a number of investigators; however, it appears to need site-specific verification and does not provide sufficient accuracy in all soils.

For example, test results presented by Powell & Uglow (1988) indicated that this method could underpredict or overpredict the normalized undrained shear strength. Similar results were presented by Lacasse & Lunne (1988) presenting correlations between predicted and measured undrained shear strength in Norwegian clays, using different reference values of strength.

6.7.2.1.2 Roque et al. Method (1988)

An alternative approach to estimating the undrained shear strength was presented by Roque et al. (1988) using a simple bearing capacity approach assuming that the blade is rectangular footing loaded in the soil horizontally. In this case, the undrained shear strength is given as follows:

$$s_u = (P_1 - \sigma_{ho})/N_C \qquad (6.14)$$

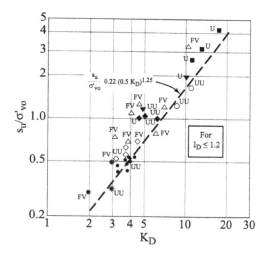

Figure 6.19 Correlation of s_u from Marchetti (1980).

where
  P_1 = DMT 1 mm expansion pressure
  σ_{ho} = *in situ* total horizontal stress = $K_o\sigma'_{vo}+u_o$
  N_C = bearing capacity factor.

Values of N_C varying from 5 to 9 were suggested by Roque et al. (1988) as given in Table 6.4.

This procedure resembles the semi-empirical technique used to predict the undrained shear strength from Menard pressuremeter using the limit pressure, P_L, as will be previously described in Chapter 7 where

$$s_u = (P_L - \sigma_{ho})/N \qquad (6.15)$$

In both Equations 6.14 and 6.15, it is assumed that a yield or limit pressure is obtained during the expansion phase of the test such that $P_1 = P_L$. For the pressuremeter, values of N from the literature are often in the range of 5–7, which compares well with values suggested by Roque et al. (1988). This technique requires an evaluation of the original (prepenetration) at rest *in situ* horizontal stress and some assumption of the soil type to estimate the bearing capacity factor, N_C.

6.7.2.1.3 Schmertmann Method (1989)

Schmertmann (1989a) presented further theoretical explanations for an expected trend between K_D and the undrained strength based on the limit pressure from cylindrical cavity expansion and suggested that a good approximation for predicting the normalized undrained strength would be as follows:

$$s_u/\sigma'_{vo} = K_D/8 \qquad (6.16)$$

6.7.2.1.4 "Effective" Lift-Off Pressure Method

Since both the normalized shear strength and K_D are normalized by σ'_{vo}, it is possible to estimate the undrained shear strength directly from $P_0 - u_o$. This eliminates the need to estimate soil unit weight to estimate σ_{vo} but still requires an estimate of u_o at the test location. In this case, the undrained shear strength would be given as follows:

$$s_u = (P_0 - u_o)/\beta \qquad (6.17)$$

where
  β = an empirical constant.

Values of β should be in the range of 6 to 10.

Table 6.4 Bearing capacity factors suggested by Roque et al. (1988)

Soil	N_c
Brittle clay and silt	5
Medium clay	7
Nonsensitive plastic clay	9

6.7.2.1.5 Larsson & Eskilson (1989) "effective" 1 mm Expansion Pressure Method

A method for estimating the undrained shear strength in clays using the value of P_1 was suggested by Larsson & Eskilson (1989) as follows:

$$s_u = (P_1 - u_o)/F \qquad (6.18)$$

This is similar to the method previously described using P_0. For both organic and inorganic clays in Sweden, Larsson & Eskilson (1989) found that F typically is in the order of 10.

6.7.2.1.6 Yu et al. (1993)

Yu et al. (1993) performed a numerical study of the undrained penetration mechanics of the DMT by modeling the blade as the expansion of a flat cavity. An elastoplastic soil model was used, and a plane strain condition was assumed so that no strain was permitted in the vertical direction. The results of this study indicated that the lift-off pressure, P_0, is a function of the initial horizontal stress, the undrained shear strength, and the rigidity index, G/s_u, of the soil. It was found that the normalized lift-off pressure, defined as

$$N_{po} = (P_0 - \sigma_{ho})/s_u \qquad (6.19)$$

was not a constant but increases with the rigidity index of the soil as follows:

$$N_{po} = -1.75 + 1.57 \ln(G/s_u). \qquad (6.20)$$

For typical values of rigidity index for clays, the normalized lift-off pressure ranges from about 3.6 to 8.3, which is similar to the range in values of N_C suggested by Roque et al. (1988). Rearranging Equation 6.18 gives

$$s_u = (P_0 - \sigma_{ho})/N_{po} \qquad (6.21)$$

The difficulty in using Equation 6.19 is that an estimate of the *in situ* horizontal stress is needed. An initial estimate of σ_{ho} could be made using the estimated K_o from the DMT results.

Finno (1993) presented results of a three-dimensional numerical study of the penetration of the DMT through saturated cohesive soils. In particular, the study evaluated the effects of DMT penetration on the lift-off, P_0, reading. It was found that DMT K_D values (and hence P_0) are sensitive to variations in K_o and OCR, which is consistent with results presented by Yu et al. (1993) but that strength and compressibility parameters do not significantly affect P_0. Since DMT-derived s_u values depend primarily on P_0 (using the Marchetti method), it was suggested that these values of strength are only as good as the empirical correlations based on OCR, and therefore, site-specific correlations are needed.

6.7.2.1.7 Kamei and Iwasaki Method (1995)

Kamei & Iwasaki (1995) suggested that for very soft clays and peat, a correlation could be established between the undrained shear strength obtained from laboratory UU triaxial compression tests and unconfined compression tests and the DMT elastic modulus, E_D, as

$$s_u = 0.018 E_D. \qquad (6.22)$$

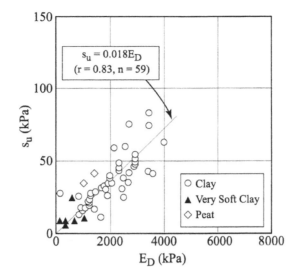

Figure 6.20 Correlation between s_u and E_D. (After Kamei & Iwasaki 1995.)

The results of tests compared on Holocene deposits are shown in Figure 6.20, all of which have undrained strengths less than 100 kPa.

It may be reasonable to expect such a correlation in very soft soils since the value of P_1 is only slightly higher than P_0, giving very low values of E_D. Therefore, since E_D reflects the difference in going from P_0 to P_1, it is reasonable to expect that as the strength increases, E_D increases. A simpler, more direct approach may be to simply take the pressure difference, i.e., $P_1 - P_0$, as an estimate of undrained strength, as was suggested by Lutenegger & Blanchard (1993).

6.7.2.1.8 Lutenegger Cavity Expansion "effective stress" Method (2006)

Lutenegger (2006a) suggested that undrained strength could be estimated by simple cavity expansion theory. From cylindrical cavity expansion theory, the installation radial effective stress of a driven cylindrical probe into a soft clay can be given as follows:

$$\sigma_r' = \left[1 + (3/M)^{0.5}\right] s_u \tag{6.23}$$

where

s_u = initial *(in situ)* undrained shear strength
M = critical state line gradient.

This prediction of σ_r' assumes that the soil adjacent to the penetrating probe (shaft of a pile, cone, DMT, etc.) is at a critical state under plane strain conditions with a radial major principal stress. The plane strain value of the critical state line gradient, M, may be obtained from

$$M = 3 \sin \varphi_{ps}' \tag{6.24}$$

where

φ_{ps}' = plane strain effective friction angle.

In terms of undrained strength, Equation 6.23 may be rewritten as

$$s_u = \sigma_r'/\alpha \qquad (6.25)$$

where

$$\alpha = \left[1 + (3/M)^{0.5}\right]. \qquad (6.26)$$

For most clays, reasonable values of φ_{ps}' range from 20° to 30°, and from Equation 6.26, it follows that the value of α has a narrow range and only varies from 2.56 to 2.72. Therefore, a reasonable estimate of the undrained strength may be obtained as

$$s_u = \sigma_r'/2.65. \qquad (6.27a)$$

As previously noted, it has been shown by several investigators that the DMT P_0 value is nearly identical to the initial penetration stress acting on the face of a cylindrical probe (e.g., full-displacement pressuremeter or lateral stress cone), and P_2 (from a C-Reading) is nearly identical to the penetration pore pressure (e.g., from piezoblade tests), which also closely match the penetration pore pressure from a CPTU. Therefore, the total and effective stress conditions acting around a cylindrical probe and the DMT do not differ that much. In terms of the data obtained from the DMT, then, Equation 6.27a may be rewritten as

$$s_u = \left(P_0 - P_2\right)/2.65. \qquad (6.27b)$$

A comparison of the results from the field vane and DMT tests from two sites is shown in Figure 6.21. Even though there is some scatter in the data, the results do not appear to be site-specific and appear to be grouped between $\alpha = 2.0$ to 3.0, which fits well with Equation 6.27b. Recently, Failmezger et al. (2015) found that this method worked well for soft clays offshore.

Table 6.5 gives a summary of the reported correlations between s_u and K_D. A comparison of several of these empirical equations is shown in Figure 6.22. Other expressions for estimating undrained shear strength from the DMT, but not using K_D, are summarized in Table 6.6.

6.7.2.2 Stress History – OCR

The DMT results may also be used to estimate the stress history of fine-grained soils, either directly through estimating the OCR or by estimating the preconsolidation stress, σ_p'. A number of comparisons have been made illustrating the usefulness of the DMT for this. It should be fairly intuitive that if the DMT can be used successfully to estimate the undrained shear strength, then because of the interrelationship between stress history and shear strength, the test can also be used successfully to estimate the stress history.

6.7.2.2.1 Marchetti Method

Based on the results of laboratory oedometer tests, Marchetti (1980) suggested a simple empirical expression to predict the OCR from DMT results as follows:

$$OCR = \left(0.5\, K_D\right)^{1.56} \qquad (6.28)$$

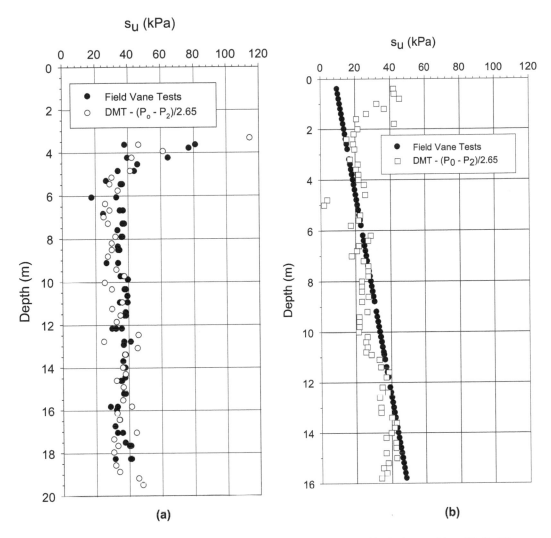

Figure 6.21 Comparison between $(P_0 - P_2)/2.65$ and s_u at two sites: (a) Amherst, Ma.; (b) Bothkennar, Scotland.

The data presented by Marchetti (1980) are shown in Figure 6.23. Again, as with estimates of the undrained shear strength, a number of investigators have shown that the use of this expression does not always lead to a reliable estimate of the stress history and that site-specific correlations using the form of Equation 6.28 are needed. A number of reported correlations between OCR and K_D are summarized in Table 6.7. A comparison of these correlations is shown in Figure 6.24.

6.7.2.2.2 Cavity Expansion Method

Chang et al. (1997) also suggested that the OCR in clays could be estimated from a cavity expansion model as follows:

$$OCR = 2\left[(P_0 - \sigma_{vo})/(4.13\,\sigma'_{vo}) \right]^{1.18}. \tag{6.29}$$

Table 6.5 Correlations for estimating undrained shear strength in clays from DMT K_D

Correlation	Soil	References
$s_u/\sigma'_{vo} = 0.22 \, (0.5K_D)^{1.25}$	Misc. Clays	Marchetti (1980)
$s_u/\sigma'_{vo} = (0.17 \text{ to } 0.21) \, (0.5K_D)^{1.25}$ (field vane) $s_u/\sigma'_{vo} = 0.14 \, (0.5K_D)^{1.25}$ (triaxial compression) $s_u/\sigma'_{vo} = 0.20 \, (0.5K_D)^{1.25}$ (simple shear)	Norwegian sensitive clays	Lacasse & Lunne (1988)
$s_u/\sigma'_{vo} = K_D/8$	Misc. Clays	Schmertmann (1989a)
$s_u/\sigma'_{vo} = 0.175 \, (0.5K_D)^{1.25}$	Singapore marine clay	Chang (1991)
$s_u/\sigma'_{vo} = 0.32 \, (0.5K_D)^{1.07}$	Taiwan marine clay	Su et al. (1993)
$s_u/\sigma'_{vo} = 0.35 \, (0.47K_D)^{1.25}$	Japan clay	Kamei & Iwasaki (1995)
$s_u/\sigma'_{vo} = (0.14K_D)^{1.55}$	Leda clay	Tanaka & Bauer (1998)
$s_u/\sigma'_{vo} = (0.17K_D)^{0.96}$	Venice lagoon	Ricceri et al. (2002)
$s_u/\sigma'_{vo} = 0.274K_D$	Pakistan	Aziz & Akbar (2017)
$s_u/\sigma'_{vo} = 0.22(0.5K_D)^{1.12}$	Stiff clay – Poland	Lechowicz et al. (2017)

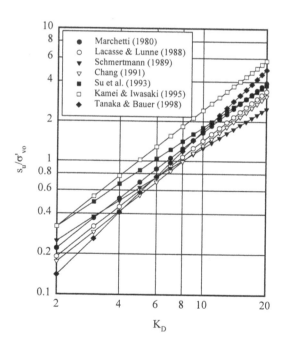

Figure 6.22 Comparison of empirical correlations between K_D and s_u.

A similar correlation was suggested by Cai et al. (2015) as follows:

$$OCR = 2\left[(P_1 - \sigma_{vo})/(4.77 \, \sigma'_{vo})\right]^{1.18}$$

6.7.2.2.3 Initial Lateral Stress Ratio Method

Lutenegger (2015) suggested an empirical correlation relating OCR to the initial lateral stress ratio, K_i. Results obtained at several clay sites in northern New York and Ontario, Canada are shown in Figure 6.25. The data fit a good trend except for results obtained in

Table 6.6 Other correlations for estimating undrained shear strength in clays from DMT

Correlation	Soil	References
$s_u = (P_1 - u_o)/9$	Organic clays – Sweden	Larsson & Eskilson (1989)
$s_u = 0.018E_D$	Soft clay – Japan	Kamei & Iwasaki (1995)
$s_u = (P_2 - P_0)/2.65$	Soft clays	Lutenegger (2006a)
$s_u = 0.12(P_0 - \sigma_{vo})$	Soft to stiff clays	Cai et al. (2015)
$s_u = 0.09(P_1 - \sigma_{vo})$		
$s_u = 0.1e^{0.28(P_0 - u_o)}$	Stiff clay – Poland	Lechowicz et al. (2017)
$s_u = 0.1e^{0.21(P_1 - u_o)}$		

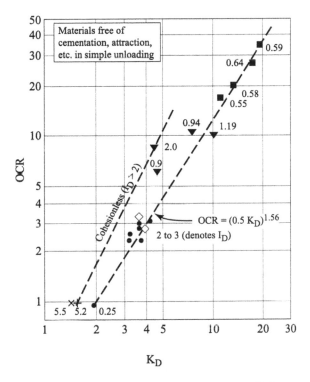

Figure 6.23 Correlation between K_D and OCR presented by Marchetti (1980).

Table 6.7 Reported correlations for estimating OCR from K_D in clays from DMT

Expression	Soil	References
$OCR = (0.5 K_D)^{1.56}$	Misc. Clays	Marchetti (1980)
$OCR = (0.372 K_D)^{1.40}$	Florida clays	Davidson & Boghrat (1983)
$OCR = 0.24 (K_D)^{1.32}$	Stiff clays of UK	Powell & Uglow (1988)
$OCR = 0.225 (K_D)^n$ (n varies from 1.35 to 1.67)	Norwegian sensitive clays	Lacasse & Lunne (1988)
$OCR = 10^{[0.16K - 2.5)]}$	Swedish clays	Larsson & Eskilson (1989)
$OCR = (0.5 K_D)^{0.84}$	Singapore marine clay	Chang (1991)
$OCR = 1.65 (0.5 K_D)^{1.13}$	Taiwan marine clay	Su et al. (1993)
$OCR = 0.34 K_D^{1.43}$	Japan soft clay	Kamei & Iwasaki (1995)
$OCR = (0.3 K_D)^{1.36}$	Leda clay	Tanaka & Bauer (1998)

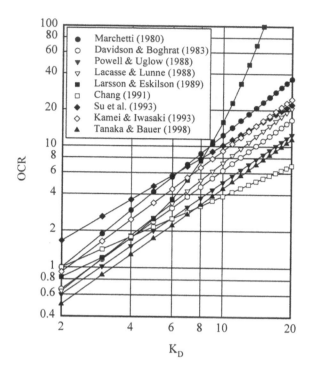

Figure 6.24 Comparison of reported empirical correlations between K_D and OCR.

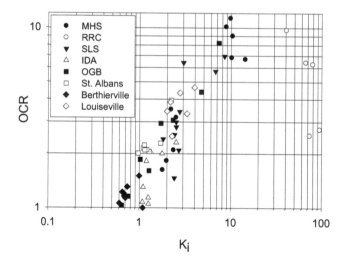

Figure 6.25 Correlation between K_i and OCR.

fissured clay near the ground surface and above the water table. The correlation may be more robust since it makes use of two of the DMT pressure measurements.

6.7.2.3 Preconsolidation Stress

The correlation proposed by Marchetti (1980) between K_D and OCR is essentially a correlation between $\sigma'p$ and $P_0 - u_o$ if the effect of the vertical effective stress is removed.

6.7.2.3.1 Mayne Method (1987)

Mayne (1987) suggested that a direct approach to estimating the stress history in clays would be to compare the DMT lift-off pressure P_0 to the preconsolidation stress σ'_p. It was suggested that the preconsolidation stress could be obtained as follows:

$$\sigma'_p = (P_0 - u_o)/\eta \qquad (6.30)$$

where:
η = an empirical factor

Based on a simple elastic-plastic model of the expansion of a cylindrical cavity, the value of η in Equation 6.30 would have a range from between 0.8 and 2.5 for a typical range of rigidity index for clays. Using a database from a number of test sites, Mayne (1987) found that the observed value of η ranged from about 1 to 3 but showed considerable scatter. This correlation was updated by Mayne (1995) for 24 intact natural clays as follows:

$$\sigma'_p = (P_0 - u_o)/2. \qquad (6.31)$$

While these results tend to substantiate the trend suggested between K_D and OCR by Marchetti, they still appear to be site-specific. This may in part be related to the fact that an implicit assumption involved with this technique is that the value of $P_0 - u_o$ represents excess pore water pressure. The author has previously shown that this assumption is very close in soft and very soft clays with OCR less than about 2.5, but in more heavily overconsolidated clays, this assumption would not be valid (Lutenegger 1988). Even in soft clay, P_0 still contains a small component of effective stress, and therefore, one should expect scatter in results using this approach. In light of more recent data in soft clays, it may be more appropriate to substitute P_2 for P_0 in Equation 6.31; however, this approach has limitations in highly overconsolidated soils.

6.7.2.3.2 Cavity Expansion "effective stress" Method

The author performed tests at eight marine clay sites to determine the relationship between stress history and DMT results. In connection with this work, DMT P_2 readings were also obtained. All oedometer tests were performed on samples obtained from thin-walled Shelby tubes or a piston sampler, and incremental loading oedometer tests were performed. The data from all sites shown in Figure 6.26 indicate a simple correlation between $(P_0 - P_2)$ and the preconsolidation stress and indicate (approximately)

$$\sigma'_p = P_0 - P_2. \qquad (6.32)$$

6.7.2.4 Lateral Stresses

Because of its geometry, the DMT largely records the horizontal response to penetration when placed vertically. Thus, a measure of horizontal total stress is obtained, which Marchetti used to define the horizontal stress index, K_D. We should expect that if K_D can be used to reasonably predict OCR, then it can also be used to predict K_o for many soils, since K_o and OCR are often related. In recent years, engineers have become increasingly aware of the influence that lateral stresses have on engineering behavior.

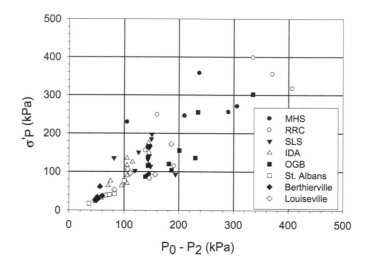

Figure 6.26 Correlation between preconsolidation stress and $(P_2 - P_0)$.

The value of K_D was directly related to K_o by Marchetti (1980) primarily using the empirical relation to OCR presented by Brooker & Ireland (1965):

$$K_o = (K_D/1.5)^{0.47} - 0.6 \qquad (6.33)$$

The initial correlation shown in Figure 6.27 appeared to be independent of soil type (excluding sands) and stress history and therefore has been used by a number of investigators. The correlation suggested by Marchetti (1980) was based on a limited number of soil types and tests and was stated to be applicable to "uncemented deposits, free of attraction, etc. Additionally, Marchetti used the observations relating OCR to K_o presented by Brooker & Ireland (1965) that were based on tests performed on reconstituted samples, not natural clay deposits.

One of the difficulties in establishing a direct relationship between K_D and K_o is that a reference value of K_o is difficult to obtain. Unlike other soil parameters such as undrained strength or compressibility, that may be reasonably determined by acceptable methods,

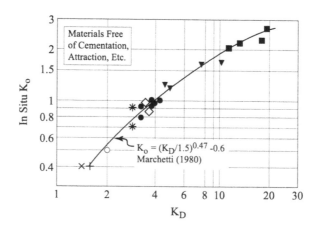

Figure 6.27 Correlation between K_D and K_o presented by Marchetti (1980).

there is no specific technique which is agreed upon by the profession as the preferred method for determining K_o. Several recent investigations have made use of other field or laboratory tests to compare the K_o value obtained using the DMT: e.g., push-in spade cells (Chan & Morgenstern 1986), K_o-Stepped Blade (Lutenegger & Timian 1986), and prebored pressuremeter (Powell & Uglow 1988). These and other studies indicate that K_o values derived from the DMT K_o correlation given by Equation 6.37 are nearly all within a factor of about 1.5 for a wide range of geologic materials.

Powell & Uglow (1988) compared estimated values of K_o using the DMT at five sites in the U.K., where previous laboratory and field work had provided a reliable database of K_o values. As shown in Figure 6.28, at several of the sites, the values of K_o predicted by the DMT were lower than previously measured values, especially for very stiff clays.

Lunne et al. (1990) suggested that it might be possible to separate clay behavior and empirical correlations between K_D and K_o by considering the difference between "young" clays and "old" clays. "Young" clays were distinguished as having normalized undrained strength, (s_u/σ'_{vo}), less than 0.7. They suggested that on the basis of previous work (e.g., Lacasse & Lunne 1983; Powell & Uglow 1988; 1989), a correlation of the following form could be used:

$$K_o = a\, K_D^{\,m} \tag{6.34}$$

For "young" clays, at least up to K_o of about 1.5, they suggested the following expression:

$$K_o = a\, K_D^{\,0.54} \tag{6.35}$$

where "a" is on average 0.34 but was observed to vary from 0.28 to 0.38.
For "old" clays on average:

$$K_o = 0.68\, K_D^{\,0.54} \tag{6.36}$$

Masood & Kibria (1991) suggested that the exponent "m" in Equation 6.34 may be related to P.I. of the soil.

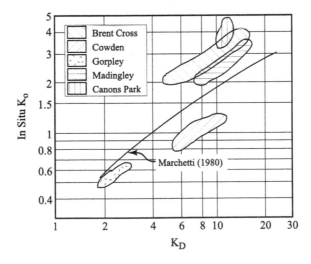

Figure 6.28 Correlation between K_D and K_o in UK stiff clays. (after Powell & Uglow 1988).

A comparison between K_o values obtained from self-boring pressuremeter tests (SBPMTs) and K_D values for a number of clays was presented by Mayne & Kulhawy (1990) as

$$K_o = 0.27 \, K_D \tag{6.37}$$

The data are shown in Figure 6.29 and show surprisingly little scatter. It might be argued that the SBPMT gives the best estimate of *in situ* K_o, and therefore, Equation 6.37 is very appealing for use in practice, especially since it is derived from a database developed using a single type of reference test.

A summary of different correlations reported between K_D and K_o for a variety of clay deposits is given in Table 6.8. These correlations are compared in Figure 6.30.

6.7.2.5 Constrained Modulus

Since the DMT expansion pressure is used to provide an indication of the deformation behavior of the soil after an initial penetration phase, it is expected that the DMT modulus may be useful at predicting "elastic" soil response. Expansion of the DMT diaphragm from P_0 to P_1 produces a known displacement that was used by Marchetti to define the Dilatometer Modulus, E_D. Marchetti (1975) had previously suggested that the DMT expansion could be used to define a lateral subgrade reaction modulus value; however, engineers are more often in need of a deformation parameter for settlement estimates.

As previously noted, Marchetti (1980) had suggested a correlation between E_D and the one-dimensional constrained modulus, M, as

$$M = R_M \, E_d \tag{6.38}$$

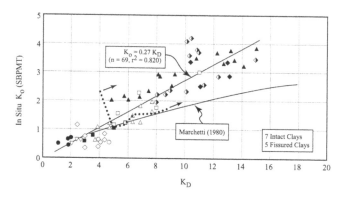

Figure 6.29 Correlation between K_D and K_o from self-boring PMT. (After Mayne & Kulhawy 1990.)

Table 6.8 Reported correlations for estimating K_o from K_D in clays

Expression	Soil	References
$K_o = (K_D/1.5)^{0.47} - 0.6$	Misc. Clays	Marchetti (1980)
$K_o = K_D/8.27$	Leda clay	Tanaka & Bauer (1998)
$K_o = 0.34 \, K_D^{0.54}$	"Young clays"	Lunne et al. (1990)
$K_o = 0.68 \, K_D^{0.54}$	"Old clays"	Lunne et al. (1990)
$K_o = 0.27 \, K_D$	Clays	Mayne & Kulhawy (1990)

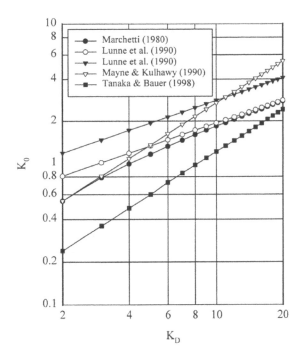

Figure 6.30 Comparison of reported correlations between K_D and K_o.

where

R_M = "modulus ratio" = $f(I_D, K_D)$.

For clays ($I_D < 0.6$), Marchetti suggested

$$R_M = 0.14 + 2.36 \log K_D \tag{6.39}$$

Marchetti thus proposed a correlation between E_D, K_D, and M and the local oedometric constrained modulus at the effective *in situ* overburden stress, σ'_{vo}. More specifically, the correlation is for the local reload modulus. Therefore, in normally consolidated soil, the DMT should not be expected to provide information about the reload modulus, and in over-consolidated soil, the DMT will provide no direct measurement of the virgin compression modulus.

Iwasaki et al. (1991) showed good agreement between constrained modulus values estimated by the Marchetti correlation (Equation 6.39) and the constrained modulus obtained from laboratory oedometer tests on high quality samples for a soft alluvial deposit in Japan. Chang (1991) suggested a simple correlation between R_m and K_D for Singapore marine clays as

$$R_m = 0.02 (K_D)^{3.5}. \tag{6.40}$$

Su et al. (1993) found that for Taiwan marine clays

$$R_m = 0.50 + \log K_D. \tag{6.41}$$

Ozer et al. (2006) found direct correlations between P_0 and P_1 and M for Lake Bonneville clay:

$$M = 0.89(P_0 - u_0)^{1.25} \tag{6.42a}$$

$$M = 0.54(P_1 - u_0)^{1.27} \tag{6.42b}$$

There is a distinct advantage in estimating M, which is particularly appealing to practitioners, since the Janbu (1963; 1967) technique for estimating settlement may be used. The relative accuracy of the estimate of M was clearly demonstrated by Schmertmann (1986) who compared DMT settlement estimates with measured settlements over a wide range in magnitude.

Constrained modulus values for comparison with the DMT should be obtained from back calculated settlement records on projects wherein an exact determination of footing or other loading stress may be made. Based on the results of measured field performance in clays, Saye & Lutenegger (1988) showed results in which approximately $M_{meausred}/M_{DMT} = 2.5$. It should also be kept in mind that the accuracy of predictions with respect to field performance is also dependent upon the variability associated with the use of the mathematical model involved. This applies to any test.

6.7.2.6 Elastic Modulus

In many design situations, the engineer may need an estimate of elastic modulus, E_S, e.g., for use in drilled shaft design or immediate settlement estimates. Depending on the design problem, a different value of E_S may be required, i.e., E_i, E_{25}, etc.

Davidson & Boghrat (1983) suggested that in highly overconsolidated clays, the value of E_i obtained from unconsolidated undrained triaxial compression tests could be related directly to E_D, using a factor of about 1.4. Robertson et al. (1988) suggested a factor of 10 for clays, for use in laterally loaded pile design. Chang (1991) found that E_u obtained as a secant at 50% yield in UU triaxial compression tests was approximately equal to E_D in marine clays.

Borden et al. (1985) suggested a relationship between the initial tangent modulus from unconfined compression tests on partially saturated compacted A-6 soil and E_D which had the form:

$$E_i = 0.142\, E_D^{1.298} \tag{6.43}$$

Su et al. (1993) showed data from UU triaxial compression tests in which the initial tangent modulus, E_i, was approximately correlated to E_D as

$$E_i = 1.13 + 0.14\, E_D. \tag{6.44}$$

where
   E_i and E_D are in MPa

The majority of these results suggest a simple correction factor of the form

$$E_S = R_E\, E_D \tag{6.45}$$

where R_E varies from about 0.4 to 10 for clays. The large range in R_E is mostly related to the reference value and definition of E_S, which ranges from the initial tangent "Young's modulus" to a secant modulus at 25%–50% of failure. The variation in R_E may be related to I_D if I_D relates to soil stiffness.

6.7.2.7 Small-Strain Shear Modulus

Kalteziotis et al. (1991) demonstrated that the results of the DMT may be useful in estimating the small-strain shear modulus of fine-grained soils. They presented correlations between shear wave velocity measurements obtained using cross-hole dynamic tests and both P_0 and P_1 pressures from six sites in Greece as

$$V_S = 3.7(P_0)^{0.63} \qquad (6.46a)$$

$$V_S = 3.0(P_1)^{0.63} \qquad (6.46b)$$

where
V_S = shear wave velocity (m/s).

These results are shown in Figure 6.31.

The shear modulus, G_{max}, may be obtained using the shear wave velocity and unit weight of the soil from

$$G_{max} = V_S^2 \rho. \qquad (6.47)$$

Tanaka et al. (1994) suggested that for soft marine clays in Japan

$$G_{max} = 7.5E_D. \qquad (6.48)$$

For soft Bangkok clay, Shibuya et al. (1998) obtained

$$G_{max} = (5.5 \text{ to } 11.3)E_D. \qquad (6.49)$$

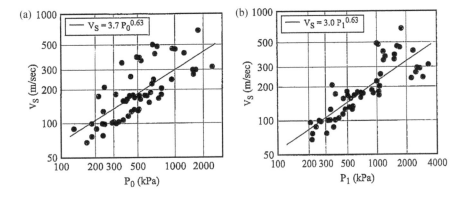

Figure 6.31 Comparison between P_0 and V_S (a) and P_1 and V_S (b) in clays. (After Kalteziotis et al. 1991.)

Table 6.9 Reported values of R_G for clays

R_G	Soil	References
4–25	Greek clays	Kalteziotis et al. (1991)
7.5	Japan marine clay	Tanaka et al. (1994)
5.5–11.3	Bangkok clay	Shibuya et al. (1998)
7.5	Marine clays	Shiwakoti et al. (2001)

These results suggest that R_g ranges from about 5 to 12 for clays, which is consistent with results previously shown. On the other hand, Kalteziotis et al. (1991) found that R_G ranged from 4 to 25 for clays although a large amount of data fell close to $R_G = 10$. Table 6.9 summarizes reported values of R_G for clays.

Howie et al. (2007) found that the value of R_G is dependent on I_D and varies from about 40 for $I_D = 0.1$ to 8 for $I_D = 1$.

6.7.2.8 Liquidity Index

The author (Lutenegger 1988) had suggested that the DMT modulus, E_D, should relate to the liquidity index of clays, since for a pure liquid, the difference between P_1 and P_0 should be zero. Redel et al. (1997) found a good correlation between liquidity index (L.I.) and E_D for a clay site in Israel for $I_D < 0.6$:

$$L.I. = -0.02\, E_D + 1.17 \tag{6.50}$$

6.7.2.9 California Bearing Ratio

Borden et al. (1985) presented results of a combined laboratory and field investigation to determine the usefulness of the DMT in evaluating pavement support characteristics. For three different soil types, they found a linear correlation between corrected unsoaked CBR and DMT modulus, E_D, as

$$CBR = 0.041\, E_D \quad \text{(Soil Type: A-5)} \tag{6.51a}$$

$$CBR = 0.052\, E_D \quad \text{(Soil Type: A-6)} \tag{6.51b}$$

$$CBR = 0.031\, E_D \quad \text{(Soil Type: A-2-4)} \tag{6.51c}$$

where
　　CBR = %
　　E_D = tsf.

6.7.2.10 Coefficient of Consolidation

As previously discussed in Chapter 4, the CPTU can be used to provide information about the flow characteristics of soil, specifically, the coefficient of consolidation and the hydraulic conductivity. The DMT is another test that can provide this information, through the use of time-pressure dissipation tests.

6.7.2.10.1 DMTC (P₂) Dissipation

Since it has already been established that the recontact pressure, P_2, may be used to give an indication of the excess pore water pressure, a procedure in which sequential pressures are obtained over time should give an indication of the change or dissipation of pore water pressure following penetration of the blade. This procedure is informally referred to as a DMTC test or a C-Reading dissipation test. The test is conducted as follows:

1. Following the penetration and release of the pushing thrust, a stopwatch is started to log the elapsed time after penetration. This then becomes time zero for the test.
2. The normal sequence of obtaining an A-, B-, and C-Reading is first performed as a routine part of the test. The elapsed time to obtain the C-Reading is noted.
3. At various time intervals, the sequence of obtaining the A-, B-, and C-Reading is repeated, but only the C-Reading and the elapsed time are recorded.
4. A sufficient number of data points are obtained over time until a minimum of 50% of the excess pore water pressure has dissipated. (This may require a test to run for a few minutes to several hours, depending on the drainage characteristics.)

Typical results obtained from a DMTC dissipation test (corrected to P_2) are shown in Figure 6.32. As can be seen, these results show the characteristic shape of time-rate of consolidation tests. How then is the coefficient of consolidation obtained?

In a manner similar to that described in Chapter 4 for the CPTU, the dissipation test can be used to obtain c_h according to

$$c_h = TH^2/t \tag{6.52}$$

where
  c_h = coefficient of consolidation in the horizontal direction
  T = theoretical time factor for a given % dissipation
  H = length of drainage path
  t = elapsed time to achieve a given % of dissipation.

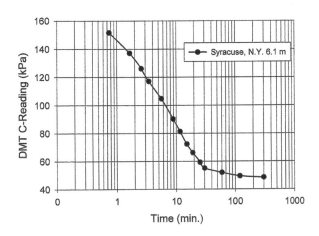

Figure 6.32 Typical DMTC dissipation test.

In the theory developed for the CPTU, H is taken as the radius of the cone (for cylindrical cavity expansion) and T is a function of the rigidity index of the soil (or E/s_u), the pore pressure parameter at failure, A_f, the cone tip geometry and the location of the porous element with respect to the cone tip. For example, Gupta & Davidson (1986) suggested that for $A_f = 0.9$, $E/s_u = 200$ that T = 0.97 for a 60° cone with the pore water pressure measured at the cone base.

Robertson et al. (1988) proposed that as an initial means of estimating c_h, an "equivalent radius" procedure could be used, i.e., back calculating the equivalent radius of a cylindrical cone given by the projected end area of the DMT blade and then using theory for a cylindrical probe. This gives a radius about 22 mm. Schmertmann (1988) suggested that an equivalent radius of 24 mm would produce approximately the same results for c_h for both CPTU dissipation and DMTC dissipation tests in the same cohesive soil. While the "equivalent radius" technique may be used to provide an initial estimate of c_h, Kabir & Lutenegger (1990) found that when compared to results from horizontal flow oedometer tests, the back calculated value of the equivalent radius ranged from 5 to 40 mm for different clays. This suggests that an alternative technique might be more appropriate.

There are several potential sources of error that may occur using this technique:

1. The apex angle of the DMT blade is only about 10°–20° in comparison to the 60° apex of most cones. This means that the initial excess pore pressure distribution will be different.
2. The point of measurement of the pore water pressure in the DMT is about 7.6 blade thicknesses behind the base of the leading wedge and about 12.7 blade thicknesses behind the wedge tip. This is considerably different than the base of the wedge where most CPTUs measure pore pressure. This may affect the rate of pore pressure dissipation observed.
3. The geometry between the two instruments is different. In some soils, e.g., soft clays, this has little or no effect and the DMT blade approximates a cylindrical cavity expansion, while in other soils, the aspect ratio of the blade may have a significant influence on the results.

An alternative procedure is to treat the DMT blade as a cone with a radius equal to the half-thickness (7.5 mm) and use the theoretical time factor for an 18° cone with pore water pressure measured 10 radii behind the tip. For 50% dissipation, Gupta & Davidson (1986) gave $T_{50} = 25$. Schmertmann (1988) suggested using the time for 50% dissipation of pore water pressure in which case Equation 6.52 may be simplified to

$$c_h \left(cm^2/s \right) = 150/t_{50}. \tag{6.53}$$

The excess pore water pressure at any time can be obtained from

$$u_e = P_2 - u_o \tag{6.54}$$

The initial excess pore water pressure would be

$$u_{e,i} = P_{2i} - u_o. \tag{6.55}$$

Since the value of P_{2i} is actually not measured in the test, because it takes about 1 min after penetration to obtain the first P_2 reading, P_{2i} must be obtained by back extrapolation.

Schmertmann (1988) suggested to simply plot the values of P_2 on a square-root-of-time plot and extrapolate the initial few data points as a straight line back to $t = 0$.

Once P_{2i} is obtained, the normalized excess pore water pressure may be determined for each P_2

reading as

$$U = u_e/u_{ei} = (P_2 - u_o)/(P_{2i} - u_o). \tag{6.56}$$

The value of t_{50} can be taken as the value corresponding to $U = 0.5$. The value of u_o must either be measured from piezometers, estimated from ground water conditions, or measured from the test if sufficient time is given in the test to allow P_2 to reach a constant value, indicating complete dissipation and u_o conditions.

6.7.2.10.2 DMTA Dissipation

A procedure using only the A-Reading dissipation to estimate the horizontal coefficient of consolidation, c_h, was presented by Marchetti & Totani (1989) and is referred to as a DMTA dissipation test. In this method, only the A-Reading is obtained from the test at selected time intervals following the penetration to the test depth. Like P_2 dissipation tests, a plot of A vs. log time shows a characteristic "S" shape as shown in Figure 6.33. Unlike the DMTC dissipation, no B-Reading or C-Reading are obtained in this procedure. Using the A-Reading vs. log time curve, the point of contraflexure is defined as T_{flex}. The coefficient of consolidation is obtained from

$$c_h = (5 \text{ to } 10 \text{ cm}^2)/T_{flex} \tag{6.57}$$

The value of c_h estimated by Equation 6.61 is the overconsolidated value and would need to be appropriately adjusted (for example, using local experience) to give an estimate of the normally consolidated value. Table 6.10 gives a qualitative rating of the rate of consolidation based on T_{flex} suggested by Marchetti & Totani (1989).

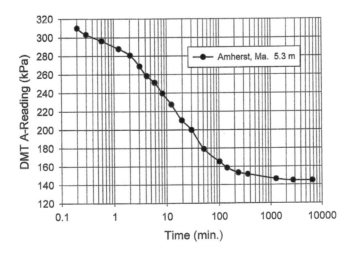

Figure 6.33 Typical DMTA dissipation test.

Table 6.10 Rating of consolidation speed based on T_{flex}

T_{flex} (min)	Rating
<10	Very fast
10–30	Fast
30–80	Medium
80–200	Slow
>200	Very slow

6.7.3 Interpretation of DMT Results in Coarse-Grained Soils

The DMT is well suited for use in granular soils such as sands because of its robust construction. While there is some uncertainty as to the appropriate interpretation, there are currently a number of available correlations for use in granular soils. Table 6.11 presents a summary of some of the reported correlations suggested between various soil properties and DMT results for granular soils.

Table 6.11 Reported correlations between soil properties and DMT in coarse-grained soils

Soil parameter	DMT index	References
D_r	K_D	Robertson & Campanella (1986) Konrad (1991)
State parameter	$(P_1 - P_0)$, K_D	Konrad (1988; 1989)
φ' (sands)	I_D, K_D, thrust or adjacent q_c	Schmertmann (1982) Marchetti (1985; 1997) Campanella & Robertson 1991)
K_o	K_D, thrust	Schmertmann (1982) Jamiolkowski et al. (1988) Lawter & Borden (1990)
OCR	K_D, thrust	Schmertmann & Crapps (1983)
M	I_D, E_D, K_D	Marchetti (1980) Baldi et al. (1986) Robertson et al. (1988) Bellotti et al. (1994; 1997)
E	E_D	Marchetti (1980) Campanella & Robertson (1983) Baldi et al. (1986)
G_{max}	E_D	Baldi et al. (1986; 1989) Hryciw & Woods (1988) Hryciw (1990)
Liquefaction potential	K_D	Marchetti (1982) Robertson & Campanella (1986) Reyna & Chameau (1991) Marchetti (2016)
k_h (subgrade reaction modulus)	P_0, K_D	Robertson et al. (1988) Gabr & Borden (1988) Schmertmann (1989b)
CBR	E_D	Borden et al. (1985)

6.7.3.1 Relative Density (D_r)

Engineers may wish to use the results of the DMT to estimate the in-place relative density of sands, for example, to make an initial assessment of liquefaction potential or to evaluate effects of ground improvement. Even though the interpretation of DMT results in sand is still very complex, some suggestions have been made to estimate D_r from the DMT lift-off pressure via K_D.

Robertson & Campanella (1986) suggested a simple correlation between K_D and D_r in normally consolidated, uncemented sand as shown in Figure 6.34. Unfortunately, it appears that the penetration mechanics and value of P_0 in sands are highly complex and, in addition to D_r, depend on a number of other variables, including stress history, initial *in situ* stresses, compressibility, aging, and cementation. Konrad (1991) presented data from a dune sand that illustrated this complexity and the importance of stress history on the correlation between K_D and D_r. The test results indicate that in comparison to normally consolidated sands, overconsolidated sands show considerable scatter, making such a correlation difficult and perhaps unreliable.

A correlation for estimating D_R of sands using K_D and stress history and normalized DMT modulus (E_D/σ'_{vo}) was presented by Lee et al. (2011) as shown in Figure 6.35.

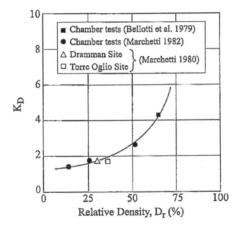

Figure 6.34 Correlation between K_D and D_R in NC uncemented sands. (After Robertson & Campanella 1986.)

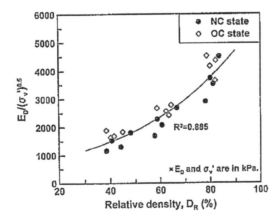

Figure 6.35 Correlation between K_D and D_R in NC and OC sands. (From Lee et al. 2011.)

6.7.3.2 State Parameter

The use of the state parameter to describe the behavior of sands was discussed in Chapter 4 as a rational approach combining the effects of void ratio and stress level (Been & Jefferies 1985). Konrad (1988; 1989) has suggested that the state parameter should also be considered to evaluate the results of DMTs in sand.

Results presented by Konrad (1989) demonstrated that for dry Ottawa sand the normalized state parameter (i.e., the state parameter divided by the value $e_{max} - e_{min}$) is related to the pressure difference $P_1 - P_0$ normalized with respect to the mean or octahedral effective stress. Konrad (1989) suggested that the value of K_D obtained from the DMT should be related to both the mean effective stress and the normalized state parameter. Engineering properties of the undistorted sand may then be obtained from existing correlations related to the state parameter.

6.7.3.3 Drained Friction Angle

It has been suggested that the penetration of the DMT blade in sands and other freely draining soils represents a drained bearing capacity failure approximating a plane-strain condition. The projected end of the DMT blade represents a long, narrow punch. Since the failure condition is controlled in part by the *in situ* state of stress in granular materials, it is reasonable to expect that the results from the DMT may be used to determine φ'.

6.7.3.3.1 Schmertmann Method

A procedure for obtaining the effective axisymmetric friction angle in sands was presented by Schmertmann (1982) using the wedge penetration theory and is attractive since it incorporates the horizontal effective stress that may be estimated from the DMT results. The method is iterative and requires measurement of the pushing thrust to advance the DMT, such that an estimate of the blade tip resistance may be obtained. This measurement may be made using a load cell located immediately behind the blade as previously described.

6.7.3.3.2 Campanella and Robertson Method

Campanella and Robertson (1991) observed that the DMT parameter K_D was directly related to both the normalized blade end bearing (q_d/σ'_{vo}), obtained from the measurement of thrust, which in turn could be related to parallel CPT-normalized tip resistance, i.e., q_c/σ'_{vo}. Based on the fact that reasonably good estimates of φ' could be obtained in sands from q_c/σ'_{vo}, they reasoned that it should be expected that a direct correlation should also exist between φ' and K_D. A chart for predicting peak friction angle was proposed by Campanella & Robertson (1991) but requires an estimate of K_o from DMT results.

6.7.3.3.3 Marchetti Method

Marchetti (1997) suggested that the method given by Campanella & Robertson (1991) could give φ' directly in terms of K_D after first assuming some values of K_o for different conditions: $K_{ONC} = 1 - \sin \varphi'$; $K_o = 1$; and $K_o = K_p^{0.5}$. The resulting correlations between φ' and K_D are shown in Figure 6.36. The three curves shown in Figure 6.36 represent three different estimates of K_o, and it can be seen that for practical purposes, the correlation is only mildly sensitive to the value of K_o for any given value of K_D (typically $\pm 1.5°$ from the $K_o = 1$ curve).

$$\Phi' = 28.2° + (K_D - 0.5)\Big/\Big[0.074 + 0.063(K_D - 0.5)^{0.92}\Big] \qquad (6.58a)$$

$$\Phi' = 27.5° + (K_D - 0.5)\Big/\Big[0.080 + 0.063(K_D - 0.5)^{0.94}\Big] \qquad (6.58b)$$

$$\Phi' = 26.8° + (K_D - 0.5)\Big/\Big[0.10 + 0.062(K_D - 0.5)^{0.95}\Big] \qquad (6.58c)$$

To provide some conservatism, Marchetti (1997) suggested that an initial estimate of φ', which is somewhat lower than all three curves of Figure 6.36, could be obtained from

$$\Phi' = 28° + 14.6\log K_D - 2.1\log^2 K_D. \qquad (6.59)$$

This is shown by the lower solid line in Figure 6.36. This approach has been shown to be very reasonable based on triaxial compression tests on undisturbed sands (Mayne 2015).

6.7.3.4 In Situ Stresses

The DMT may be used to provide an estimate of the *in situ* horizontal stress at the location where the test is performed. Provided that the blade is advanced vertically into the ground, the membrane faces horizontally and provides a response, at least partially controlled by the initial, i.e, preinsertion, horizontal stress. This has been confirmed by the calibration tests previously shown. Marchetti (1980) proposed a correlation between the DMT horizontal stress index K_D and the at-rest earth pressure coefficient, K_o, as

$$K_o = (K_D/1.5)^{0.47} - 0.6. \qquad (6.33)$$

The basis for this correlation was related in part to a correlation proposed linking K_D first to OCR and then using an assumed relationship between K_o and OCR. Most of the test data presented by Marchetti were for fine-grained soils, and only a limited amount of data represented sands or other granular soils.

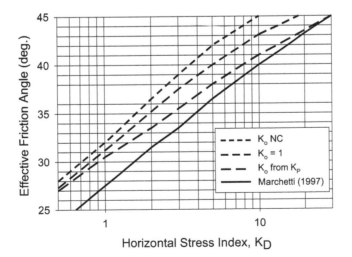

Figure 6.36 Estimates of φ' from K_D. (After Marchetti 1997.)

On the basis of some carefully controlled calibration chamber tests on two sands and on some field experience, Baldi et al. (1986) and Jamiolkowski et al. (1988) suggested that values of K_o obtained by the empirical correlation given by Equation 6.37 tended to be too high in the case of dense and very dense sands and somewhat too low in loose sands. As previously indicated, the penetration of the DMT blade into sands represents a complex deformation mechanism involving both *in situ* stress conditions and stiffness properties of the sands.

Jamiolkowski et al. (1988) suggested that the correlation of K_D/K_o, referred to as the amplification factor, could be made to the state parameter as a rational approach to interpreting test results. They suggested that the amplification factor could be related to the state parameter as

$$K_D/K_o = ae^m \tag{6.60}$$

where
  a and m are empirical coefficients.

Values of a =1.35 and m = −8.08 were given by Jamiolkowski et al. (1988) for Hokksund and Ticino sands. Carriglio et al. (1990) later gave values of a = 1.65 and m = −7.60. Lawter & Borden (1990) confirmed this approach using test results from Cape Fear sand and suggested that a = 1.73 and m = −6.07. For NC sands, the ratio K_o^{DMT}/K_o has a limiting value of about 0.6 at low relative density and increases as D_r increases. Results from more recent calibration chamber tests (Lee et al. 2011) show that the ratio K_o/K_D is dependent on the normalized value of E_D and stress history (OCR) as shown in Figure 6.37a. However, as seen in Figure 6.37b, if the mean effective stress is taken into account, the data follow as single trend.

6.7.3.5 Stress History

Marchetti (1980) recommended an empirical correlation between stress history and K_D as

$$OCR = (0.5\, K_D)^{1.56} \tag{6.28}$$

6.7.3.6 Constrained Modulus

Marchetti (1980) had suggested a relationship between the tangent drained constrained modulus, M, and the Dilatometer Modulus, E_d, as

$$M = R_M\, E_d \tag{6.38}$$

or

$$R_M = M/E_d. \tag{6.61}$$

Values of R_M were suggested by Marchetti (1980) as being related to both the Material Index, I_D and the Lateral Stress Index, K_D. For coarse-grained soils, i.e., predominantly sands and gravels with $I_D > 3.0$, the values of R_M for use in Equation 6.61 were given as

$$R_M = 0.50 + 2.00 \log K_D \, (\text{for } K_D < 10) \tag{6.62a}$$

$$R_M = 0.32 + 2.18 \log K_D \, (\text{for } K_D > 10). \tag{6.62b}$$

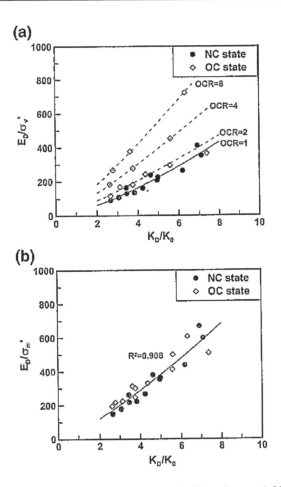

Figure 6.37 (a and b) Relationship between K_D and K_o in sands. (From Lee et al. 2011.)

6.7.3.7 Elastic Modulus

For different sands, the correlation between E_S and E_D varies with stress history and the reference value of E_S. In general,

$$E = E_D R_E \tag{6.63}$$

or

$$R_E = E_{25}/E_D. \tag{6.64}$$

Suggested values of R_E for different granular soils are summarized in Table 6.12.
   Viana de Fonseca et al. (1998) found that the soil modulus was related to P_0 as

$$E_S/E_D = 2.35 - 2.21 \log P_0 \ (P_0 \text{ in MPa}) \tag{6.65}$$

where
   E_S = soil modulus at 10% axial strain.

Table 6.12 Suggested values of R_E for granular soils

	R_E	Reference E	References
NC sand	1	E_{25}	Campanella et al. (1985)
NC Ticino sand	0.88 ± 0.27	E_{25}	Baldi et al. (1986)
NC Ticino sand	1.05 ± 0.25	$E_{0.1}$	Jamiolkowski et al. (1988)
OC Ticino sand	4.29 ± 0.62	E_{25}	Baldi et al. (1986)
OC Hokksund sand	2.49 ± 0.74	E_{25}	Baldi et al. (1986)
OC Ticino sand	3.66 ± 0.80	$E_{0.1}$	Jamiolkowski et al. (1988)
Sand	2	E_i	Robertson et al. (1989)

6.7.3.8 Small-Strain Shear Modulus

Baldi et al. (1986) had noted a correlation between the maximum shear modulus, G_{max}, obtained from resonant column tests on normally consolidated Ticino sand and the DMT modulus, E_D, as

$$R_G = G_{max}/E_D = 2.72 \pm 0.59 \tag{6.66}$$

Bellotti et al. (1986) found similar results for P_0 river sand, with G_{max} obtained from cross-hole tests:

$$R_G = G_{max}/E_D = 2.2 \pm 0.7. \tag{6.67}$$

This compares well with more recent results presented by Bellotti et al. (1994) who showed that for both NC and OC Toyoura sand,

$$R_G = G_{max}/E_D = 2.96 - (0.02 \, D_r)(D_r \text{ in } \%). \tag{6.68}$$

Viana de Fonseca et al. (1998) suggested that R_G was related to the lift-off pressure as

$$R_G = 16.7 - 16.3 \log P_0 \, (P_0 \text{ in MPa}) \tag{6.69}$$

A comparison between small-strain shear modulus, G_{max}, and DMT modulus, E_D, was presented by Hryciw & Woods (1988) for a series of tests performed in an overconsolidated silty sand. They found that the value of R_G ranged from about 1.5 to 6.0. Data for a wide range of relative densities for different sands presented by Sully & Campanella (1989) indicate that the ratio of G_{max}/E_D varies from about 1 to 4.5.

A global correlation for R_G was presented by Cruz et al. (2013), which shows that the value of R_G ($= G_{max}/E_D$) is dependent on ID for sedimentary and residual coarse-grained soils as shown in Figure 6.38. The upper bound global correlation was

$$R_G = 7.0 (I_D)^{-1.1}. \tag{6.70}$$

Rivera-Cruz et al. (2013) also showed that the value of R_G was dependent on U_D, with R_G decreasing as U_D increases.

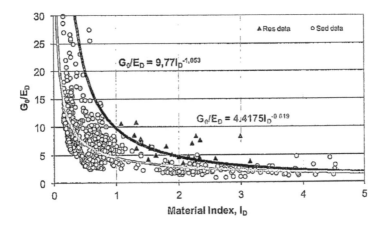

Figure 6.38 Correlation between I_D and R_G for sedimentary and residual coarse-grained soils. (From Cruz et al. 2013.)

6.7.3.9 Coefficient of Subgrade Reaction

Schmertmann (1981; 1989b) suggested that a prediction of the vertical coefficient of subgrade reaction could be obtained from the DMT as

$$k_s = (0.5/K_o)((B+1)/(2B))^2 ((K_D - K_o)/0.5\,t)(P_0 - u_o) \tag{6.71}$$

where
   B = width of the foundation (ft)
   t = blade thickness.

According to Schmertmann (1989b), Equation 6.71 expresses the average pressure increase required to wedge apart the soil one half of the blade thickness plus a size and direction correction.

Gabr & Borden (1988) suggested that a more direct method of determining the coefficient of horizontal subgrade reaction for predicting the load deflection behavior of drilled piers in cohesionless could be taken as

$$k_s = (P_0 - \sigma_{ho})/(0.5\,t) \tag{6.72}$$

where
   σ_{ho} = initial total horizontal stress.

6.7.3.10 Liquefaction Potential

The DMT may have application for evaluating the liquefaction potential of granular soils, especially those most susceptible to liquefaction, which predominantly include uncemented loose saturated sands and silty sands. These materials typically exist in a low horizontal stress environment, which the DMT may detect and also typically have little significant stress history (Marchetti 1982). The estimation of liquefaction potential using the DMT is essentially based on the detection of low relative density layers using the Horizontal Stress Index, K_D (Marchetti 1982; Robertson & Campanella 1986; Reyna & Chameau 1991;

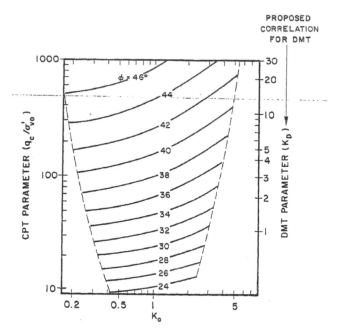

Figure 6.39 Liquefaction potential from DMT K_D. (Modified after Monaco et al. 2005.)

Tsai et al. 2009; Marchetti 2016). Marchetti (1997) suggested the following tentative categories for uncemented saturated sands:

1. $K_D > 1.7$: liquefaction is not a problem.
2. $K_D < 1.3$: liquefaction is a problem.
3. $1.3 < K_D < 1.7$: additional study is necessary.

Correlations have also been suggested between the cyclic stress ratio to cause liquefaction and either E_D or K_D as shown in Figure 6.39.

6.8 SEISMIC DILATOMETER

The seismic dilatometer (SDMT) may represent the fastest growing *in situ* test being used in site characterization worldwide. The SDMT combines the standard DMT equipment with a seismic module attached behind the DMT blade for measuring the shear wave velocity, V_S (Martin & Mayne 1997; Marchetti et al 2008; Amoroso et al. 2013). The SDMT was initially used in the field in the late 1990s (Martin & Mayne 1997; 1998) but has quickly gained wide popularity as a simple and rapid method for obtaining V_S as a result of further development of the equipment.

The seismic module is placed in a drill rod immediately behind the blade and is equipped with two receivers spaced 0.5 m apart. V_S is obtained as a true interval as the ratio between the difference in distance between the source and the receivers $(s_2 - s_1)$ and the delay of the arrival of the impulse from the first to the second receiver, Δt. The pulse is created at the surface in the same way that shear waves are generated for measuring shear wave velocity with the seismic cone or piezocone (SCPT/SCPTU) using a hammer to strike against a fixed plate on the ground surface.

Marchetti et al. (2008) found that the shear modulus, G_{max}, obtained from the SDMT V_S values were related to the DMT Modulus, E_D, and Lateral Stress Index, K_D, but that the relationship was dependent on the soil type, which could be expressed by the DMT Material Index, I_D, Figure 6.40. Calculated values of the DMT Constrained Modulus, M_{DMT}, from E_D, were then used to correlate G_0/M_{DMT} for different soils. Figure 6.41 shows ratios of G_0/M_{DMT} as a function of K_D for different soils identified from the DMT Material Index, I_D. The following correlations were presented by Marchetti et al. (2008):

$$\text{Clay}\left(I_D < 0.6\right): G_0/M_{DMT} = 26.18 K_D^{-1.0066} \tag{6.73a}$$

$$\text{Silt}\left(0.6 < I_D < 1.8\right): G_0/M_{DMT} = 15.686 K_D^{-0.921} \tag{6.73b}$$

$$\text{Sand}\left(I_D > 1.8\right): G_0/M_{DMT} = 4.5613 K_D^{-0.7967}. \tag{6.73c}$$

The correlations may be used to estimate G_0 from standard DMT results when shear wave velocity measurements are not available.

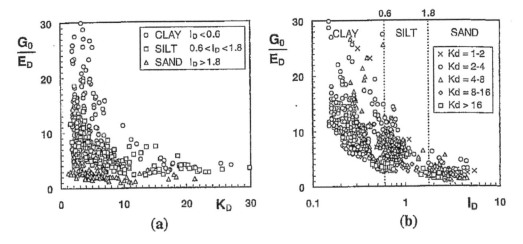

Figure 6.40 Dependence of G_{max}/E_D on K_D and I_D. (From Marchetti et al. 2008.)

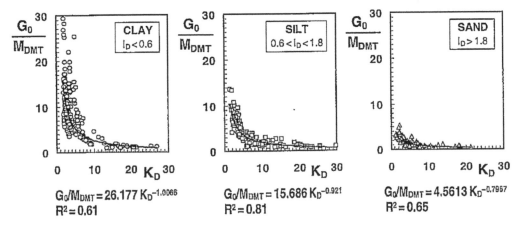

Figure 6.41 Correlations between G_{max}/M_{DMT} and K_D for different soils. (From Marchetti et al. 2008.)

6.9 DESIGN APPLICATIONS

The DMT has a wide range of applicability in different geologic materials as previously noted in Table 6.1. Most of the current applications of the DMT to engineering design problems make use of conventional soil parameters predicted by the DMT. Therefore, the accuracy of the predictions generally indicates the ability of the DMT to accurately predict properties. The real value of any soil tests is in their ability to accurately predict field performance. In addition to obvious uses as a site-profiling tool and in obtaining predictions of

Table 6.13 Some reported design applications of DMT for design

Application	References
Settlement prediction	Schmertmann et al. (1986)
	Hayes (1986)
	Saye & Lutenegger (1988b)
	Skiles & Townsend (1994)
	Steiner (1994)
	Monaco et al. (2006)
	Failmezger et al. (2015)
Bearing capacity of shallow foundations	Lutenegger (2006c)
Laterally loaded drilled shafts and driven piles	Gabr & Borden (1988)
	Robertson et al. (1989)
	Marchetti et al. (1991)
	Lutenegger & Miller (1993)
	Gabr et al. (1994)
	Ruesta & Townsend (1997)
	Anderson et al. (2006b)
Axially loaded piles	Marchetti et al. (1986)
	Gabr et al. (1994)
	Lutenegger & Miller (1993)
	Togliani & Reuter (2015)
Liquefaction potential of sands	Marchetti (1982)
	Robertson & Campanella (1986)
	Reyna & Chameau (1991)
	Totani et al. (1997)
	Passos et al. (2004)
	Grasso & Maugeri (2006)
	Maugeri & Monaco (2006)
	Marchetti (2016)
Ground improvement	Schmertmann (1982)
Verification and compaction control	Schmertmann et al. (1986)
	Lutenegger (1986)
	Lacasse & Lunne (1986)
	Sawada & Sugawara (1995)
	Miller et al. (2006)
	Amoroso et al. (2015b)
	Kurek & Balachowski (2015)
Uplift of anchor foundations	Lutenegger et al. (1988)
Slope stability	Rankka (1990)
	Totani et al. (1997)
	Peiffer (2015)
Retaining structures	Anderson et al. (2006a)
	Cunha & Reyes (2015)
	Deb & Konai (2015)
Changes in lateral stress	Peiffer et al. (1994)

conventional soil properties, several more direct applications of the DMT to specific engineering problems have been reported. Table 6.13 presents a summary. The list is no doubt larger since the application of the DMT is rapidly expanding, and new uses are continually being investigated.

The success of the DMT to accurately predict performance has generally been linked to a design approach based on an accepted engineering practice. Thus, at the current time, conventional design procedures using DMT-derived conventional soil engineering properties are generally being used. This is in contrast to a hybrid design approach, which in often used with other *in situ* tests; for example, the pressuremeter approach to foundation design based on E_m and P_L.

6.10 SUMMARY OF DMT

The DMT has quickly earned a place in the geotechnical profession as a cost-effective tool for conducting routine site investigations. The device has developed to a point where there is now substantial theoretical justification for many of the empirical correlations. The DMT is a simple and efficient test for estimating a number of soil engineering properties in a wide range of earth materials and may also be used as an extremely useful logging device since it allows closely spaced vertical test points. In specific applications, the test may be even more efficient if the engineer chooses to test at a few preselected depths. The recontact pressure, P_2, has shown to be a valuable component of the test and should be a routine part of every test.

REFERENCES

Akbar, A. and Clarke, B., 2001. A Flat Dilatometer to Operate in Glacial Till. *Geotechnical Testing Journal, ASTM*, Vol. 24, No. 1, pp. 51–60.

Akbar, A., Kibria, S., and Clarke, B., 2005. The Newcastle Dilatometer Testing in Lahore Cohesive Soils. *Proceedings of the 16th International Conference on Soil Mechanics and Foundation Engineering*, Vol. 2, pp. 651–654.

Amoroso, S., Monaco, P., and Marchetti, D., 2013. Use of the Seismic Dilatometer to Estimate In Situ G-G_{max} Decay Curves in Various Soil Types. *Proceedings of the 4th International Symposium on Geotechnical and Geotechnical Site Characterization*, pp. 489–497.

Amoroso, S., Rodrigues, C., Viana de Fonseca, A., and Cruz, N., 2015a. Liquefaction Evaluation of Aveiro Sands from SCPTU and SDMT. *Proceedings of the 3rd International Conference on the Flat Dilatometer*, pp. 293–300.

Amoroso, S., Rollins, K., Monaco, P., and Thorp, A., 2015b. Use of SDMT Testing for Measuring Sol Densification by Ground Improvement in Christchurch, New Zealand. *Proceedings of the 3rd International Conference on the Flat Dilatometer*, pp. 177–184.

Anderson, J., Townsend, F., and Grajales, B., 2006a. Prediction of P-y Curves from Dilatometer Tests Case Histories and Results. *Proceedings of the 2nd International Conference on the Flat Dilatometer*, pp. 50–61.

Anderson, J., Ogunro, V., Detwiler, J., and Starnes, J., 2006b. DMT Testing for the Estimation of Lateral Earth Pressures in Piedmont Residual Sols. . *Proceedings of the 2nd International Conference on the Flat Dilatometer*, pp. 5184-189.

Arulrajah, A., Nikraz, H., and Bo, M.W., 2004. Characterization of Soft Marine Clay Using the Flat Dilatometer. *Proceedings of the 2nd International Conference on Geotechnical and Geophysical Site Characterization*, Vol. 1, pp. 287–292.

Aziz, M. and Akbar, A., 2017. Interrelationships of Flat Rigid Dilatometer Parameters with Unconfined Compression Test Results. *Geotechnical Testing Journal, ASTM*, Vol. 40, No. 2, pp. 258–268.

Baldi, G., Bellotti, R., Ghionna, V., Jamiolkowski, M., Marchetti, S., and Pasqualini, E., 1986. Flat Dilatometer Tests in Calibration Chambers. *Use of In Situ Tests in Geotechnical Engineering, ASCE*, pp. 431–446.

Baldi, G., Bellotti, R., Ghionna, V., Jamiolkowski, M., and Lo Presti, D.C.F., 1989. Modulus of Sands from CPT's and DMT's. *Proceedings of the 12th International Conference on Soil Mechanics and Foundation Engineering*, Vol. 1, pp. 165–170.

Bechai, M., Law, K.T., Cragg, C.B.H., and Konard, J.M., 1986. In Situ Testing of Marine Clay for Towerline Foundations. *Proceedings of the 39th Canadian Geotechnical Conference*, pp. 115–119.

Been, K. and Jefferies, M.G., 1985. A State Parameter for Sands. *Geotechnique*, Vol. 35, No. 2, pp. 99–112.

Bellotti, R., Ghionna, V., Jamiolkowski, M., Lancellota, R., Manfredini, G., 1986. Deformation Characteristics of Cohesionless Soils from In Situ Tests. *Use of In Situ Tests in Geotechnical Engineering, ASCE*, pp. 47–73.

Bellotti, R., Fretti, C., Jamiolkowski, M., and Tanizawa, F., 1994. Flat Dilatometer Tests in Toyoura Sand. *Proceedings of the 13th International Conference on Soil Mechanics and Foundation Engineering*, Vol. 4, pp. 1779–1782.

Bellotti, R., Benoit, J., Fetti, C., and Jamiolkowski, M., 1997. Stiffness of Toyoura Sand from Dilatometer Tests. *Journal of Geotechnical and Geoenvironmental Engineering, ASCE*, Vol. 123, No. 9, pp. 836–846.

Benoit, J., NeJame, L, Atwood, M., and Findlay, R.C., 1990. Dilatometer Lateral Stress Measurements in Soft Sensitive Clays. Transportation Research Record No. 1278, pp. 150–155.

Boghrat, A., 1987. Dilatometer Testing in Highly Overconsolidated Soils. *Journal of Geotechnical Engineering, ASCE*, Vol. 113, No. 5, pp. 516–519.

Borden, R.H., Aziz, C.N., Lowder, W.M., and Khosla, N.P., 1985. Evaluation of Pavement Subgrade Support Characteristics by Dilatometer Test. Transportation Research Record No. 1022, pp. 120–127.

Borden, R.H., Saliba, R.E., and Lowder, W.M., 1986. Compressibility of Compacted Fills Evaluated by the Dilatometer. Transportation Research Record No. 1089, pp. 1–10.

Borden, R.H., Sullivan, W.J., and Lien, W., 1988. Settlement Predictions in Residual Soils by Dilatometer, Pressuremeter, and One-Dimensional Compression Tests. Comparison With Measured Field Response. *Proceedings of the 2nd International Conference on Case Histories in Geotechnical Engineering*, Vol. 2, pp. 1149–1154.

Brooker, E. and Ireland, H., 1965. Earth Pressure at Rest Related to Stress History. *Canadian Geotechnical Journal*, Vol. 1, No. 1, pp. 1–15.

Brown, D.A. and Vinson, J., 1998. Comparison of Strength and Stiffness Parameters for a Piedmont Residual Soil. *Proceedings of the 1st International Conference on Site Characterization*, Vol. 2, pp. 1229–1234.

Burgess, N., 1983. Use of the Flat Dilatometer in the Beaufort Sea. *Proceedings of the 1st International Conference on the Flat Dilatometer*, Edmonton, Mobile Augers and Research Ltd.

Cai, L., Chang, M., and The, C., 2015. Analysis of Dilatometer Test in Clay. *Proceedings of the 3rd International Conference on the Flat Dilatometer*, pp. 385–392.

Campanella, R.G. and Robertson, P.K., 1983. Flat Plate DMT: Research at UBC. *Proceedings of the 1st International Conference on Flat Dilatometer*, Edmonton, Mobile Augers and Research Ltd.

Campanella, R.G. and Robertson, P.K., 1991. Use and Interpretation of a Research Dilatometer. *Canadian Geotechnical Journal*, Vol. 28, No. 1, pp. 113–126.

Campanella, R.G., Robertson, P.K., Gillespie, D, and Grieg, J., 1985. Recent Developments in In Situ Testing of Soils. *Proceedings of the 11th International Conference on Soil Mechanics and Foundation Engineering*, Vol. 2, pp. 849–854.

Carriglio, F., Ghionna, V.N., Jamiolkowski, M., and Lancellotta, R., 1990. Stiffness and Penetration Resistance of Sands Versus State Parameter. *Journal of Geotechnical Engineering, ASCE*, Vol. 116, No. 8, pp. 1015–1020.

Chan, A.C.Y. and Morgenstern, N.R., 1986. Measurement of Lateral Stresses in a Lacustrine Clay Deposit. *Proceedings of the 39th Canadian Geotechnical Conference*, pp. 285–290.

Chang, M.F., 1986. The Flat Dilatometer Test and Its Application to Two Singapore Clays. *Proceedings of the 4th International Geotechnical Seminar on Field Instrumentations and In Situ Measurements*, Singapore, pp. 85–101.

Chang, M.F., 1988. In-Situ Testing on Residual Soils in Singapore. *Proceedings of the 2nd International Conference Geomechanics in Tropical Soils*, Vol. 1, pp. 97–107.

Chang, M.F., 1991. Interpretation of Overconsolidation Ratio from In Situ Tests in Recent Clay Deposits in Singapore and Malaysia. *Canadian Geotechnical Journal*, Vol. 28, No. 1, pp. 210–225.

Chang, M.F., Choa, V., Cao, L.F., and Myintwin, B., 1997. Overconsolidation Ratio of a Seabed Clay from In Situ Tests. *Proceedings of the 14th International Conference on Soil Mechanics and Foundation Engineering*, Vol. 1, pp. 453–456.

Chu, J., Bo, M., Chang, M., and Choa, V., 2002. Consolidation and Permeability Properties of Singapore Marine Clay. *Journal of Geotechnical and Geoenvironmental Engineering, ASCE*, Vol. 128, No. 9, pp. 724–732.

Clough, G.W. and Goeke, P.M., 1986. In Situ Testing for Lock and Dam 26 Cellular Cofferdam. *Use of In Situ Tests in Geotechnical Engineering, ASCE*, pp. 131–145.

Cruz, N. and Viana de Fonseca, A., 2006. Portuguese Experience in Residual Soil Characteristics by DMT Tests. *Proceedings of the 2nd International Conference on the Flat Dilatometer*, pp. 359–364.

Cruz, N., Rodrigues, C., and Viana de Fonseca, A., 2013. Detecting the Presence of Cementation Structures in Soils Based on DMT Interpreted Charts. *Proceedings of the 4th International Conference on Geotechnical and Geophysical Site Characterization*, Vol. 2, pp. 1723–1728.

Cunha, R. and Reyes, A., 2015. Design of Retaining structures in a Tropical Soil with the Use of the Marchetti Dilatometer. *Proceedings of the 3rd International Conference on the Flat Dilatometer*, pp. 77–82.

Davidson, J.L. and Boghrat, A., 1983. Flat Dilatometer Testing in Florida. *Proceedings of the International Symposium on In Situ Testing of Soil and Rock*, Paris, Vol. 2, pp. 251–255.

Deb, K. and Konai, S., 2015. Estimation of Lateral Earth Pressure on Cantilever Sheet Pile Using Flat Dilatometer Test (DMT) Data: Numerical Study. *Proceedings of the 3rd International Conference on the Flat Dilatometer*, pp. 401–408.

Fabius, M., 1985. Experience with the Dilatometer in Routine Geotechnical Design. *Proceedings of the 38th Canadian Geotechnical Conference* pp. 163–170.

Failmezger, R., Till, P., Frizzell, J., and Kight, S., 2015. Redesign of Shallow Foundations Using Dilatometer Tests – More Case Studies After DMT '06 Conference. *Proceedings of the 3rd International Conference on the Flat Dilatometer*, pp. 83–91.

Finno, R.J., 1993. Analytical Interpretation of Dilatometer Penetration Through Saturated Cohesive Soils. *Geotechnique*, Vol. 43, No. 2, pp. 241–254.

Gabr, M.A. and Borden, R.H., 1988. Analysis of Load Deflection Response of Laterally Loaded Piers Using DMT. *Proceedings of the 1st International Symposium on Penetration Testing*, Vol. 2, pp. 513–520.

Gabr, M.A., Lunne, T., and Powell, J.J.M., 1994. P-y Analysis of Laterally Loaded Piles in Clay Using DMT. *Journal of Geotechnical Engineering, ASCE*, Vol. 120, No. 5, pp. 816–837.

Giacheti, H., Piexoto, A., De Mio, G., and de Carvalho, D., 2006. Flat Dilatometer Testing in Brazilian Tropical Soils. *Proceedings of the 2nd International Conference on the Flat Dilatometer*, pp. 103–110.

Grasso, S. and Maugeri, M., 2006. Using K_D and VS from Seismic Dilatometer (DSDMT) for Evaluating Soil Liquefaction. *Proceedings of the 2nd International Conference on the Flat Dilatometer*, pp. 281–288.

Gupta, R.C. and Davidson, J.L., 1986. Piezoprobe Determined Coefficient of Consolidation. *Soils and Foundations*, Vol. 26, No. 1, pp. 12–22.

Hammandshiev, K.B. and Lutenegger, A.J., 1985. Study of OCR of Loess by Flat Dilatometer. *Proceedings of the 12th International Conference on Soil Mechanics and Foundation Engineering*, Vol. 4, pp. 2409–2414.

Hamouche, K.K., Leroueil, S., Roy, M., and Lutenegger, A.J., 1995. In Situ Evaluation of K_o in Eastern Canada Clays. *Canadian Geotechnical Journal*, Vol. 32, No. 4, pp. 677–688.

Hayes, J.A., 1983. Case Histories Involving the Flat Dilatometer. *Proceedings of the 1st International Conference on Flat Dilatometer*, Edmonton, Mobile Augers and Research Ltd.

Hayes, J.A., 1986. Comparison of Dilatometer Test Results with Observed Settlement of Structures and Earthwork. *Proceedings of the 39th Canadian Geotechnical Conference*, pp. 311–316.

Howie, J., Rivera, I., Weemes, I., and lbanna, M., 2007. Seismic Dilatometer Testing for Deformation and Strength Parameters. *Proceedings of the 6th Canadian Geotechnical Conference*, pp. 1794–1801.

Hryciw, R.D., 1990. Small Strain Shear Modulus of Soil by Dilatometer. *Journal of Geotechnical Engineering, ASCE*, Vol. 116, No. 11, pp. 1700–1716.

Hryciw, R.D. and Woods, R.D., 1988. DMT-Cross Hole Shear Correlations. *Proceedings of the 1st International Symposium on Penetration Testing*, Vol. 1, pp. 527–532.

Iwasaki, K., Tsuchiya, H., Sakai, Y., and Yamamoto, Y., 1991. Applicability of the Marchetti Dilatometer Test to Soft Ground in Japan. *Proceedings of Geo-Coast '91*, pp. 29–32.

Jamiolkowski, M., Ghionna, V.N., Lancellotta, R., and Pasqualini, E., 1988. New Correlations of Penetration Tests for Design Practice. *Proceedings of the 1st International Symposium on Penetration Testing*, Vol. 1, pp. 263–296.

Janbu, N., 1963. Soil Compressibility as Determined by Oedometer and Triaxial Tests. *Proceedings of the 3rd European Conference on Soil Mechanics and Foundation Engineering*, pp. 19–25.

Janbu, N., 1967. Settlement and Calculations Based on the Tangent Modulus Concept. Three guest lectures at Moscow State University, Bulletin No. 2, *Soil Mechanics, Norwegian Institute of Technology*, pp. 1–57.

Kabir, M.G. and Lutenegger, A.J., 1990. In Situ Estimation of the Coefficient of Consolidation in Clays. *Canadian Geotechnical Journal*, Vol. 27, No.1, pp. 58–67.

Kaderabek, T.J., Barreiro, D., and Call, M.A., 1986. In Situ Tests on a Florida Peat. *Use of In Situ Tests in Geotechnical Engineering, ASCE*, pp. 649–667.

Kaggwa, W.S., Jaksa, M.B., and Jha, R.K., 1995. Development of Automated Dilatometer and Comparison with Cone Penetration Tests at University of Adelaide, Australia. *Proceedings of the International Conference on Advances in Site Investigation Practice*, pp. 372–382.

Kalteziotis, N.A., Pachakis, M.D., and Zervogiannis, H.S., 1991. Applications of the Flat Dilatometer Test (DMT) in Cohesive Soils. *Proceedings of the 10th European Conference on Soil Mechanics and Foundation Engineering*, Vol. 1, pp. 125–128.

Kamei, T. and Iwasaki, K., 1995. Evaluation of Undrained Shear Strength of Cohesive Soils Using a Flat Dilatometer. *Soils and Foundation*, Vol. 35, No. 2, pp. 111–116.

Kay, J. and Chiu, C., 1993. A Modified Dilatometer for Small Strain Stiffness Characterization. *Proceedings of the 11the Southeast Asian Geotechnical Conference*, pp. 125–128.la

Kim, S., Jeong, S., Lee, S., Kim, D., and Kim, Y., 1997. Characterization of In Situ Properties of Korean Marine Clays Using CPTU and DMT. *Proceedings of the 14th International Conference on Soil Mechanics and Foundation Engineering*, Vol. 1, pp. 519–522.

Konrad, J.M., 1988. The Interpretation of Flat Plate Dilatometer Tests in Sands in Terms of the State Parameter. *Geotechnique*, Vol. 38, No. 2, pp. 263–277.

Konrad, J.M., 1989. The Dilatometer Test in Sands: Use and Limitation. *Proceedings of the 12th International Conference on Soil Mechanics and Foundation Engineering*, Vol. 1, pp. 247–250.

Konrad, J.M., 1991. In Situ Tests in a Dune Sand. *Canadian Geotechnical Journal*, Vol. 28, No. 2, pp. 304–309.

Konrad, J.M., Bozozuk, M., and Law, K.T., 1985. Study of In-Situ Test Methods in Deltaic Silt. *Proceedings of the 11th International Conference on Soil Mechanics and Foundation Engineering*, Vol. 2, pp. 879–886.

Kurek, N. and Balachowski, L., 2015. CPTU/DMT Control of Heavy Tamping Compaction of Sands. *Proceedings of the 3rd International Conference on the Flat Dilatometer*, pp. 191–196.

Lacasse, S. and Lunne, T., 1983. Dilatometer Test in Two Soft Marine Clays. Norwegian Geotechnical Institute Publication No. 146, pp. 1–8.

Lacasse, S. and Lunne, T., 1986. Dilatometer Tests in Sand. *Use of In Situ Tests in Geotechnical Engineering, ASCE*, pp. 686–699.

Lacasse, S. and Lunne, T., 1988. Calibration of Dilatometer Correlations. *Proceedings of the 1st International Symposium Penetration Testing*, Vol. 1, pp. 539–548.

Larsson, R. and Eskilson, S., 1989. Dilatometerforsok i Organisk Jord. Swedish Geotechnical Institute Report No. 258, 78pp.

Lawter, R.S. and Borden, R.H., 1990. Determination of Horizontal Stress in Normally Consolidated Sands by Using the Dilatometer Test: A Calibration Chamber Study. Transportation Research Record No. 1278, pp. 135–140.

Lechowicz, Z., Rabarijoely, S., and Kutia, T., 2017. Determination of Undrained Shear Strength and Constrained Modulus from DMT for Stiff Overconsolidated Clays. *Annals of Warsaw University of Lif Sciences*, Vol. 49, No. 2, pp. 107–116.

Lee, M., Choi, S., Kim, M., and Lee, W., 2011. Effects of Stress History on CPT and DMT results in Sand. *Engineering Geology*, Vol. 117, Nos. 3–4, pp. 259–265.

Lunne, T., Eidsmoen, T., Gillespie, D., and Howland, J.D., 1987. The Offshore Dilatometer. *Proceedings of the 6th International Symposium of Offshore Engineering*.

Lunne, T., Powell, J.J.M., Hauge, E.A., Mokkelbost, K.H., and Uglow, I.M., 1990. Correlation of Dilatometer Readings with Lateral Stress in Clays. Transportation Research Record No. 1278, pp. 183–193.

Lutenegger, A.J., 1986. Application of Dynamic Compaction in Friable Loess. *Journal of Geotechnical Engineering, ASCE*, Vol. 112, No. 6, pp. 663–667.

Lutenegger, A.J., 1988. Current Status of the Marchetti Dilatometer Test. *Proceedings of the 1st International Symposium on Penetration Testing*, Vol. 1, pp. 137–155.

Lutenegger, A., 1990. Determination of In Situ Lateral Stresses in a Dense Glacial Till. Transportation Research Record No. 1278, pp. 190–203.

Lutenegger, A., 2000. National Geotechnical Experimentation Site - University of Massachusetts. *National Geotechnical Experimentation Sites, ASCE*, pp. 102–129.

Lutenegger, A.J., 2006a. Cavity Expansion Model to Estimate Undrained Shear Strength in Clay from Dilatometer. *Proceedings of the 2nd International Conference on the Flat Dilatometer*, pp. 319–326.

Lutenegger, A.J., 2006b. Consolidation Lateral Stress Ratios in Clay from Flat Dilatometer. *Proceedings of the 2nd International Conference on the Flat Dilatometer*, pp. 327–333.

Lutenegger, A.J., 2006c. Flat Dilatometer Method for Estimating Bearing Capacity of Shallow Foundations on Sand. *Proceedings of the 2nd International Conference on the Flat Dilatometer*, pp. 334–340.

Lutenegger, A., 2015. Dilatometer Tests in Sensitive Champlain Sea Clay: Stress History and Shear Strength. *Proceedings of the 3rd International Conference on the Dilatometer*, pp. 473–480.

Lutenegger, A.J. and Blanchard, J.D., 1993. A Comparison Between Full Displacement Pressuremeter Tests and Dilatometer Tests in Clay. *Proceedings of the 3rd International Symposium on Pressuremeters*, pp. 309–320.

Lutenegger, A.J. and Donchev, P., 1983. Flat Dilatometer Testing in Some Metastable Loess Soils. *Proceedings of the International Symposium on In Situ Testing of Soil and Rock*, Paris, Vol. 2, pp. 337–340.

Lutenegger, A.J., and Kabir, M., 1988. Use of Dilatometer C-Reading for Site Stratigraphy. *Proceedings of the 1st International Symposium on Penetration Testing*, Vol. 1, pp. 549–554.

Lutenegger, A. and Miller, G., 1993. Behavior of Laterally Loaded Drilled Shafts in Stiff Soil. *Proceedings of the 3rd International Conference on Case Histories in Geotechnical Engineering*, pp. 147–152.

Lutenegger, A. and Timian, D.A., 1986. Flat-Plate Penetrometer Tests in Marine Clays. *Proceedings of the 39th Canadian Geotechnical Conference*, pp. 301–309.

Lutenegger, A.J., Smith, B.L., and Kabir, M., 1988. Use of In Situ Tests to Predict Uplift Performance of Multihelix Anchors. *Special Topics in Foundations, ASCE*, pp. 93–109.

Marchetti, S., 1975. An In Situ Test for the Measurement of Horizontal Soil Deformability. *In Situ Measurement of Soil Properties, ASCE*, Vol. 2, pp. 255–259.

Marchetti, S., 1980. In Situ Test by Flat Dilatometer. *Journal of the Geotechnical Engineering Division, ASCE*, Vol. 106, No. GT3, pp. 299–321.

Marchetti, S., 1982. Detection of Liquefiable Sand Layers by Means of Quasi-Static Penetration Tests. *Proceedings of the 2nd European Conference on Penetration Testing*, pp. 689–695.

Marchetti, S., 1985. On the Field Determination of K_0 in Sand. Panel Presentation Session: In Situ Testing Techniques. *Proceedings of the 11th International Conference on Soil Mechanics and Foundation Engineering*.

Marchetti, S., 1997. The Flat Dilatometer Design Applications. *Proceedings of the 3rd Geotechnical Engineering Conference Cairo University*.

Marchetti, S., 2016. Incorporating the Stress History Parameter KD of DMT into Liquefaction Correlations in Clean Uncemented Sands. *Journal of Geotechnical and Geoenvironmental Engineering, ASCE*, Vol. 142, No. 2, 4 pp.

Marchetti, S. and Totani, G., 1989. c_h Evaluations from DMTA Dissipation Curves. *Proceedings of the 12th International Conference on Soil Mechanics and Foundation Engineering*, Vol. 1, pp. 281–286.

Marchetti, S., Monaco, P., Totani, G., and Marchetti, D., 2008. In Situ Tests by Seismic Dilatometer (SDMT). *From Research to Practice in Geotechnical Engineering, ASCE*, pp. 292–311.

Marchetti, S., Totani, G., Campanella, R.G., Robertson, P.K. and Taddei, B., 1986. The DMT-σ_{HC} Method for Piles Driven in Clay. *Use of In Situ Tests in Geotechnical Engineering, ASCE*, pp. 765–779.

Marchetti, S., Totani, G., Calabrese, M., and Monaco, P., 1991. P-y Curves from DMT Data for Piles Driven in Clay. *Proceedings of the 4th International Conference on Piling and Deep Foundations*, Vol. 1, pp. 263–272.

Martin, G.K. and Mayne, P.W., 1997. Seismic Flat Dilatometer Tests in Connecticut Valley Varved Clay. *Geotechnical Testing Journal, ASTM*, Vol. 20, No. 3, pp. 357–361.

Martin, G.K. and Mayne, P.W., 1998. Seismic Flat Dilatometer Tests in Piedmont Residual Soils. *Proceeding of the 1st International Conference on Site Characterization*, Vol. 2, pp. 837–843.

Masood, T. and Kibria, S., 1991. Experience with Dilatometer Testing in Cohesive Soils. *Proceedings of the 9th Asian Regional Conference on Soil Mechanics and Foundation Engineering*, Vol. 1, pp. 51–54.

Maugeri, M. and Monaco, P., 2006. Liquefaction Potential evaluated by SDMT. *Proceedings of the 3rd International Conference on the Flat Dilatometer*, pp. 295–305.

Mayne, P.W., 1987. Determining Preconsolidation Stress and Penetration Pore Pressures from DMT Contact Pressures. *Geotechnical Testing Journal, ASTM*, Vol. 10, pp. 146–150.

Mayne, P.W., 1995. Profiling Yield Stress in Clays by In Situ Tests. Transportation Research Record No. 1479, pp. 43–50.

Mayne, P.W., 2015. Peak Friction Angle of Undisturbed Sands Using DMT. *Proceedings of the 3rd International Conference on the Flat Dilkatometer*, pp. 237–242.

Mayne, P.W. and Frost, D.D., 1991. Dilatometer Experience in Washington, D.C., and Vicinity. Transportation Research Record No. 1169, pp. 16–23.

Mayne, P.W. and Kulhawy, F.H., 1990. Direct and Indirect Determinations of In Situ K_0 in Clays. Transportation Research Record No. 1278, pp. 141–149.

Miller, H., Stetson, K., Benoit, J., and Connors, P., 2006. Comparison of DMT and CPTU Testing on a Deep Dynamic Compaction Project. *Proceedings of the 2nd International Conference on the Flat Dilatometer*, pp. 140–147.

Minkov, M., Karachorov, P, Donchev, P., and Genov, R., 1984. Field Tests of Soft Saturated Soils. *Proceedings of the 6th Conference on Soil Mechanics and Foundation Engineering*, Budapest, pp. 205–212.

Mlynarek, Z., Wierzbicki, J., and Manka, M., 2015. Geotechnical Parameters of Loess Soils from CPTU and SDMT. *Proceedings of the 3rdd International Conference on the Flat Dilatometer*, pp. 481–488.

Monaco, P., Marchetti, S., Totani, G., and Calabrese, M., 2005. Sand Liquefiability Assessment by Flat Dilatometer Tests (DMT). *Proceedings of the 16th International Conference on Soil Mechanics and Geotechnical Engineering*, Vol. 4, pp. 2693–2697.

Monaco, P., Totani, G., and Calabreses, M., 2006. DMT-Predicted vs. Observed Settlement: A Review of the Available Experience. *Proceedings of the 2nd International Conference on the Flat Dilatometer*, pp. 244–252.

Nichols, N.J., Benoit, J., and Prior, F.E., 1989. In Situ Testing of Peaty Organic Soils: A Case History. Transportation Research Record No. 1235, pp. 10–16.

Ortigao, J.A.R., Cunha, R.P., and Alves, L.S., 1996. In Situ Tests in Brasilia Porous Clay. *Canadian Geotechnical Journal*, Vol. 33, No. 1, pp. 189–198.

Ozer, A., Bartlett, S., and Lawton, E., 2006. DMT Testing for Consolidation Properties of the Lake Bonneville Clay. *Proceedings of the 2nd International Conference on the Flat Dilatometer*, pp. 154–161.

Passos, P., Farias, M., and Comha, R., 2004. Use of the DMT and DPL Tests to Evaluate Ground Improvement in Sand Deposits. *Proceedings of the 2nd International Conference on Geotechnical and Geophysical Site Characterization*, Vol. 2, pp. 1709–1716.

Peiffer, H., 2015. The Use of the DMT to Monitor the Stability of the Slopes of a Clay Exploitation Pit in the Boom Clay in Belgium. *Proceedings of the 3rd International Conference on the Flat Dilatometer*, pp. 127–133.

Peiffer, H., Van Impe, W., Cortvrindt, G., and Bottiau, M., 1994. DMT-Measurements Around PCS-Piles in Belgium. *Proceedings of the 13th International Conference on Soil Mechanics and Foundation Engineering*, Vol. 2, pp. 469–472.

Powell, J.J.M. and Uglow, I.M., 1986. Dilatometer Testing in Stiff Overconsolidated Clays. *Proceedings of the 39th Canadian Geotechnical Conference*, pp. 317–326.

Powell, J.J.M. and Uglow, I.M., 1988. Marchetti Dilatometer Testing in UK Soils. *Proceedings of the International Symposium on Penetration Testing*, Vol. 1, pp. 555–562.

Powell, J.J.M. and Uglow, I.M., 1989. The Interpretation of the Marchetti Dilatometer Test in UK Clays. *Proceedings of the Conference on Penetration Testing in the U.K.*, pp. 269–273.

Rankka, K., 1990. Measuring and Predicting Lateral Earth Pressures in Slopes in Soft Clays in Sweden. *Transportation Research Record*, No. 1278, pp. 172–182.

Redel, C., Blechman, D., and Feferbaum, S., 1997. Flat Dilatometer Testing in Israel. *Proceedings of the 14th International Conference on Soil Mechanics and Foundation Engineering*, Vol. 1, pp. 581–584.

Reyna, F. and Chameau, J.L., 1991. Dilatometer Based Liquefaction Potential of Sites in the Imperial Valley. *Proceedings of the 2nd International Conference on Recent Advances in Geotechnical Earthquake Engineering and Soil Dynamics*.

Ricceri, G., Simonini, P., and Cola, S., 2002. Applicability of Piezocone and Dilatometer to Characterize the Soils of the Venice Lagoon. *Geotechnical and Geological Engineering*, Vol. 20, pp. 89–91.

Rivera-Cruz, I., Howie, J., Vargas-Herrera, Cuto-Loria, M., and Luna-Gonzalez, O., 2013. A New Approach for Identification of Soil Behaviour Type from Seismic Dilatometer (SDMT) Data. *Proceedings of the 4th International Conference on Geotechnical and Geophysical Site Characterization*, Vol. 2, pp. 947–954.

Robertson, P.K., 1990. Soil Classification Using the Cone Penetration Test. *Canadian Geotechnical Journal*, Vol. 27, No. 1, pp. 151–158.

Robertson, P., 2009. CPT-DMT Correlations. *Journal of Geotechnical and Geoenvironmental Engineering, ASCE*, Vol. 135, No. 11, pp. 1762–1771.

Robertson, P. and Campanella, R.G., 1986. Estimating Liquefaction Potential of Sands Using the Flat Plate Dilatometer. *Geotechnical Testing Journal ASTM*, Vol. 9, No. 1, pp. 38–40.

Robertson, P.K., Campanella, R.G., Gillespie, D., and By, T., 1988. Excess Pore Pressures and the Flat Dilatometer. *Proceedings of the 1st International Symposium on Penetration Testing*, Vol. 1, pp. 567–576.

Robertson, P.K., Davies, M.P. and Campanella, R.G., 1989. Design of Laterally Loaded Driven Piles Using the Flat Plate Dilatometer. *Geotechnical Testing Journal ASTM*, Vol. 12, No. 1, pp. 30–38.

Roque, R., Janbu, N., and Senneset, K., 1988. Basic Interpretation Procedures of Flat Dilatometer Tests. *Proceedings of the 1st International Symposium on Penetration Testing*, Vol. 1, pp. 577–587.

Ruesta, P.F. and Townsend, F.C., 1997. Prediction of Lateral Load Response for a Pile Group. Transportation Research Record No. 1569, pp. 36–46.

Sawada, S. and Sugawara, N., 1995. Evaluation of Densification of Loose Sand by SBP and DMT. *Proceedings of the 4th International Symposium on Pressuremeters*, pp. 101–107.

Saye, S.R. and Lutenegger, A. J., 1988a. Performance of Two Metal Grain Bins on Compressible Alluvium in Western Iowa. *Measured Performance of Shallow Foundations, ASCE*, p. 27–45.

Saye, S.R. and Lutenegger, A.J., 1988b. Site Assessment and Stress History of Stiff Alluvium with the Marchetti Dilatometer. *Proceedings of the 1st International Symposium on Penetration Testing*, Vol. 1, pp. 589–593.

Schmertmann, J.H., 1981. Discussion of In Situ Tests by Flat Dilatometer. *Journal of the Geotechnical Engineering Division*, ASCE, Vol. 107, No. GT6, pp. 831–832.

Schmertmann, J.H., 1982. A Method for Determining the Friction Angle in Sands from the Marchetti Dilatometer Test. *Proceedings of the 2nd European Symposium on Penetration Testing*, Vol. 2, pp. 853–861.

Schmertmann, J.H., 1988. Dilatometer Digest No. 10, *GPE Inc*, 22 pp.

Schmertmann, J.H., 1989a. Dilatometer Digest No. 11, GPE Inc., 18 pp.

Schmertmann, J.H., 1989b. Discussion of Performance of a Raft Foundation Supporting a Multistory Structure. *Canadian Geotechnical Journal*, Vol. 26, No. 1, pp. 185–186.

Schmertmann, J.H. and Crapps, D.K., 1983. Use of In Situ Penetration Tests to Aid Pile Design and Installation. *Geopile '83*, pp. 27–47.

Schmertmann, J.H., Baker, W., Gupta, R., and Kessler, K., 1986. CPT/DMT QC of Ground Modification at a Power Plant. *Use of In Situ Tests in Geotechnical Engineering, ASCE*, pp. 985–1001.

Shibuya, S., Hanh, L., Wilailak, K., Lohani, T., Tanaka, H., and Hamouche, K., 1998. Characterizing Stiffness and Strength of Soft Bangkok Clay from In Situ and Laboratory Tests. *Proceeding of the 1st International Conference on Site Characterization*, Vol. 2, pp. 1361–1366.

Shiwakoti, D., Tanaka, H., and Tanaka, M., 2001. A Study of Small Strain Shear Modulus of Undisturbed Soft Marine Clays and its Correlations to Other Soil Parameters. *Proceedings of the 15th International Conference on Soil Mechanics and Geotechnical Engineering*, Vol. 3, pp. 2247–2252.

Skiles, D.L. and Townsend, F.C., 1994. Predicting Shallow Foundation Settlement in Sands from DMT. *Vertical and Horizontal Deformations of Foundations and Embankments, ASCE*, Vol. 1, pp. 132–142.

Sonnenfeld, S., Schmertmann, J., and Williams, R., 1985. A Bridge Site Investigation Using SPT's, MPMT's and DMT's from Barges. ASTM Special Technical Publication 883, pp. 515–535.

Steiner, W., 1994. Settlement Behavior of an Avalanche Protection Gallery Founded on Loose Sandy Silt. *Vertical and Horizontal Deformations of Foundations and Embankments, ASCE*, Vol. 1, pp. 207–221.

Stetson, K., Benoit, J., and Carter, M., 2003. Design of an Instrumented Flat Dilatometer. *Geotechnical Testing Journal, ASTM*, Vol. 26, No. 3, pp. 302–309.

Su, P.C., Chen, Y.C., Sun, C.Y., and Wang, G.S., 1993. The Flat Dilatometer Tests in Clay. *Proceedings of the 11th Southeast Asian Geotechnical Conference*, pp. 205–210.

Sully, J.P. and Campanella, R.G., 1989. Correlation of Maximum Shear Modulus with DMT Test Results in Sand. *Proceedings of the 12th International Conference on Soil Mechanics and Foundation Engineering*, Vol. 1, pp. 339–343.

Tanaka, A. and Bauer, G.E., 1998. Dilatometer Tests in a Leda Clay Crust. *Proceedings of the 1st International Conference on Site Characterization*, Vol. 1, pp. 877–882.

Tanaka, H., Tanaka, M., Iguchi, H., and Nishida, K., 1994. Shear Modulus of Soft Clays Measured by Various Kinds of Tests. *Proceedings of the International Symposium on Pre-Failure Deformation of Geomaterials*, Vol. 1, pp. 235–240.

Togliani, G. and Reuter, G., 2015. Pile Capacity Prediction (Class C): DMT vs. CPTU. *Proceedings of the 3rd International Conference on the Flat Dilatometer*, pp. 271–281.

Totani, G., Calabreses, M., Marchetti, S., and Monaco, P., 1997. Use of In-Situ Flat Dilatometer (DMT) for Ground Characterization in the Stability Analysis of Slopes. *Proceedings of the 4th International Conference on Soil Mechanics and Geotechnical Engineering*, Vol. 1, pp. 607–610.

Tsai, P., Lee, D., Kung, G., and Juang, C., 2009. Simplified DMT-Based Methods for Evaluating Liquefaction Resistance of Soils. *Engineering Geology*, Vol. 103, pp. 13–22.

Viana de Fonseca, A., Fernandes, M., and Cardoso, A., 1998. Characterization of a Saprolitic Soil from Porto Granite Using In Situ Testing. *Proceedings of the 1st International Conference on Site Characterization*, Vol. 2, pp. 1381–1387.

Wang, C. and Borden, R., 1996. Deformation Characteristics of Piedmont Residual Soils. *Journal of Geotechnical Engineering, ASCE*, Vol. 122, No. 10, pp. 822–830.

Wong, J., Wong, M., and Kassim, K., 1993. Comparison Between Dilatometer and Other In Situ and Laboratory Tests in Malaysian Alluvial Clay. *Proceedings of the 11th Southeast Asian Geotechnical Conference*, pp. 275–3279.

Yu, H.S., Carter, J.P., and Booker, J.R., 1993. Analysis of the Dilatometer Test in Undrained Clay. *Predictive Soil Mechanics: Proceedings of the Wroth Memorial Symposium*, pp. 782–795.

Chapter 7

Pressuremeter Test (PMT)

7.1 INTRODUCTION

The Pressuremeter Test (PMT) falls into the class of *in situ* tests, which are generally intended for specific property measurement. Tests are performed at a known location in the subsurface in order to provide a direct determination of soil properties. However, results from the test may be used to provide a direct input for design, as is generally the case with the Menard Pressuremeter Test (MPMT), wherein a set of empirical design rules are used for foundation design. The basic principle of the PMT is to install a cylindrical probe into the ground; expand the probe laterally against the surrounding soil or rock using water or gas; and obtain measurements of the applied pressure and probe volume or deformation. The equipment and test procedures for prebored pressuremeters are standardized by ASTM D4719 *Standard Test Method for Pressuremeter Testing in Soil* and ISO 22476-4:2009E *Geotechnical Investigation and Testing – Filed Testing-Part 4: Menard Pressuremeter Test*.

Pressuremeters are used to determine *in situ* stress conditions, elastic soil properties, limit equilibrium conditions, and consolidation behavior. Test results are used as input parameters for the bearing capacity and settlement design of shallow and deep foundations in compression as well as the behavior of deep foundations under the lateral load. The test is attractive because in theory the boundary conditions are controlled and well defined as are the stress and strain conditions. The test provides a direct measurement of the pressure-displacement behavior of soil and soft rock *in situ*. Results from the prebored pressuremeter can be used to determine a number of soil properties, such as undrained shear strength and preconsolidation stress in fine-grained soils, and the pressuremeter may also be used in geotechnical design, using specific rules based on parameters interpreted from the pressure-expansion curve.

Credit for the development of the prebored pressuremeter is generally given to Louis Menard who began work in the 1950s. Several complete books have been written that are devoted entirely to the pressuremeter: Baguelin et al (1978), Mair & Wood (1987), Briaud (1992), and Clarke (1995). In addition, proceedings have been published from several international symposia organized and dedicated solely to the pressuremeter: Paris, France, 1982; College Station, Texas, 1986; Oxford, UK, 1990; Sherbrooke, Canada, 1995; Paris, France, 2005; Paris, France, 2013; and Tunis, Tunisia, 2015. Interested readers may find a detailed and specific information regarding various aspects of the pressuremeter in these publications. The focus of this chapter is primarily on prebored PMTs since it is considered the most common test available to the profession. Other types of pressuremeters are only briefly discussed.

7.2 MECHANICS OF THE TEST

There are a number of different configurations of the PMT and a variety of installation or deployment methods available, but the basic principle of all pressuremeters is to perform a cylindrical cavity expansion test in soil or rock in order to obtain a measure of the mechanical behavior. The primary variables measured during the test are the cavity size (through either radial or diametric changes or cavity volume changes) and the internal cavity pressure (which is usually input at the ground surface and is required to cause the expansion of the cavity). The basic principle of the pressuremeter, as originally developed by Menard, is illustrated in Figure 7.1.

In operation, once the probe is installed to the test depth, the membrane is pressured by a fluid and the change in cavity size is measured as a function of cavity pressure. In simple fluid systems, compressed gas is used to force a liquid into the probe, and the liquid volume is measured for each increment of pressure. In more complex systems, compressed gas is used to inflate the membrane and the change in cavity size is detected using some form of displacement-measuring system, usually strain-gaged feeler arms.

7.3 PRESSUREMETER EQUIPMENT

Pressuremeters can generically be defined as cylindrical probes that apply a uniform pressure to a soil or rock cavity through the use of a flexible membrane. The devices currently available to geotechnical engineers vary from simple to highly sophisticated instruments. Probes are available from diameters of 32 to 102 mm. Depending on the method of installation, pressuremeters can be classified into four categories, as shown in Figure 7.2:

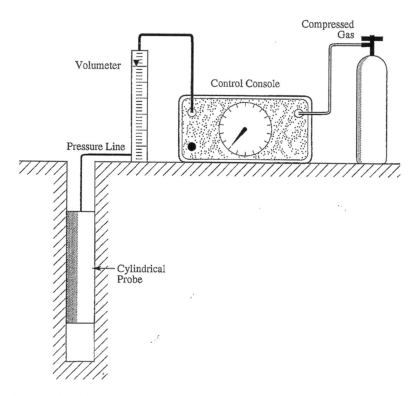

Figure 7.1 Basic principle of the Pressuremeter Test.

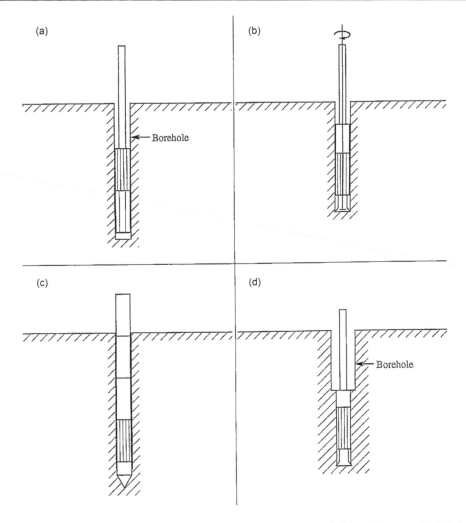

Figure 7.2 Schematic of different types of pressuremeters: (a) Prebored, (b) self-boring, (c) full displacement, and (d) push-in.

1. Prebored pressuremeter (PPMT/PMT);
2. Self-boring pressuremeter (SBPMT);
3. Full-displacement (cone) pressuremeter (FDPMT);
4. Push-in pressuremeter (PIPMT).

A brief summary of these different types of pressuremeters is presented in Table 7.1.

7.3.1 Prebored Pressuremeters

A PPMT or simply PMT is any type of pressuremeter designed to be used in a preformed borehole in the ground. Much of the current practice in the use and interpretation of PPMT data relies on the work by Menard who developed the first commercial PPMT while working at the University of Illinois in the early 1950s. Menard suggested a design in which the probe consists of a central measurement cell filled with water or other appropriate fluid which is pressurized by applying gas pressure through a console at the surface.

Table 7.1 Comparison of different types of pressuremeters

Type	Comments
Prebored "Menard"	Installed into a predrilled borehole. May be of three-cell or single (mono)-cell design. Test equipment (mono-cell) is simple, and tests are easy to perform.
Self-boring	Drills its own testing cavity by flushing soils upward using drilling fluid. Test is the most complex and requires a specialized expertise.
Full displacement	Installed by pushing without predrilling. Usually mounted behind the body of a CPT or CPTU. Displaces soil completely during installation. Test is very simple to perform.
Push-in	Installed at the bottom of a drilled hole. A thin-walled cutting shoe is attached at the front, and soil is partially displaced inside the body of the instrument.

The change in borehole size is indirectly determined by measuring the total volume displacement of the cell throughout the test. Surrounding the central cell are two "guard" cells, which are also inflated and which were designed to prevent end expansion of the central cell, thereby keeping all of the expansion radial. This type of instrument is sometimes referred to as a "three-cell" or "tri-cell probe". The probe is assumed to expand as a right cylinder so that the soil is subject to plane strain loading. The purpose of the guard cells in a tri-cell probe is to maintain this condition. For a mono-cell probe, if the length/diameter ratio is greater than about 6, this condition is close to the plane strain (Laier et al. 1975; Borsetto et al. 1983; Houlsby & Carter 1993).

7.3.1.1 Tri-Cell Probe

A tri-cell probe is historically the conventional device used to conduct tests in prebored holes. As indicated in Figure 7.1, guard cells are located on either side of a central cell. The purpose of the guard cells is to retain the expansion of the central cell so that most of the cavity expansion remains cylindrical in shape. This means that the pressure-expansion curve is obtained only from the central cell, which becomes the measuring cell. In order to perform the test, the pressure in the guard cells is kept slightly higher than the pressure in the central measuring cell. Normally, air has been used for maintaining the pressure in the guard cells, whereas liquid is used in the central measuring cell. The two pressures (i.e., pressure in the guard and measuring cells) may be independently controlled using two different pressure regulators, or a differential pressure regulator, preset to maintain a constant pressure difference, may be used.

Ideally, the pressure in the guard cells should be equal to that in the central measuring cell, and therefore, two separate pressure lines are needed with the instrument. Usually, a coaxial tubing is used so that a single line runs down the borehole. Because of the design, the actual mechanics of the instrument are somewhat complex and can present some difficulties. While the original tri-cell probe was actually fabricated with three separate inflatable membranes, the design was modified to allow a single membrane to encapsulate the central measuring cell. A schematic of this design is shown in Figure 7.3.

For example, replacing membranes may be difficult since the system must be free of leaks to insure correct volume measurements. The central cell fluid line must be de-aired again to insure correct volume measurements. Often, these measurements are difficult to check without actually performing a test. The manufacturer of the Menard pressuremeter makes other models of pressuremeters, but most of them are similar in concept to the tri-cell configuration shown in Figure 7.3. The standardized testing procedure adapted by ASTM (D4719) allows the use of a tri-cell probe.

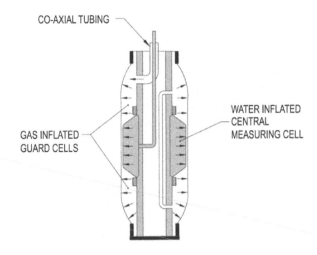

CO-AXIAL TUBING

GAS INFLATED
GUARD CELLS

WATER INFLATED
CENTRAL
MEASURING CELL

Figure 7.3 Schematic of tri-cell PMT with single outer membrane over central measuring cell.

7.3.1.2 Mono-Cell Probe

An alternative approach, which is more simple, is to use a single-cell or mono-cell probe in lieu of the tri-cell design. This type of device offers two distinct advantages: (1) the probe design is less complex and therefore easier to repair or maintain, and (2) since only one cell is down the hole, the control console and the measuring system are much simpler. Different styles of mono-cell probes and expansion equipment are available but they all essentially operate in the same manner. A liquid may still be used to measure the volume change by using a graduated voltmeter to determine the cavity expansion, or compressed gas may be used directly and the change in the borehole diameter may be directly measured electrically.

Figure 7.4 shows a schematic of a gas-operated mono-cell PMT that uses spring-loaded internal feeler arms to measure the change in diameter of the probe as expansion occurs. The control console consists of a simple gas regulator and an analog pressure gauge to control and determine the pressure for each increment of loading, and a simple electronic digital

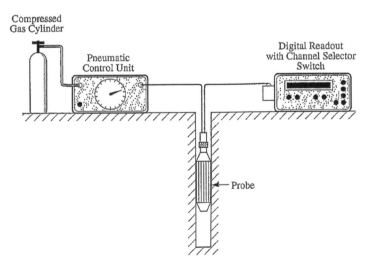

Compressed
Gas Cylinder

Pneumatic
Control Unit

Digital Readout
with Channel Selector
Switch

Probe

Figure 7.4 Schematic of electronic gas-operated mono-cell PMT.

readout to measure the diameter of the probe. Feeler arms are located around the center of the probe at 120° in order to measure the change in diameter in different directions. This may be advantageous if the pressure-diameter response is desired in a specific direction, say perpendicular to a retaining wall. Figure 7.5 shows a diagram of the probe with and without the outer membrane.

The devices that have been discussed so far and shown in Figures 7.4 and 7.5 are generally performed as pressure-controlled tests in which the probe pressure is increased in steps. An alternative is to use a mono-cell liquid system with a simple volume screw pump to perform a volume-controlled test. One such device is called the TEXAM probe, as shown in Figure 7.6. As with other liquid-filled devices, the probe and the entire system must be properly de-aired in order to obtain accurate volume change and pressure readings. This system uses a fluid-filled cylinder that is operated using a hand crank. Each rotation of the crank injects a constant amount of volume into the probe. The pressure is then applied using the inline pressure gauge. Figure 7.6 shows that an alternative is to use an optional gas pressure to perform the test. A photograph of the equipment is shown in Figure 7.7.

Even though they are more complex and more expensive, mono-cell devices that are operated by compressed gas alone offer at least two distinct advantages over liquid-filled devices. Since gas is used entirely, there is no need to de-air the system, saving a potentially large operation. More importantly, unlike liquid-filled probes that can only provide an indirect and average measurement of the ground response, gas-operated devices can be equipped to provide diametric change measurements at 120° around the probe. Therefore, any differences in directional response in a particular direction, e.g., perpendicular to a retaining structure. Prebored gas-operated devices with electrical strain-gaged feeler arms are manufactured by Roctest and OYO Inc.

Pressuremeters that use liquid to determine the volume change also have another disadvantage over gas-operated probes. The head of fluid in the tubing increases as the probe is lowered into the borehole and tends to expand the rubber membrane. A partial vacuum may be used to minimize this expansion, but at test depths greater than about 30 ft, it becomes difficult to hold this pressure head.

Results presented by Faugeras et al. (1983) and Briaud (1986; 1992) indicate that there is very little difference in the interpreted test results between a tri-cell probe and a mono-cell probe,

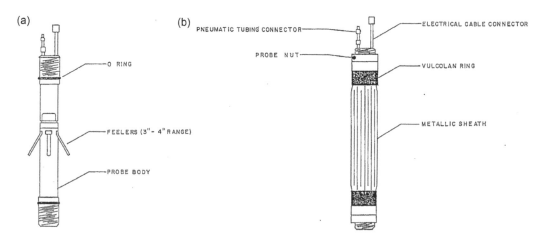

Figure 7.5 Schematic of gas-operated feeler arm mono-cell PMT probe: (a) Sheath removed and (b) fully assembled.

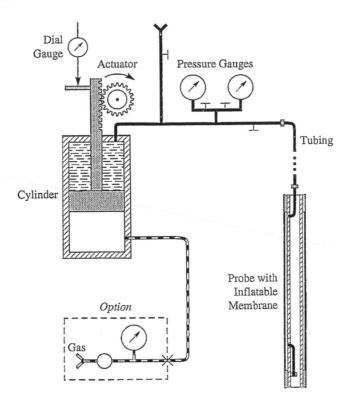

Figure 7.6 Schematic of TEXAM manual PMT.

Figure 7.7 TEXAM pressuremeter. (Courtesy Dr. J.-L. Briaud.)

provided the length/diameter ratio of the mono-cell probe is greater than about 6. The pressure-expansion curves and the interpreted results are nearly identical. Based on these observations, it appears that even though the tri-cell style of PMT has great historical background, it presents an unnecessary complication in the test. The author initially started using a tri-cell probe to perform tests in the early 1980s but quickly converted to a mono-cell probe in the late1980s and has since only used this type of equipment.

7.3.2 Self-Boring Pressuremeters

The SBPMT was designed as a method of inserting the instrument into the ground with the intention of creating a "zero-disturbance" condition. In effect, the SBPMT was specifically designed to allow the determination of the *in situ* horizontal stress and strength and stiffness characteristics. The SBPMT was developed in the early 1970s almost simultaneously in England (Wroth & Hughes 1974; Windle & Wroth 1977) and in France (Baguelin et al. 1974). SBPMTs have been used in a wide range of materials, including clays (e.g., Canou & Tumay 1986; Jefferies 1988; Finno et al. 1990; Benoit et al. 1990) and sands (e.g., Robertson & Hughes 1985; Bruzzi et al. 1986; Fahey 1991).

As originally developed, the SBPMT was designed with an internal cutting mechanism that is used to drill a cavity in the ground. The probe is advanced with a hydraulic push while the cutter rotates. Drilling fluid is used to wash soil cuttings upward through the center of the probe body and outward through the drill string. The cutter is operated by a hydraulic motor attached to the cutter rods at the surface. A schematic of the probe is shown in Figure 7.8.

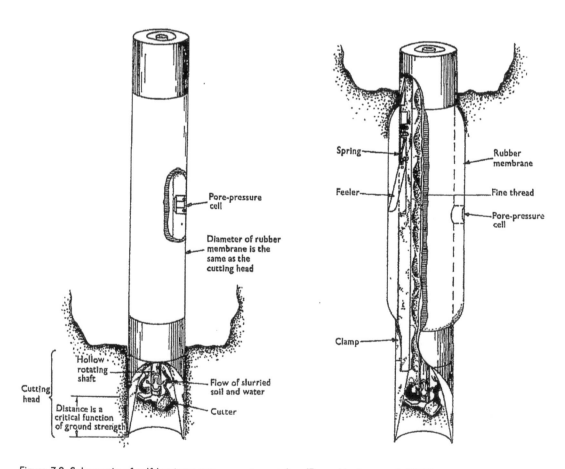

Figure 7.8 Schematic of self-boring pressuremeter probe. (From Hughes et al. 1977.)

The SBPMT is designed as a mono-cell probe and measures the expansion of the membrane using strain-gaged feeler arms at the center of the probe body. Typically, three independent strain arms are located at 120° around the probe body. A pore pressure cell is often located on the face of the membrane. A typical length/diameter ratio of 6 is used in the SBPMT. The membrane is typically protected by a segmental metallic sheath or "Chinese lantern". Most SBPMTs use an automatic data acquisition system to collect the data. A pressure transducer placed inside the probe is used to measure the expansion pressure. Nitrogen gas is typically used to produce the expansion, and the tests are usually performed as continuous loading tests.

The lower end of the probe is equipped with a sharp inward beveled cutting shoe as shown in Figure 7.8. The position of the rotating cutting is adjustable relative to the end of the cutting shoe. Drilling fluid exits through one or two holes on the cutter bit.

It is important that the proper setup is established for proper insertion of the probe to minimize disturbance. The rate of advance, the pushing force on the probe, the position of the cutter bit, the rate of rotary drilling, and the fluid pressure may all influence the quality of the test results since they may affect the insertion of the probe. The parameters may be different for different soils.

The use of the mechanical cutting system and the drilling fluid presents a rather formidable and cumbersome method of inserting the probe. In addition to requiring a considerable amount of extraneous equipment, the technique requires a large mud mixing tank and a circulating system which produces a considerable cleanup chore. The major drawback to the self-boring insertion is that the rate of testing is often very slow and can be delayed by small stones encountered by the cutter bit, which may create jamming.

An alternative insertion technique that uses a jetting system has been developed for production testing (Benoit et al. 1990). In this system, the cutter mechanism is replaced with a central jetting nozzle. The nozzle is attached to a small rod that runs through the probe body to the top of a perforated pipe attached to the top of the probe. The pipe is smaller in diameter than the probe and has discharge ports to allow the cutting slurry to escape. Standard drill rods are used to lower the probe into the ground.

The insertion procedure consists of simultaneously pushing the probe into the ground while pumping water or drilling mud down to the jetting tip. As drilling advances, the soil cuttings are flushed up inside the probe and flushed out through the discharge ports of the perforated pipe.

7.3.3 Full-Displacement (Cone) Pressuremeters

Unlike an MPMT that requires a prebored hole or an SBPMT that drills its own hole, a FDPMT does neither, but is instead advanced to the test depth by pushing the probe quasi-statically, much in the same manner as a CPT or CPTU. The FDPMT is usually fitted with a conical tip to displace the soil. A special FDPMT that has a 15-cm^2 CPT attached to the front combines the attributes of the CPT to obtain tip and sleeve resistance (and occasionally pore water or pressure) during penetration with the ability to obtain the cavity pressure-expansion behavior of the soil at the test location.

This combined tool is often referred to as a "cone pressuremeter" (CPMT). In either case, the FDPMT is used much in the same way as other PMTs in that it provides a pressure-expansion test, which may also include unload-reload loops, to obtain estimates of *in situ* stress conditions, soil strength, and soil stiffness.

It appears that the first suggestion of combining a CPT with a PMT cell was made by Baguelin & Jezequel (1983). Almost any PMT could be converted to a FDPMT by the simple addition of a conical tip to the front of the probe. The majority of advances in FDPMT work have occurred in the U.K. through the development of a 15-cm^2 probe, which is attached behind

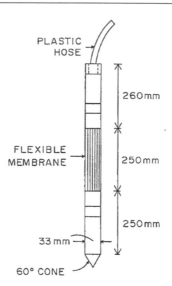

Figure 7.9 Schematic of Pencel FDPMT.

a 15-cm^2 CPTU (e.g., Withers et al. 1986; Houlsby & Withers 1988). The length of the probe is around 705 mm, which gives an L/D ratio of about 16. Cone pressuremeters can be used in both clays (e.g., Houlsby & Withers 1988; Lutenegger & Blanchard 1990; Powell 1990; Campenella et al. 1990; Rehman et al. 2011) and sands (e.g., Ghionna et al. 1995; Withers et al. 1990).

One advantage of the FDPMT is that like a CPT or DMT, the device creates a repeatable disruption to the soil each time. A simple version of the FDPMT is also available and is marketed under the name "Pencel pressuremeter", shown in Figure 7.9. This device is a modification of the "Pavement pressuremeter" introduced by Briaud & Shields (1979), and has a diameter of 33 mm with a probe length of about 250 mm, giving an L/D ratio of about 7.5. The test may be operated by either a hand crank or motor-driven mechanism to force fluid into the probe from a screw pump. Pressure and volume are recorded at the ground surface. The probe can be inserted very quickly, and the tests can be performed to develop soil properties throughout a subsurface profile without drilling a borehole.

7.3.4 Push-in Pressuremeter

PIPMT is a very special type of pressuremeter and is somewhat a hybrid in between a prebored pressuremeter and a full-displacement pressuremeter. The probe resembles a thick-walled sampling tube. The test is performed at the bottom of a borehole by pushing the probe ahead of the borehole. The probe is hollow, and the leading edge of the probe has an internally sharpened cutting shoe that cuts the soil which then enters the hollow probe. The test was initially developed for use offshore (Fyffe et al. 1986) but has also been used on land (Huang & Haefele 1988).

7.4 CREATING A BOREHOLE FOR THE PMT

The prebored PMT must be performed in a hole or cavity created by some form of test drilling. The test drilling for a prebored PMT is an important step in the process. The borehole needs to be sized correctly for the probe size in order to obtain a complete pressure-expansion curve from the test. Table 7.2 gives some typical PMT probe sizes and nominal tolerances for the test cavity.

Table 7.2 Typical PMT probe diameters and borehole sizes (after ASTM D4719)

Probe	Probe diameter (mm)	Borehole diameters: Nominal (mm)	Borehole diameters: Maximum (mm)
EX	33	34	40
AX	44	45	53
BX	58	60	70
NX	74	76	89

According to Briaud (2013), making a quality borehole is the most important step in obtaining a high-quality test. ASTM D4719 provides some guidelines for different methods of creating a borehole in different soils, which are summarized in Table 7.3.

Briaud (2013) summarized the basic differences in drilling a borehole for a PMT and Drilling a borehole for sampling. These are summarized in Table 7.4. Recommended drilling practices for creating a quality PMT borehole by the rotary drilling are presented in Table 7.5.

Table 7.3 Recommended Method of Creating Borehole for Prebored PMT (after ASTM D4719)

		Borehole Method									
Soil	Type	Rotary drilling with bottom discharge bit	Pushed thin-walled tube	Pilot hole drilling & pushed sampler	Pilot hole drilling & shaved hole	Flight auger	Hand auger above GWT	Driven or vibrated sample tube	Core barrel	Rotary percussion	Driven, pushed, or vibrated slotted tube
Clayey	Soft	2	2	2	2	NR	NR	NR	NR	NR	NR
	Firm to stiff	1	1	2	2	1	1	NR	NR	NR	NR
	Stiff to hard	1	2	1	1	1	NA	NA	1	2	NR
Silty	Above GWL	1	2	2	2	1	1	2	NR	NR	NR
	Below GWL	1	NR	NR	2	NR	NR	NR	NR	NR	NR
Sandy	Loose above GWL	1	NR	NR	2	2	2	2	NA	NR	NR
	Loose below GWL	2	NR	NR	2	NR	NR	NR	NA	NR	NR
	Medium to dense	NR	NR	NR	2	1	1	2	NR	2	NR
Gravelly	Loose	2	NA	NA	NA	NA	NA	NR	NA	2	2
	Dense	NR	NA	NA	NA	NR	NA	NR	NA	2	1
Soft rock		1	NA	2	NA	1	NA	1	2	2	NR

Note: 1 is the first choice; 2 is the second choice; NR = not recommended; NA = not applicable

Table 7.4 Differences between drilling for PMT testing and drilling for soil sampling (after Briaud 2013)

Drilling for PMT testing	Drilling for sampling
Slow rotation to minimize an enlargement of borehole diameter	Fast rotation to get to the sampling depth faster
Care about undisturbed borehole walls left behind the bit	Don't care about borehole walls left behind the bit
Don't care about soil in front of the bit	Care about undisturbed soil in front of the bit
Advance borehole beyond testing depth for soil cuttings to settle in	Stop drilling at sampling depth
Do not clean the borehole by running the bit up and down in the open hole; this will increase the hole diameter	Clean borehole by running bit fast with fast mud flow up and down in open hole; avoid unwanted cuttings in sampling tube
Care about the borehole diameter	Don't care about the borehole diameter

Table 7.5 Recommended practices for creating a quality PMT borehole using rotary drilling (after Briaud 2013)

Diameter of the drilling bit should be equal to the diameter of the probe

Three-wing bit for silts and clays (carving); roller bit for sands and gravels (chopping and washing)

Diameter of rods should be small enough to allow flush cuttings to go by

Slow drilling mud circulation to minimize erosion of the borehole

Slow rotation of the drill bit (< 60 rpm)

Drill I m past the testing depth for cuttings to settle

One drilling pass down and one withdrawal; no flushing or cleaning of the borehole

One test at a timew

7.5 TEST PROCEDURES

A detailed test method is described in ASTM D4719 *Standard Test Method for Pressuremeter Testing in Soil*. ASTM D4719 allows for two types of test procedures. Whether a tri-cell or mono-cell probe is used, the test procedure is essentially the same. There are two common methods for performing a prebored PMT: (1) incremental stress-controlled test and (2) incremental volume-controlled test. Other procedures, such as a constant rate of stress increase or a constant rate of strain increase, are also possible but more complex and may require the use of automatic data acquisition systems.

7.5.1 Test Procedure A – Equal-Pressure Increment Method

In a stress-controlled test, the pressure in the probe is increased incrementally after the probe is set into the borehole to the desired test depth. After an initial volume or strain-arm displacement reading is obtained, the pressure is set with the regulator on the console and a stop watch is started. Volume or strain-arm readings are then obtained after an elapsed time of 15 s, 30 s, and 1 min. After the 1-min. reading, the pressure is increased to the next level and another set of volume or displacement readings are obtained. The stepwise loading sequence is continued until the end of the test.

This procedure is the most common and the easiest test method to perform since the pressure is increased in a stepwise manner using a pressure regulator on the control console and the response of the soil to that pressure in terms of volume change or diameter/radius change is obtained. It is very common that one or more unload-reload cycles are performed in the pseudo-elastic portion of the curve to obtain the unload-reload modulus, E_{UR}. There is no set rule on where this should be done.

7.5.2 Test Procedure B – Equal-Volume Increment Method

In the equal-volume method, the pressure at the control; console is adjusted so that the injected fluid volume is the same for each increment. The pressure required to maintain this volume is then recorded along with the total volume to that increment. This test procedure is much more difficult to perform if a gas regulator is used to control the test since the pressure must be continuously adjusted to keep the volume increment constant. An alternative used in the TEXAM pressuremeter is to use a hand crank fluid piston system that injects a fixed volume of fluid with each rotation of the crank.

7.5.3 Continuous Loading Tests

It is also possible to perform the test under continuous loading. In continuous loading tests, the test pressure is applied to the probe in a continuously increasing manner. In this way, more data points are obtained, the test usually takes less time to complete, and the test equipment may be automated.

If compressed gas is being used to expand the membrane (either directly or using a gas/liquid interface), the gas regulator is controlled by a small motor that increases the gas pressure at a preset rate. As the gas pressure increases, measurements are taken of the probe volume or radial arm displacement. In continuous loading tests, there is no measurement of the creep as in an incremental loading test since there is no waiting period for successive load increments. The results of continuous loading tests provide a much smoother and continuous test curve in comparison with that of an incremental test. Continuous loading tests are most often used with self-boring and full-displacement tests, while incremental loading is more common for prebored tests.

An alternative method for performing continuous loading tests may be used if a liquid-filled probe is used. Instead of using a gas/liquid interface and compressed gas to perform the test, a small-liquid screw pump mechanism may be attached to the probe. A small electric motor is then attached to a screw drive to advance the pump and force liquid into the probe at a preset rate. A pressure gage or transducer placed in line provides a measure of the probe pressure, while a simple counter or linear distance transformer (LVDT) or digital dial gage provides a measure of the pump travel which can be converted to volume through a simple calibration.

7.5.4 Holding Tests

A special type of PMT test procedure may be used to measure creep effects by maintaining a constant pressure over a long period of time and monitoring the change in volume/diameter. This is typically referred to as "a holding test" since the pressure is held constant. The time for a holding test can be as short as 10 min or over an hour depending on the creep characteristics of the soil. Holding tests may be performed at different pressure levels to determine differences in creep behavior relative to the limit pressure.

7.6 DATA REDUCTION

7.6.1 Corrected Pressure-Volume Curve

The reduction of prebored PMT data involves the manipulation of the field data 1-min pressure and volume/diameter data to obtain a corrected cavity pressure vs. cavity volume or diameter/ radius curve by using the membrane and system calibrations as previously discussed. Use of the Menard rules for the design of foundation requires that specific parameters be obtained from

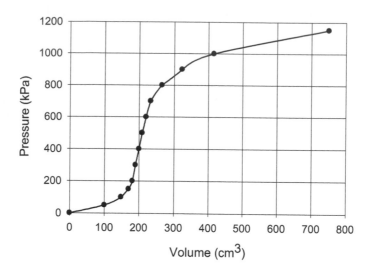

Figure 7.10 Idealized corrected PMT pressure-volume expansion curve.

the results of the test. Primarily, these parameters include: P_O, P_f, P_L, E_{UR}, and E_m, as defined in Figure 7.10. As can be seen, the PMT curve has three distinct zones: Zone 1 – pressure to inflate the membrane against the sides of the borehole and reinstate lateral stress from unloading by drilling; Zone 2 – pseudo-elastic straight-line pressure-volume response; and Zone 3 – plastic response after the pseudo-elastic response as the soil approaches a failure condition.

7.6.1.1 Initial Pressure, P_O

The pressure P_O is designated as the pressure that identifies the beginning of the elastic portion of the pressuremeter curve. Effectively, it is the point of tangency at the end of Zone 1. In many soils, this stress corresponds to the final point along the curve at which the ground stresses have been reinstated in the cavity and therefore closely represents the *in situ* total horizontal stress. Stresses up to this point represent reloading to account for borehole stress relief, and deformations represent recompression to account for disturbance and relaxations.

Since the value of P_O coincides with the beginning of elastic behavior, it may be determined from the point of tangency with the elastic portion of the curve as shown in Figure 7.11. Note that it may be necessary to enhance or enlarge the initial portion of the curve to obtain an accurate determination of P_O. Note also that a sufficient number of data points are required both at the beginning of the test, i.e., expanding against the borehole and reloading, and in the pseudo-elastic portion (Zone 2), in order to allow a sufficient accuracy in evaluating P_O. The initial pressure may also be interpreted from the creep curve as will bediscussed..

7.6.1.2 Creep Pressure, P_f

At the end of the elastic portion of the test (Zone 2), the soil has reached an initial yielding point, which signifies the beginning of plastic failure. The pressure at which this change in behavior occurs is called the creep pressure, P_f. It is effectively the point of tangency of the pressuremeter curve where the linear pseudo-elastic behavior ends. The value of P_f may also be estimated from the creep curve as will be shown in Figure 7.14. Both methods should be used to provide some redundancy in the estimate of P_f. Figure 7.12 shows PMT data on an expanded scale used to determine the creep pressure.

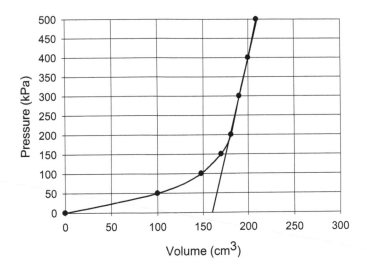

Figure 7.11 Identification of POT for initial pressure, $P_O = 200$ kPa, using expanded scale.

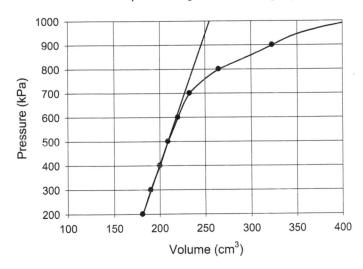

Figure 7.12 Identification of POT for creep pressure, $P_f = 550$ kPa, using expanded scale.

7.6.1.3 Limit Pressure, P_L

As pressure is increased above the initial yield point, an expanding annulus of plastic soil develops around the cylindrical cavity. When a cylindrical cavity is expanded in an elastic-perfectly plastic soil, the limit pressure P_L is theoretically reached when the infinite expansion of the cavity occurs, i.e., $\Delta V/V = 1$. That is, the change in cavity volume is equal to the current cavity volume. Because of mechanical limitations, pressuremeters are not capable of expanding sufficiently to reach the limit pressure, and therefore, the estimation of P_L requires the extrapolation of data from lower values of $\Delta V/V$.

In a prebored pressuremeter, the limit pressure is defined as the pressure where the probe volume reaches twice the original soil cavity, defined from volume measurements as $V_o + V_i$. V_i is the corrected volume reading at the pressure where P_O is obtained, i.e., where the probe membrane made contact with the borehole wall. Therefore, the limit pressure, P_L, would be

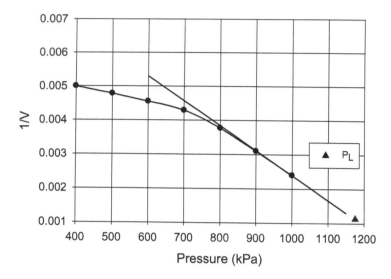

Figure 7.13 ASTM method of extrapolating to obtain P_L.

obtained as the pressure where the volume is equal to 2 ($V_o + V_i$). ASTM D4719 suggests that the limit pressure may be obtained by extrapolation to this volume by making a plot of $1/V$ vs. P, as shown in Figure 7.13. Only the data points obtained after the linear portion of the pressuremeter curve, i.e., after P_f, should be used in this extrapolation.

7.6.1.4 Net Limit Pressure, P_L*

The net limit pressure $P*_L$ combines the extrapolated value of P_L and the interpreted value of P_O as:

$$P*_L = P_L - P_O \tag{7.1}$$

The net limit pressure is often used in design calculations for bearing capacity.

7.6.1.5 Pressuremeter Modulus, E_m

The pressuremeter modulus may be determined using the slope of the pressure-volume curve in the pseudo-elastic (linear) portion of the corrected PMT curve. According to ASTM D4719, the pressuremeter modulus is determined from:

$$E_m = 2(1 + \upsilon)(V_O + V_M)(\Delta P/\Delta V) \tag{7.2}$$

where
   υ = Poisson's ratio
   Vo = volume of the measuring portion of the uninflated (at rest) probe at 0 volume reading at the ground surface (cm^3)
   V = corrected volume reading of the measuring portion of the probe
   ΔP = corrected pressure increase in the center part of the straight-line portion of the pressure-volume curve

ΔV = corrected volume increase in the center part of the straight-line portion of the pressure-volume curve corresponding to ΔP pressure increase

V_M = corrected volume reading in the center portion of the ΔV volume increase

$Vo + V$ = current volume of inflated probe

If the diameter of the probe is measured during the test, the pressuremeter modulus may be determined from:

$$E_m = 2(1 + \upsilon)(R_P + \Delta R_M)(\Delta P/d\Delta R) \tag{7.3}$$

where

R_p = radius of probe in uninflated condition (mm)

ΔR_M = increase in radius of probe up to the point corresponding to the pressure where E_m is measured (mm)

$d\Delta R$ = increase in radius of the probe corresponding to ΔP pressure increase (mm)

ΔR = increase in probe radius (mm)

$RP + \Delta R$ = current radius of inflated probe (mm)

7.6.1.6 Unload-Reload Modulus, E_{UR}

If an unload-reload loop is performed (recommended on most tests), Equations 7.2 or 7.3 may be used to determine the unload-reload modulus, E_{UR}. Briaud (2013) noted that the problem with the unload-reload modulus is that it is not necessarily a unique value for a given test. It depends on the amplitude of the unload-reload loop and also on the stress level where the unload-reload is performed. As a result, the value of E_{UR} may vary from operator to operator depending on the specific test procedures. Briaud (2013) recommended that the unload-reload loop be performed at the end of the pseudo-elastic part of the curve and that unload be taken to about ½ of the estimated pressure at this point.

7.6.2 Creep Curve

During the test, volume or diameter readings are taken at 30 s and 1 min. after applying each pressure increment. The difference between these values is called the creep. In addition to plotting the full corrected pressure vs. volume/diameter curve, as shown in Figure 7.10, the creep curve is also plotted, as shown in Figure 7.14. Ideally, the creep volume will decrease as the pressure P_O is approached, and then the pseudo-elastic portion of the curve will be very small and more or less constant up to the creep pressure, P_f, provided equal-pressure increments have been used in the test. After P_f, the creep volume will increase with each pressure increment as the limit pressure P_L is approached. The intersection of straight lines drawn through each section of the creep curve can also be used to interpret the initial horizontal stress, P_O, and the creep pressure, P_f, as shown in Figure 7.14.

Figure 7.15 shows the results obtained using an NX-sized mono-cell PMT in medium stiff clay in Leona, Kansas. Even though the ASTM standard suggests obtaining the PMT curve in 7–10 pressure increments, the author has found that more data points, especially at the beginning of the test, are very useful in interpreting the value of P_O, and it is essential to have at least 4–5 data points after the creep pressure P_f to accurately extrapolate the limit pressure P_L.

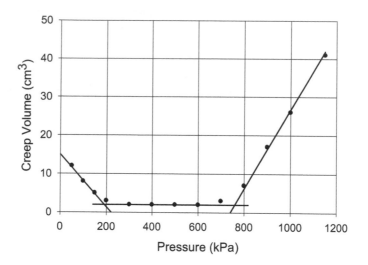

Figure 7.14 Idealized pressuremeter creep curve.

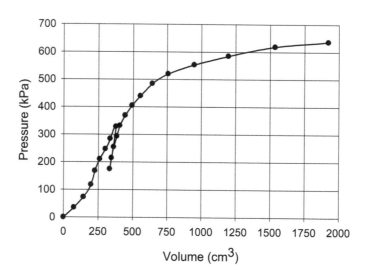

Figure 7.15 Test results obtained using a mono-cell PMT curve in Leona, Kansas.

7.6.3 Relationships Between PMT Parameters

Typical ranges in values of P^*_L and E_m are given in Table 7.6 for both clay and sand. Even though the ranges are high in each category, it can easily be seen that clay values will generally be lower than sand values.

Results of the PMT may be used to give an indication of soil type based on the relationships between PMT parameters interpreted from the corrected PMT pressure-expansion curve. For example, Briaud (1992) suggested that clays show $E_m/P^*_L > 12$ and sands show $7 < E_m/P^*_L < 12$.

Similarly, according to Briaud (1992; 2013), in clays the P_f/P_L ratio is typically around 0.5, while for sands, it is around 0.33. Walker (1979) had shown that $P_f/P_L = 0.59$ for the weathered rock at several sites in Canada and Australia. Bahar (1998) found that $P_f/P_L = 0.56$ for

Table 7.6 Typical range of expected values of P*$_L$ and E$_M$

	Clay				
	Soft	*Medium*	*Stiff*	*Very Stiff*	*Hard*
P*$_L$ (kPa)	0–200	200–400	400–800	800–1600	>1600
E$_m$ (MPa)	0–2.5	2.5–5.0	5.0–12	12–25	>25

	Sand			
	Loose	*Compact*	*Dense*	*Very Dense*
P*$_L$ (kPa)	0–500	500–1500	1500–2500	>2500
E$_m$ (MPa)	0–3.5	3.5–12	12–22.5	>22.5

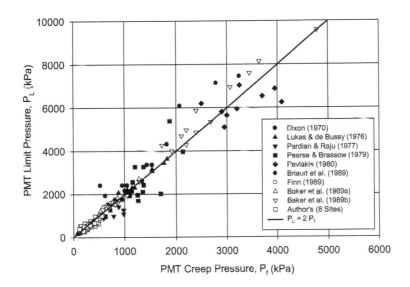

Figure 7.16 Observed relationship between P$_f$ and P$_L$.

three stiff clay sites in Algiers. These interrelationships may be useful to check the quality of the test results but may also be used to make an approximate estimate of P$_L$ in cases where the test does not give sufficient data points beyond the initial yield pressure.

Test results collected by the author from several published sources representing a large number of tests in a wide range of fine-grained soils are shown in Figure 7.16. Without any accounting for differences in test equipment or procedures and only considering the interpreted (or cited) values of P$_f$ and P$_L$ shows a very strong linear relationship over a wide range of stress. Results obtained by the author at eight test sites consisting of mostly fine-grained soils are also shown. Figure 7.17 shows the same data as that of Figure 7.16 but on an expanded scale for lower values in softer materials. All of the data fall reasonably close to the trend line of P$_L$ = 2 P$_f$.

7.7 FACTORS AFFECTING TEST RESULTS

Even though the basic PMT is a relatively simple device, other types of pressuremeters are more complex and there are a number of factors that can influence the results obtained from the test.

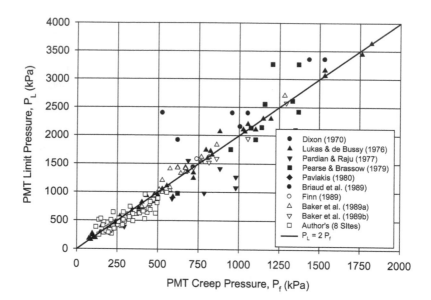

Figure 7.17 Observed relationship between P_f and P_L on an expanded scale.

7.7.1 Method of Installation

As previously noted, an important factor influencing the test results is the method used to create the cavity for testing. Some drilling methods will disturb the soil more than others. In moisture-sensitive soils, e.g., unsaturated soils above the water table, dry drilling methods are preferred. If the cavity is too small, some initial pressure may develop while inserting the probe, as shown in Figure 7.18, and it will not be able to interpret P_O. If the cavity is too large, the probe may expand to its full capacity before the creep pressure or limit pressure is reached, as shown in Figure 7.19.

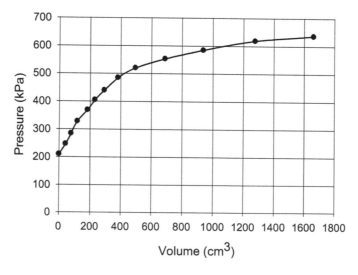

Figure 7.18 PMT curve – borehole too small.

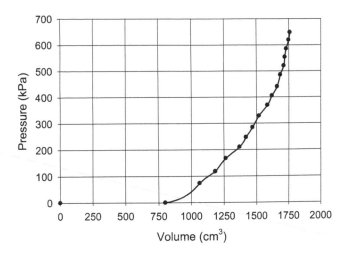

Figure 7.19 PMT curve – borehole too large.

7.7.2 Calibration of Membrane

Different membranes are available for different soils. A thick rubber membrane may be used in fine-grained soils where there are no gravel particles that could puncture the membrane during expansion. Membranes are also available with an outer metallic sheath made of individual overlapping strips of thin stainless steel, as shown in Figure 7.20. During expansion, the metal strips expand with the underlying rubber membrane and protect the membrane from damage.

Calibration of the membranes should be done carefully so that pressure losses from the stiffness of the membrane can be subtracted from field expansion pressures to give the correct soil pressures. Membranes are usually "exercised" a number of times, i.e., expanded

Figure 7.20 NX-sized rubber membrane with stainless steel metallic strips.

and deflated repeatedly in air before being used in a test. Some experience is required for membrane calibration, and small increments of pressure should be used to develop the membrane pressure calibration curve.

7.7.3 Volume Losses

Volume losses can occur in a liquid PMT system because of expansion of the tubing inside the control console and in the line between the console and the probe. Typically, pressure losses are calibrated by placing the probe in a thick-walled steel pipe with an inside diameter close to the outside diameter of the probe and incrementally expanding the probe while taking volume measurements. The calibration for volume loss is especially important when testing stiff soils. If a gas system is used and the change in probe diameter is measured using instrumented electronic arms inside the probe, there is no need for volume loss calibration.

7.7.4 Geometry of Cutter (SBPMT)

The geometry of the cutter blade inside the body of the SBPMT may influence the test results by disturbing the soil surrounding the probe as the probe advances into the ground. Most configurations of cutters have now been standardized by users of the SBPMT for different ground conditions.

7.7.5 Rate of Installation (SBPMT)

The rate of advance of the SBPMT probe, the fluid pressure, and the cutter speed are all factors that can influence the quality of the installation and the test results obtained. The force applied to the drilling rods and the pumping pressure to remove cuttings from the cutting face and drill rods may also affect the test results.

7.8 INTERPRETATION OF TESTS RESULTS IN FINE-GRAINED SOILS

The interpretation of prebored PMTs to obtain estimates of specific soil properties in fine-grained soils is largely empirical. However, like most other *in situ* tests, considerable experience has been gained in the past 25 years, and a large database exists in a wide variety of soils. Primarily, PMTs in clays may be used to estimate *in situ* horizontal stress, undrained shear strength, deformation modulus, and coefficient of consolidation.

At the present time, it can be considered that there are two distinctly different approaches to the interpretation and application of PMT results. In general terms, these can be considered as (1) empirical approach and (2) theoretical approach.

The empirical approach is largely associated with prebored PMTs and essentially originated by Louis Menard, and the expanded use of the PMT in France. This approach makes use of the different PMT parameters obtained from the test, i.e., P_O, P_f, P_L, and E_m, to estimate conventional soil parameters. Additionally, empirical design rules have been developed, largely from practical experience, to use the PMT results directly in foundation design.

The theoretical approach originated largely with the development of the self-boring pressuremeter in the U.K. The premise with this approach is that the PMT directly provides a measure of the stress-strain characteristics of the soil under known loading conditions, i.e., cylindrical cavity expansion. This approach uses the test results to directly determine soil properties through analysis of the PMT curve.

7.8.1 *In Situ* **Horizontal Stress**

At the beginning of the test, as pressure is applied to the probe, the membrane moves more or less freely out to engage the borehole walls. During the initial loading, any disturbed soil is easily compressed, and even up to the beginning of the elastic portion, the soil compresses easily. During these processes, the creep measurements will generally show a progressive decrease as soil resistance to deformation builds. As soon as the elastic portion of the test is reached, the creep measurements become very small and are generally uniform throughout this phase.

Using the creep curve, it is possible to identify the beginning of the elastic response by determining the intersection of these two portions of the curve and therefore identify P_O, as previously shown in Figure 7.16. This procedure should be checked against the previous point-of-tangency procedure as an independent estimation of P_O.

Marsland & Randolph (1977) proposed a technique to estimate P_O, primarily applicable to tests performed in stiff clays, but in principle applicable to other soils. They proposed that, in the proximity of σ_{ho}, the pressure-cavity strain relationship should be linear (i.e., the surrounding soil behaves elastically). For tests performed in a borehole, where the soil is completely unloaded before the cavity is expanded, the reference cavity pressure, P_O, should lie within, but not necessarily at, the start of the approximately linear section of the pressuremeter curve. The elastic response of the soil should cease (and hence the curve should cease to the linear) when the undrained strength of the soil is reached the wall of the cavity.

Marsland & Randolph (1977) suggested that an iterative procedure to estimate P_O should provide a similar response in the borehole strain. Therefore, P_O should lie in the linear portion of the curve but closer to the middle rather than at the beginning, since an increase in cavity pressure above P_O causes plastic yield and would be equal to the shear strength, i.e.,

$$P_O + s_u = P_f \tag{7.4}$$

It should be possible to then estimate P_O by trial and error.

The procedure is as follows: (1) make an initial estimate of P_O based on other information, say stress history or assumed K_o; (2) using P_O, determine V_o or R_o and plot R vs. ln $\Delta V/V$ to establish s_u; (3) compare $P_O + s_u$ with P_f; and (4) perform additional iteration until the solution converges.

7.8.2 **Undrained Shear Strength**

In saturated clays, the PMT is essentially an undrained test and a measure of the undrained shear strength from PMT results. There are essentially two approaches to evaluate the undrained shear strength from PMT results. In the first approach, a theoretical evaluation is made from the PMT curve, using the test results in the plastic range, after yielding, i.e., after P_f. An additional estimate may be made based on the limit pressure, P_L. In the second approach, which is used primarily with prebored PMT results, the limit pressure, P_L, is empirically correlated with s_u. Most methods use the PMT limit pressure P_L, which must first be obtained by a graphical solution, and require data points beyond the pseudo-elastic portion of the test.

7.8.2.1 *Theoretical Evaluation*

s_u from P_L
Based on the theoretical undrained cylindrical cavity expansion theory and the assumption of ideal elastic-plastic behavior presented by Gibson & Anderson (1961), the undrained shear strength may be obtained as:

$$P = \sigma_{HO} + s_u \left[1 + \ln(G/s_u) + \ln(\Delta V/V) \right] \tag{7.5}$$

where

P = pressure

σ_{HO} = *in situ* total horizontal stress (obtained from the initial portion of the PMT curve or estimated by other means)

G = shear modulus

ΔV = increase in cavity volume

V = current cavity volume

A plot of P vs. ln ($\Delta V/V$) in the plastic pressure range, i.e., for points after the yield stress, P_f, gives a straight line, the slope of which is s_u.

For an infinite cavity expansion, Equation 7.5 may be rewritten in terms of the limit pressure, P_L, as:

$$P_L = \sigma_{HO} + s_u \left[1 + \ln G/s_u \right] \tag{7.6}$$

Equation 7.6 may be reduced to:

$$P_L - \sigma_{HO} = s_u N_P \tag{7.7}$$

or

$$s_u = \left(P_L - \sigma_{HO} \right)/N_P \tag{7.8}$$

where

$$N_P = \left[1 + \ln E/3 \, s_u \right]$$

Equation 7.6 assumes that Poisson's ratio of the soil is equal to 0.5 for undrained loading. The value of N_P varies from about 3.2 to 8.0 for $I_r = G/s_u$ values ranging from 10 to 1000. The value of P_L in Equation 7.8 must first be obtained by graphical means using the points past the initial yield pressure, P_f, to obtain the limiting cavity stress at infinite expansion. For many clays and reasonable values of G/s_u, the value of N_P only varies from about 5.5 to 6.8. Marsland & Randolph (1977) suggested that based on simple bearing capacity theory, a reasonable value of N_P would be about 6.2. Borsetto et al. (1983) have shown that using a pressuremeter of finite length leads to an overprediction of S_u when the theory for an infinite cavity expansion is used.

7.8.2.2 Empirical Approach

Alternatively, a simple empirical relationship has been used to relate s_u to the net limit pressure, P^*_L, making use of the interpreted initial pressure, P_O, instead of the horizontal stress, σ_{HO}:

$$s_u = P^*_L/N_{Pm} \tag{7.9}$$

Table 7.7 indicates the value of N_{Pm} that has been reported by a number of investigators. The comparisons in Table 7.7 are primarily for stiff clays where the reference s_u has been obtained from unconfined compression tests, triaxial compressions tests, plate load tests, or other techniques.

Table 7.7 Reported values of N_{Pm} for different clays

N_{Pm}	Soil Type	References
4.6	London Clay	Gibson & Anderson (1961)
5.5	Stiff clay	Centres d'Etudes Menard (1967)
5.1	Stiff till and hardpan	Lukas & LeClerc de Bussy (1976)
6.2	Stiff glacial clays	Marsland & Randolph (1977)
4.5	Medium clay	Nayak (1979)
4.5	Soft marine clay	Bechai et al. (1986)
5.2–7.0	Very stiff to stiff clays	Davidson & Bodine (1986)
10	Stiff clay	Martin & Drahos (1986)
8.3	Greek clays	Kalteziotis et al. (1990)
5.2	Stiff clay – Algiers	Bahar (1998)
8	Clayey soils –Turkey	Bozbey & Togrol (2010)
4.1	Clays – Algeria	Ramdane et al. (2013)

An alternative form of Equation 7.8 has been suggested in which the total vertical stress at the test location is substituted in place of the *in situ* horizontal stress. This is convenient since the vertical stress is usually easier to estimate than the horizontal stress. Equation 7.9 then becomes:

$$s_u = (P_L - \sigma_{vO})/N^*_P \tag{7.10}$$

The values of N^*_P have been reported as 4 for medium to stiff clay (Komornik et al. 1970) to as high as 12.5 for marine and lacustrine clays of Canada (Leroueil 1983).

The undrained shear strength has also been directly correlated with the value of P_f. For example, Bergado et al. (1986) found that for undrained shear strength determined by field vane tests in Bangkok marine clay:

$$s_{uv} = P_f/3.15 \tag{7.11}$$

A number of other empirical correlations have been suggested for estimating the undrained shear strength from either the limit pressure, P_L, or the net limit pressure, P^*_L. Several of the correlations are presented in Table 7.8.

Table 7.8 Other Empirical Correlations for Estimating Undrained Shear Strength

Correlation	Soil Type	References
$s_u = P^*_L/10 + 25\,kPa$	Clays	Amar & Jezequel (1972)
$s_u = P_L/7.5$	Clays	Briaud et al. (1986)
$s_u = P_L/5.9$ (s_u in kPa)	Bangkok clay	Bergado et al. (1986)
$s_u/pa = 0.21\,(P^*_L/pa)^{0.75}$ (s_u in tsf) (pa = atmospheric pressure)	Clays	Briaud (1992)
$s_u = (P_L - 10)/3.57$ (s_u in MPa)	Gault Clay	Pound & Varley (1993)

7.8.3 Preconsolidation Stress

A number of investigations have shown that there is an approximate linear relationship between the pressuremeter creep stress, P_f, and the vertical one-dimensional consolidation yield or preconsolidation stress, σ'_p, over a stress range of 20–2000 kPa (e.g., Mori, 1965; Lukas & LeClerc de Bussy, 1976; Ohya et al. 1983; Davidson & Bodine, 1986; Bergado et al. 1986). Similar results were presented by Martin & Drahos (1986) for clay in the Richmond, Virginia area. Although considerable scatter was presented in these data, most of the points fell within the range $\sigma'_p = P_f$ to $\sigma'_p = 0.6\,P_f$.

7.8.4 Small-Strain Shear Modulus

A correlation between the small-strain shear modulus and P_L was suggested by Kalteziotis et al. (1990) for clays in Greece as:

$$G_{max} = 138\,P_L^{1.42} \tag{7.12}$$

G_{max} and P_L are expressed in MPa.

Additionally, the value of G_{max} was also strongly correlated with the value of G_m according to the simple expression:

$$G_{max} = 45\,G_m, \tag{7.13}$$

where the value of G_m was defined by the authors as:

$$G_m = V(\Delta P / \Delta V)$$

where
   V = volume of the cavity at the midpoint of the linear portion of a PMT curve
   ΔV = volume increase for a pressure increment ΔP in this region

7.9 INTERPRETATION OF TEST RESULTS IN COARSE-GRAINED SOILS

PMT results obtained in coarse-grained soils are similar to those obtained in fine-grained soils, by employing PMT parameters, P_O, P_f, P_L, etc. and then applying rules for design. PMT modulus values may be used for settlement estimates for shallow foundations. Several methods have been suggested for estimating some properties of coarse-grained soils from the PMT curve, e.g., drained friction angle (Hughes et al. 1977; Baguelin et al. 1978; Manaserro 1989); however, they are not particularly reliable and require a detailed interpretation of the PMT curve. The author suggests that results from PMT tests in sands be used along with the recommended design rules developed specifically for the prebored PMT.

7.10 PRESSUREMETER TESTING IN ROCK

In weak or soft rocks, there is a practical problem of obtaining high-quality samples for laboratory testing or for that matter performing most *in situ* tests. The PMT presents a solution to this problem provided that a high-quality borehole cavity can be obtained. Stiffness and limit pressures can be usually obtained but a high-pressure apparatus is used in order to fully define the rock mass behavior. Several cases have reported the successful application of PMT in weaker rocks (shales and mudstones) for the design of drilled shaft rock sockets (Freeman et al., 1972; Jubenville & Hepworth, 1981; Briaud 1985). Table 7.9 gives a summary of some reported uses of the PMT in rock.

7.11 CORRELATIONS WITH OTHER *IN SITU* TESTS

Correlations between various PMT parameters and other *in situ* tests, especially the SPT, have been noted by a number of investigators (e.g., Chiang & Ho 1980; Tsuchiya & Toyooka 1982; Bozbey & Togrol 2010). Tables 7.10 and 7.11 present the reported correlations between SPT N-values and P_f and P^*_{LM}. Table 2.13 gives some reported correlations between N and the PMT modulus, E_m, also listed in Table 7.12.

7.12 APPLICATIONS TO DESIGN

Most of the design applications of the PMT follow the rules established based on the values of limit pressure or modulus to estimate bearing capacity or settlement of both shallow and deep foundations (Baguelin et al. 1978; 1986; Briaud 1986; 1992; Gambin & Frank

Table 7.9 Some reported uses of PMT in rock

Location	Rock type	Range of P_L (kPa)	References
California	Siltstone Sandstone Shale	1450–4300	Dixon (1970)
Illinois	Shale		Hendron et al. (1970)
India	Weathered granite	150–950	Pandian & Raju (1977)
South Africa	Weathered siltstone	5100–7050	Pavlakis (1980)
Australia	Siltstone	8000–8700	Pells & Turner (1980)
South Dakota	Shale	NR	Nichols et al (1986)
United Arab Emirates	Siltstone	NR	Mahmoud et al. (1990)
Oklahoma	Shale Sandstone	2000–20,000	Miller & Smith (2004)
Canada	Shale	NR	Cao et al. (2013)
Turkey	Sandstone Siltstone	2500–4000	Tezel et al. (2013)
India	Weathered basalt and tuff	2000–10,000	Birid (2015)
Iran	Claystone Marlstone	1000–6000	Asghari-Kaljahi et al. (2016)

NR = not reported

Table 7.10 Reported correlations between P_f and SPT N-value

Correlation	Soil	References
$P_f = 0.669N^{0.792}$ (P_f in bar)	Miscellaneous soils	Tsuchiya & Toyooka (1982)
$P_f = 0.33$ to $0.50N$	Clayey and sandy soils	Ohya et al. (1983)
$P_f = N/6.86$	Stiff clay	Bergado et al. (1986)

P_f in kg/cm^2 unless noted

Table 7.11 Reported correlations between P_L and P^*_L and SPT N-value

Correlation	Soil	References
$P^*_L = N/1.37$ $P^*_L = N/1.03$ (P^*_L in kg/cm^2)	Stiff clay Dense to very dense sand	Bergado et al. (1986)
$P_L = 75N$ (P_L in kPa)	Residual soils of Singapore	Chang (1988)
$\log P^*_L = 0.0073N + 1.1194$ (P^*_L in tsf)	Clay and shale	Nevels & Laguros (1993)
$P_L = 29.45(N_{60}) + 219.7$ (P_L in kPa)	Sandy silty clay – Turkey	Yagiz et al. (2008)
$P_L = 0.33(N_{60})^{0.51}$ $P_L = 0.26(N_{60})^{0.57}$ (P_L in MPa)	Clayey soils – Turkey Sandy soils – Turkey	Bozbey & Togrol (2010)
$P_L = 0.425(N_{60})^{1.2}$ (P_L in MPa)	Clayey soils – Turkey	Kayabasi (2012)
$P_L = -0.872 + 0.067(N_{60})$ (P_L in MPA)	Clayey soils – Istanbul	Agan & Algin (2014)

1995; Frank 2009). One of the most direct applications of the PMT is the estimation of the behavior of laterally loaded drilled shafts or driven piles using the pressure-expansion curve. Methods for evaluating the lateral behavior of drilled shafts using the prebored PMT and driven piles using the FDMPT have been developed. Cyclic PMT procedures and results have been given by Briaud et al. (1983a) and Failmezger & Sedran (2013). Table 7.13 gives some typical applications of the PMT in design.

7.12.1 Design of Shallow Foundations

7.12.1.1 Bearing Capacity

The bearing capacity of shallow foundations may be determined from:

$$q_{ult} = KP^*_L + \gamma D_f \tag{7.14}$$

where

  q_{ult} = ultimate unit bearing capacity
  K = bearing capacity factor
  P^*_L = equivalent net limit pressure within the zone of influence of the footing
  γ = total unit weight of soil
  D_f = depth of embedment of footing

The bearing capacity factor K is a function of the footing depth/width ratio and soil type, i.e., clay, silt, and sand. Typical values of K range from about 0.8 for a surface footing to about 2.0 for $D_f/B = 3$ for sand and about 1.2 for clay. Briaud (1992) presented a summary of the accuracy of this approach using a database of published cases and found it to be sufficiently reliable.

7.12.1.2 Settlement

The settlement of shallow foundations may be estimated using the calculated PMT modulus as:

$$s = [(2/9E_d)(qB_o)(\lambda_d B/B_o)^\alpha + (\alpha qB\lambda_c/9E_c)] \tag{7.15}$$

Table 7.12 Reported correlations between SPT N-value and pressuremeter modulus[a] (reference provided in Chapter 2)

Correlation	Soil	References
$E_m = \text{Log}^{-1}[0.65180 \text{ Log } N + 1.33355]$ (E_m in tsf)	Piedmont residual soil	Martin (1977)
$E_m = 7.7N$	Clay	Nayak (1979)
$E_m = 15N$	Clayey soil	Ohya et al. (1983)
$E_m = 4N$	Sandy soil	
$E_m = 6.84 N^{0.986}$ (E_p in bar)	Miscellaneous soil types[b]	Tsuchiya & Toyooka (1982)
$E_{m.} = 22N + 160$	Gneissic saprolite ($20 < N < 30$)	Rocha Filho et al. (1985)
$E_m = 26N + 120$	Gneissic saprolite ($30 < N < 60$)	
$\ln E_m = 3.509 + 0.712 \ln N$ (E_p in ksf)	Residual soil	Barksdale et al. (1986)
$E_m = 15N + 240$	Lateritic or mature Gneissic residual soil ($7 < N < 15$)	Toledo (1986)
$E_m = \text{Log}^{-1}[0.70437 \text{ Log } N + 1.17627]$ (E_m in tsf)	Piedmont residual soil	Martin (1987)
$E_m = 1.6N$ (E_m in MPa)	Residual soil	Jones & Rust (1989)
$\log E_m = 1.0156 \log N + 1.1129$ (E_p in tsf)	Clay and clay shale	Nevels & Laguros (1993)
$E_m \text{ (kPa)} = 388.7 N_{60} + 4554$	Sandy silty clay	Yagiz et al. (2008)
$E_m \text{ (MPa)} = 1.33(N_{60})^{0.77}$	Sandy soils – Istanbul	Bozbey & Togrol (2010)
$E_m \text{ (MPa)} = 1.61(N_{60})^{0.71}$	Clayey soils – Istanbul	
$E_m = 0.285(N_{60})^{1.4}$ (E_m in MPa)	Clayey soils – Turkey	Kayabasi (2012)
$Em \text{ (MPa)} = 2.22 + 0.0029(N_{60})^{2.5}$	Clayey soils – Turkey	Agan & Algin (2014)

[a] E_p in kg/cm² unless noted.
[b] Individual equations given by authors for eight different soil types ranging from very soft organic soil to mudstone.

where
    s = settlement
    E_d = average PMT modulus within the zone of significant influence below the footing
    q = footing net bearing stress
    B_o = reference footing width (60 cm or 2 ft)
    λ_d = deviatoric shape factor
    B = footing width
    α = soil rheological factor
    E_c = average PMT modulus just below the footing
    λ_c = spherical shape factor

Values of shape factors and rheological factors depend on the length/width ratio of the footing and soil type and stiffness, and may be obtained from Briaud (1992) or other sources.

Table 7.13 Design applications of PMT

Application	References
Shallow foundation bearing capacity	Briaud et al. (1986)
Shallow foundation settlement	Briaud et al. (1986)
	Barksdale et al. (1986)
	Lukas (1986)
	Briaud (2007)
	Baguelin et al. (2009)
	Ouabel et al. (2020)
Axially loaded deep foundations	Briaud (1985)
	Gambi & Frank (2009)
Laterally loaded deep foundations	Briaud et al. (1983b)
	Briaud (1985)
	Robertson et al. (1986)
	Meyerhof & Sastry (1987)
	Huang et al. (1989)
	Briaud (1997)
	Anderson et al. (2003)
	Failmezger et al. (2005)
	Bouafia (2013)
	Farid et al. (2013)
Earth retaining structures	Grant & Hughes (1986)
	Finno et al. (1990)
	Aoyagi et al. (1995)

7.12.2 Deep Foundations

7.12.2.1 Ultimate Axial Load of Deep Foundations

Design rules for estimating the end bearing and side resistance of deep foundations under axial loading are also available and primarily make use of the limit pressure or net limit pressure. The approach is similar to that used for the bearing capacity of shallow foundations and takes into account geometry and soil characteristics to give design coefficients to obtain unit end bearing and unit side resistance (Baguelin et al. 1978; 1986; Briaud 1986; 1992; Gambin & Frank 1995; Frank 2009).

7.12.2.2 Laterally Loaded Shafts and Piles

Results from the PMT have been extensively used to predict the lateral load behavior of deep foundations. The methods are predominantly based on developing pile p-y curves determined from the pressure-expansion curves from the PMT. The primary difference is in the deployment of the PMT to simulate pile installation, i.e., predrilled or displacement.

7.13 SUMMARY OF PMT

The prebored pressuremeter is a very useful test for determining specific soil properties, such as shear strength and stiffness. It can also be used to design foundations using a set of well-established rules for both shallow and deep foundations. The test is not complicated, and the equipment is reasonably economical. Test results can be evaluated quickly. The test is especially useful for estimating the settlement of shallow foundations and the lateral load behavior of drilled deep foundations. Other types of pressuremeters, especially the SBPMT, may be expensive and require expert experience in order to perform the test and interpret results. The Pencel PMT can be deployed quickly and may be used where other types of PMT are not available.

REFERENCES

Agan, C. and Algin, H., 2014. Determination of Relationships Between Menard Pressuremeter Test and Standard Penetration Test Data by Using ANN Model: A Case Study on the Clayey Soil in Sivas, Turkey. *Geotechnical Testing Journal, ASTM*, Vol. 37, No. 3, pp. 1–12.

Amar, S. and Jezequel, J., 1972. Essais en Place et an Laboratoire sul Sols Coherents: Comparison des Results. *Bulettin de Liaison des Ponts et Chausses*, No. 58, pp. 97–108.

Anderson, J., Townsend, F., and Grajales, B., 2003. Case History Evaluation of Laterally Loaded Piles. *Journal of Geotechnical and Geoenvironmental Engineering, ASCE*, Vol. 129, No. 3, pp. 187–196.

Aoyagi, T., Honda, T., Morita, Y., and Fukagawa, R., 1995. Deformation of Diaphragm Walls Estimated from Pressuremeter. The Pressuremeter and its New Avenues, pp. 405–410.

Asghari-Kaljahi, E., Khalili, Z., and Yasrobi, S., 2016. Pressuremeter Tests in the Hard Soils and Soft Rocks of Arak Aluminum Plant Site, Iran. *Proceedings of the 5th Symposium on Geotechnical and Geophysical Site Characterization*, pp. 743–747.

Baguelin, F.J., Bustamante, M. and Frank, R., 1986. The Pressuremeter for Foundations: French Experience. *Use of In Situ Tests in Geotechnical Engineering, ASCE*, pp. 31–46.

Baguelin, F., Bustamante, M., and Frank, R., 1974. The Pressuremeter for Foundations: French Experience. *Use of In Situ Tests in Geotechnical Engineering, ASCE*, pp. 48–73.

Baguelin, F.J. and Jezequel, J.-F., 1983. The LPC Pressiopenetrometer. *Geotechnical Practice in Offshore Engineering, ASCE*, pp. 203–219.

Baguelin, F., Jezequel, J., and Shield, D., 1978. *The Pressuremeter and Foundation Design*. Trans Tech Publications, 617 pp.

Baguelin, F., Lay, L., Ung, S., and Sanfratello, J., 2009. Pressuremeter, Consolidation State and Settlement in Fine-Grained Soils. *Proceedings of the 17th International Conference on Soil Mechanics and Geotechnical Engineering*, Vol. 2, pp. 961–964.

Bahar, R., 1998. Interpretation of Pressuremeter Tests Carried Out in Stiff Clays. *Proceedings of the 2nd International Symposium on the Geotechnics of Hard Soils – Soft Rocks*, Vol. 1, pp. 413–422.

Barksdale, R., Ferry, C., and Lawrence, J., 1986. Residual Soil Settlement from Pressuremeter Moduli. *Use of In Situ Tests in Geotechnical Engineering, ASCE*, pp. 447–461.

Bechai, M., Law, K.T., Cragg, C.B.H., and Konard, J.M., 1986. In Situ Testing of Marine Clay for Towerline Foundations. *Proceedings of the 39th Canadian Geotechnical Conference*, pp. 115–119.

Benoit, J., Oweis, I., and Leung, A., 1990. Self-Boring Pressuremeter Testing of the Hackensack Meadows Varved Clays. *Proceedings of the 3rd International Symposium on the Pressuremeter*, pp. 95–94.

Bergado, D., Khaleqoe, M., Neeyapan, R., and Chang, C., 1986. Correlations of In Situ Tests in Bangkok Subsoils. *Geotechnical Engineering*, Vol. 17, pp. 1–37.

Birid, K., 2015. Interpretation of Pressuremeter Tests in Rock. *Proceedings of the 7th International Conference on the Pressuremeter*, pp. 289–299.

Borsetto, M., Imperato, L., Nova, R., and Peano, A., 1983. Effects of Pressuremeters of Finite Length in Soft Clay. *Proceedings of the International Symposium on In Situ Testing*, Vol. 2, pp. 211–215.

Bouafia, A., 2013. P-Y Curves from the Prebored Pressuremeter Test for Laterally Loaded Single Piles. *Proceedings of the 18th International Conference on Soil Mechanics and Foundation Engineering*, pp. 2695–2698.

Bozbey, I. and Togrol, E., 2010. Correlation of Standard Penetration Teat and Pressuremeter Data: A Case Study from Istanbul, Turkey. *Bulletin of Engineering Geology and the Environment*, Vol. 69, pp. 505–515.

Briaud, J., 1985. Pressuremeter and Deep Foundation Design. ASTM STP 950, pp. 376–405.

Briaud, J. 1986. Pressuremeter and Foundation Design. *Use of In Situ Tests in Geotechnical Engineering, ASCE*, pp. 74–115.

Briaud, J., 1992. *The Pressuremeter*. A.A. Balkema Publishers, Rotterdam, 322 pp.

Briaud, J., 1997. SALLOP: Simple Approach for Lateral Loads on Piles. *Journal of Geotechnical and Geoenvironmental Engineering, ASCE*, Vol. 123, No. 10, pp. 958–964.

Briaud, J., 2007. Spread Footings in Sand: Load Settlement Curve Approach. *Journal of Geotechnical and Geoenvironmental Engineering, ASCE*, Vol. 133, No. 8, pp. 905–920.

Briaud, J., 2013. The Pressuremeter Test: Expanding its Use. *Proceedings of the 18th International Conference on Soil Mechanics and Geotechnical Engineering*, Vol. 1, pp. 107–126.

Briaud, J. and Shields, D., 1979. A Special Pressuremeter and Pressuremeter Tests for Pavement Evaluation and Design. *ASTM Geotechnical Testing Journal*, Vol. 2, No. 3, pp. 143–151.

Briaud, J., Lytton, R., and Hung, J., 1983a. Obtaining Moduli from Cyclic Pressuremeter Tests. *Journal of Geotechnical Engineering, ASCE*, Vol. 109, No. 5, pp. 657–665.

Briaud, J., Smith, T., and Meyer, B., 1983b. Laterally Loaded Piles and the Pressuremeter: Comparison of Existing Methods. ASTM STP 835, pp. 97–111.

Briaud, J., Tand, K., and Funegard, E., 1986. Pressuremeter and Shallow Foundations on Clay. Transportation Research Record No. 1105, pp. 1–14.

Bruzzi, D., Ghionna, V., Jamiolkowski, M., Lancellotta, R., and Manfredini, G., 1986. Sel-Boring Pressuremeter Tests in Po River Sand. ASTM STP 950, pp. 57–74.

Cao, L., Peaker, S., and Sirati, A., 2013. Rock Modulus from In Situ Pressuremeter and Laboratory Tests. *Proceedings of the 18th International Conference on Soil Mechanics and Foundation Engineering*, 4 pp.

Campenella, R., Howie, J., Sully, J., Hers, I., and Robertson, P., 1990. Evaluation of Cone Presssuremeter Tests in Soft Cohesive Soils. *Proceedings of the 3rd International Symposium on the Pressuremeter*, pp. 125–136.

Canou, J. and Tumay, M., 1986. Field Evaluation of French Self-Boring Pressuremeter PAF 76 in a Soft Deltaic Louisiana Clay. ASTM STP 950, pp. 97–118.

Centre d'Etudes Menard, 1967. Interpretation d'un Essai Pressiometrique, Publication D 31/67.

Chang, M.F., 1988. In-Situ Testing on Residual Soils in Singapore. *Proceedings of the 2nd International Conference Geomechanics in Tropical Soils*, Vol. 1, pp. 97–107.

Chiang, Y. and Ho, Y., 1980. Pressuremeter Method for Foundation Design in Hong Kong. *Proceedings of the 6th S.E. Asian Conference on Soil Mechanics*, pp. 32–42.

Clarke, B., 1995. *Pressuremeters in Geotechnical Design*. Taylor & Francis Publishers, 363 pp.

Davidson, R.R. and Bodine, D.G., 1986. Analysis and Verification of Louisiana Pile Foundation Designed Based on Pressuremeter Results, *ASTM Special Technical Publication 950*, pp. 423–439.

Dixon, S., 1970. Pressure Meter Testing of Soft Bedrock. ASTM STP 477, pp. 126–136.

Fahey, M., 1991. Measuring Shear Mdulus in sand with the Self-Boring Pressuremeter. *ASTM Geotechnical Testing Journal*, Vol. 11, No. 3, pp. 187–194.

Failmezger, R. and Sedran, G., 2013. New Method to Compute Reload and Unload Pressuremeter Moduli. *Proceedings of the 18th International Conference on Soil Mechanics and Foundation Engineering*, 4pp.

Failmezger, R., Zdinak, A., Darden, J., and Fahs, R., 2005. Use of Rock Pressuremeter for Deep Foundation Design. *Proceedings of the 5th International Symposium on Pressuremeter*, 10 pp.

Farid, M., Slah, N., and Cosentino, P., 2013. Pencel Pressuremeter Testing for Determining P-Y Curves for Laterally Loaded Deep Foundations. *Procedia Engineering*, Vol. 54, pp. 491–504.

Faugeras, J.C., Gourves, R., Meunier, P., Nagura, M., Matsubara, L. and Saguawara, N., 1983. On the Various Factors Affecting Pressuremeter Test Results. *Proceedings of the International Symposium on In Situ Testing in Soil and Rock*, Vol. 2, pp. 275–281.

Finno, R., Benot, J., and Chumg, C., 1990. Filed and Laboratory Measurement of K_O in Chicago Clay. *Proceedings of the 3rd International Symposium on the Pressuremeter*, pp. 331–340.

Frank, R., 2009. Design of Foundations in France with the use of Menard Pressuremeter Tests (MPM). *Soil Mechanics and Foundation Engineering*, Vol. 46, pp. 219–231.

Freeman, C.F., Klanjnerman, D. and Prasad, G.D., 1972. Design of Deep Socketed Caissons into Shale Bedrock. *Canadian Geotechnical Journal*, Vol. 9, pp. 105–118.

Fyffe, S., Reid, W., and Summers, J., 1986. The Push-In Pressuremeter: 5 Years Offshore Experience. ASTM STP 950, pp. 22–37.

Gambin, M. and Frank, R., 1995. The Present Design Rules for Foundations Based on Menard PMT Results. *Proceedings of the 4the International Symposium on the Pressuremeter*, pp. 425–432.

Ghionna, V., Jamiolkowski, M., Pedroni, S., and Piccoli, S., 1995. Cone Pressuremeter Tests in Po River Sand. *Proceedings of the 4th International Symposium on the Pressuremeter*, pp. 471–480.

Gibson, R and Anderson, W., 1961. Measurement of Soil Properties with the Pressuremeter. *Civil Engineering and Public Works Review*, Vol. 3, No. 6, pp. 615–618.

Grant, W. and Hughes, J., 1986. Pressuremeter Tess and Shoring Wall Design. *Use of In Situ Tests in Geotechnical Engineering, ASCE*, pp. 588–601.

Hendron, A., Mesri, G., and Way, G., 1970. Compressibility Characteristics of Shales Measured by Laboratory and In situ Tests. ASTM STP 477, pp. 137–153.

Houlsby, G. and Carter, J., 1993. The Effects of Pressuremeter Geometry on the Results of Tests in Clay. *Geotechnique*, Vol. 43, No. 4, pp. 567–576.

Houlsby, G. and Withers, N., 1988. Analysis of the Cone Pressuremeter Test in Clay. *Geotechnicque*, Vol. 36, No. 4, pp. 575–587.

Huang, A. and Haefele, K., 1988. A Push-In Pressuremeter/Sampler. *Proceedings of the 1st International Symposium on Penetration Testing*, Vol. 1, pp. 533–538.

Huang, A., Lutenegger, A., Islam, M., and Miller, G., 1989. Analyses of Laterally Loaded Drilled Shafts Using In Situ Tests. Transportation Research Record No. 1235, pp. 60–67.

Hughes, J., Wroth, C., and Windle, D., 1977. Pressuremeter Tests in Sands. *Geotechnique*, Vol. 27, No. 4, pp. 455–477.

Jefferies, M., 1988. Determination of Horizontal In Situ Stress in Clay with Self-Bored Pressuremeter. *Canadian Geotechnical Journal*, Vol. 25, No. 3, pp. 559–573.

Jones, G.A. and Rust, E., 1989. Foundations on Residual Soil Using Pressuremeter Moduli. *Proceedings of the 12th International Conference on Soil Mechanics and Foundations Engineering*, Vol. 1, pp. 519–523.

Jubenville, D.M. and Hepworth, R.C., 1981. Drilled Pier Foundations in Shale, Denver, Colorado Area. Drilled Piers and Caissons, *ASCE*, pp. 66–81.

Kalteziotis, N., Tsiambaos, G., Sabatakakis, N., and Zerogiannis, H., 1990. Prediction of Soil Dynamic Parameters from Pressuremeter and Other In Situ Tests. *Proceedings of the 3rd International Symposium on Pressuremeters*, pp. 391–400.

Kayabasi, A., 2012. Prediction of Pressuremeter Modulus and Limit Pressure of Clayey Soils by Simple and Non-Linear Multiple Regression Techniques: A Case Study from Mersin, Turkey. *Environmental Earth Sciences*, Vol. 66, pp. 2171–2183.

Komornik, A., Wiseman, G., and Frydman, S., 1970. A Study of In Situ Testing with the Pressuremeter. *Proceedings of the Conference on In Situ Investigations of Soil and Rock*, London, pp. 145–153.

Laier, J., Schmertmann, J., and Schaub, J., 1975. Effect of Finite Pressuremeter Length in Dry Sand. *In Situ Measurement of Soil Properties, ASCE*, Vol. 1, pp. 241–259.

Leroueil, S., Tavenas, F., and LeBitian, J.-P., 1983. Properties Caracteristiques des Argiles de l'est du Canada. *Canadian Geotechnical Journal*, Vol. 20, No. 4, pp. 681–705.

Lukas, R., 1986. Settlement Prediction Using the Pressuremeter. AST STP 950, pp. 406–422.

Lukas, R. and LeClerc de Bussy, B., 1976. Pressuremeter and Laboratory Test Correlations for Clays. *Journal of the Geotechnical Engineering Division, ASCE*, Vol. 102, No. GT9, pp. 945–962.

Lutenegger, A. and Blanchard, J., 1990. A Comparison Between Full Displacement Pressuremeter Tests and Dilatometer Tests in Clay. *Proceedings of the 3rd International Symposium on the Pressuremeter*, pp. 309–320.

Mahmoud, M., Evans, J., and Trenter, N., 1990. The Use of the Pressuremeter Test in Weak Rocks at the Port of Jebel Ali. *Proceedings of the 3rd International Symposium on the Pressuremeter*, pp. 341–350.

288 *In Situ* Testing Methods in Geotechnical Engineering

Mair, R. and Wood, D., 1987. *Pressuremeter Testing: Methods and Interpretation.* Butterworth Publishers.

Manaserro, F., 1989. Stress-Strain Relationships from Self-Boring Pressuremeter Tests in Sand. *Geotechnicque,* Vol. 39, No. 2, pp. 293–308.

Marsland, A. and Randolph, M., 1977. Comparisons of the Results from Pressuremeter Tests and Large In Situ Plate Tests in London Clay. *Geotechnicque,* Vol. 27, No. 2, pp. 217–243.

Martin, R.E., 1987. Settlement of Residual Soils. *Foundations and Excavation in Decomposed Rocks of the Piedmont Province, ASCE,* pp. 1–14.

Martin, R. and Drahos, E., 1986. Pressuremeter Correlations for Preconsolidated Clay. *Use of In Situ Tests in Geotechnical Engineering, ASCE,* pp. 206–220.

Meyerhof, G. and Sastry, V., 1987. Full-Displacement Pressuremeter Method for Rigid Piles Under Lateral Loads and Moments. *Canadian Geotechnical Journal,* Vol. 24, NO. 4, pp. 471–478.

Miller, G. and Smith, J., 2004. Texas Cone Penetrometer-Pressuremeter Correlations for Soft Rock. GeoSupport 2004 – SCE, pp. 441–449.

Mori, H., 1965. Discussion, *Proceedings of the 6th International Conference on Soil Mechanics and Foundation Engineering,* Vol. 3, pp. 502–504.

Nayak, N.V., 1979. Use of the Pressuremeter in Geotechnical Design Practice. *Proceedings of the International Symposium on In Situ Testing of Soils and Rocks and Performance of Structures,* Roorkee pp. 432–436.

Nevels, J.B. and Laguros, J.G., 1993. Correlation of Engineering Properties of the Hennessey Formation Clays and Shales. *Geotechnical Engineering of Hard Soils - Soft Rocks,* Vol. 1, pp. 215–220.

Nichols, T., Collins, D., and Davidson, R., 1986. In Situ and Laboratory Geotechnical Tests of the Pierre Shale near Hays, South Dakota – A Characterization of Engineering Behavior. *Canadian Geotechnical Journal,* Vol. 23, No. 2, pp. 181–194.

Ohya, S., Imai, T., and Nagura, M., 1983. Recent Developments in Pressuremeter Testing, In Situ Measurement and S-Wave Velocity Measurement. Recent Developments in Laboratory and Field Tests and Analysis of Geotechnical Problems, pp. 189–209.

Ouabel, H., Bendiouis, A., and Zadjaoui, A., 2020. Numerical Estimation of Settlement Under Shallow Foundation by the Pressuremeter Method. *Civil Engineering Journal,* Vol. 6, No. 1, pp. 156–163.

Pandian, N. and Raju, A., 1977. In Situ Testing of Soft Rocks with Pressuremeter. *Indian Geotechnical Journal,* Vol. 7, No. 4, pp. 305–314.

Pavlakis, M., 1980. Pressuremeter Testing of Weathered Karroo Siltstone. *Proceedings of the 7th Regional Conference for Africa on Soil Mechanics and Foundation Engineering,* pp. 147–155.

Pells, P.J. and Turner, R.M., 1980. End Bearing on Rock with Particular Reference to Sandstone. *Proceedings of the International Conference on Structural Foundations on Rock,* Sydney, 12 pp.

Pound, C. and Varley, P., 1993. The Application of the High Pressure Dilatometer on the Channel Tunnel Project. The Engineering Geology of Weak Rock, pp. 251–258.

Powell, J., 1990. A Comparison of Four Different Pressuremeters and Their Methods of Interpretation in a Stiff Heavily Overconsolidated Clay. *Proceedings of the 3rd International Symposium on the Pressuremeter,* pp. 287–298.

Ramdane, B., Nassima, A. and Ouarda, B., 2013. Interpretation of a Pressuremeter Test in Cohesive Soils. *Proceedings of the International Conference on Geotechnical Engineering,* Tunisia, 10 pp.

Rehman, Z., Akbar, A., and Clarke, B., 2011. Characterization of a Cohesive Soil Bed Using Cone Pressuremeter. *Soils and Foundations,* Vol. 51, No. 5, pp. 823–833.

Robertson, P. and Hughes, J., 1985. Determination of Properties of Sand from Self-Boring Pressuremeter Tests. ASTM STP 950, pp. 283–302.

Robertson, P., Hughes, J., Campanella, R., Brown, P., and McKeown, S., 1986. Design of Laterally Loaded Piles Using the Pressuremeter. ASTM STP 950, pp. 443–457.

Rocha Filho, P., Antunes, F., and Falcao, M.F.Q., 1985. Qualitative Influence of the Weathering Degree Upon the Mechanical Properties of Young Gneiss Residual Soil. *Proceedings of the 1st International Conference on Geomechanics in Tropical Lateritic and Saprolitic Soils,* Vol. 1, pp. 281–294.

Tezel, G., Hacialioglu, E., Onal, F., and Ozmen, G., 2013. Comparison of High Pressure Pressuremeter (HyperPac) and Pre-Bored Pressuremeter Tests – A Case Study. *Proceedings of the 18th International Conference on Soil Mechanics and Foundation Engineering*, 5 pp.

Toledo, R.D., 1986. Field Study of the Stiffness of a Gneissic Residual Soil Using Pressuremeter and Instrumented Pile Load Test. *MS Thesis, Civil Engineering Department*, Pontifical Catholic University, Rio de Janeiro.

Tsuchiya, H. and Toyooka, Y., 1982. Comparison Between N-Values and Pressuremeter Parameters. *Proceedings of the 2nd European Symposium on Penetration Testing*, Vol. 1, pp. 169–174.

Walker, L., 1979. The Selection of Design Parameters in Weathered Rocks. *Proceedings of the 7th European Conference on Soil Mechanics and Foundation Engineering*, Vol. 2, pp. 287–294.

Windle, D. and Wroth, C.P., 1977. In Situ Measurement of the Properties of Stiff Clays. *Proceedings of the 9th International Conference on Soil Mechanics and Foundation Engineering*, Vol. 1, pp. 347–352.

Withers, N., Schaap, L., and Dalton, C., 1986. The Development of the Full Displacement Pressuremeter. ASTM STP 950, pp. 38–56.

Withers, N., Howie, J., Hughes, J., and Robertson, P., 1989. Performance and analysis of Cone Pressuremeter Tests in Sands. *Geotechnique*, Vol. 39, No. 3, pp. 433–454.

Wroth, C.P. and Hughes, J., 1974. The Development of a Special Instrument for the In Situ Measurement of Strength and Stiffness of Soils. *Proceedings of the Conference on Subsurface Exploration for Underground Excavation and Heavy Construction, ASCE*, pp. 295–311.

Yagiz, S., Akyol, E., and Sem, G., 2008. Relationship Between the Standard Penetration Test and the Pressuremeter Test on Sandy Silty Clays: A Case Study from Denizli. *Bulletin of Engineering Geology and the Environment*, Vol. 67, pp. 405–410

Chapter 8

Borehole Shear Test (BST)

8.1 INTRODUCTION

The Borehole Shear Test (BST) was developed as a method of measuring the drained *in situ* shear strength of soils by performing a series of direct shear tests on the sides of a borehole (Handy & Fox 1967). The test is intended to provide independent measurements of the drained friction angle, φ', and cohesion, c', of soils. It is the only *in situ* test to give a direct measurement of these parameters *in situ*.

The test represents a simple concept and gives results that can be directly used in geotechnical design. The mode of testing may be suited to evaluating side friction resistance of axially loaded piles and drilled shafts or grouted anchors. The equipment is relatively rugged and portable; the test is easy to conduct and essentially operator independent, and can be used in a wide range of soils.

8.2 MECHANICS

The BST is conducted by first advancing a borehole into the soil to a desired test depth and then lowering an expandable shear head into the borehole to engage soil along the sides of the hole. The shear head is equipped with shear plates that have a sharp "teeth" that grip the soil along the sides of the hole. Once the shear head has been expanded and sufficient time for consolidation is given, the shear head is pulled upward slowly to induce a shear failure in the soil. This concept is illustrated in Figure 8.1.

This procedure gives a single measurement of the normal stress and shear stress acting on the soil at failure. In this way, the BST simulates, at least in part, the procedure used in laboratory direct shear box tests where normal stress is applied to a sample and then failure is produced by shearing the soil. In the BST, like in the laboratory shear box, this procedure is repeated a number of times using different values of normal stress until sufficient pairs of data points (normal and shear stress) are obtained to define the failure envelope of the soil as shown in Figure 8.2.

8.3 EQUIPMENT

The current configuration of the BST apparatus is shown schematically in Figure 8.3, and apparatus consists of three basic components: (1) the shear head, (2) the control console, and (3) the shear force reaction base plate.

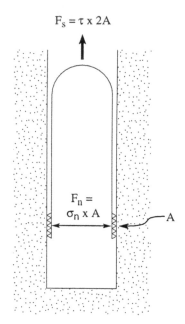

$$F_s = \tau \times 2A$$

$$F_n = \sigma_n \times A$$

A

Figure 8.1 Concept of Borehole Shear Test (BST).

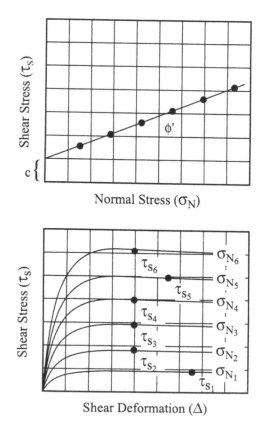

Figure 8.2 Idealized test results from BST.

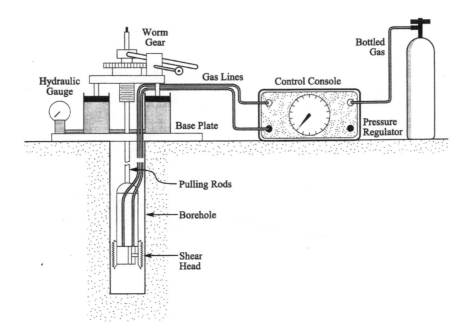

Figure 8.3 Schematic of BST equipment.

8.3.1 Shear Head

The shear head consists of a double-acting pneumatic cylinder that is used to expand the shear plates outward, to apply a normal force to the side of the borehole, and to engage the soil for shearing. Serrated shear plates mounted on opposite sides of the shear head are used to "bite" into the soil, helping insure that failure will occur within the soil and not just at the surface of the borehole. The shear head is lowered into the hole using a series of small-diameter rods attached to the top of the shear head. The rods also act to transfer shear stress from the shear head to the reaction base during the test. The shear head is double-acting so that while it is being lowered into the hole, the plates can be in the fully retracted position. At the end of the test, the plates can be retracted for ease in removing the shear head from the hole.

8.3.2 Control Console

The normal stress against the sides of the borehole is held constant during the test by pressurizing the pneumatic shear head cylinder using a regulated compressed gas applied to the shear head through the control console. Gas pressure is usually supplied by a small tank of carbon dioxide or nitrogen. The actual gas pressure applied to the shear head is registered on a pressure gage mounted in the console and is held constant using a pressure regulator. During the test, constant pressure on the shear head is maintained in order to keep the normal stress constant, like in a direct shear box test.

8.3.3 Shear Force Reaction Base Plate

Once the normal stress has been applied and sufficient consolidation time has been allowed, the soil is failed by slowly pulling the shear head upward. This is accomplished using the shear force reaction base plate located at the ground surface and centered over the borehole.

Figure 8.4 BST in progress.

The rods extending from the shear head pass through the center of the base plate. A simple worm-gear mechanism attached to the pulling rods by a rod clamp is used to pull the shear head upward by means of a hand crank. A closed hydraulic system on the base plate is used to measure the pulling force, as shown in Figure 8.3. The apparatus is self-contained and can be operated by hand. The equipment is fully portable and can be used in remote or limited access locations. Figure 8.4 shows a photograph of a test in progress.

8.4 TEST PROCEDURES

As noted, the BST is performed in two separate phases: (1) a *consolidation* phase, during which some time is allowed to elapse between the application of normal stress and that of shearing stress, and (2) a *shearing* phase, during which the shear head is pulled to create the soil shear failure. This procedure is directly analogous to performing a laboratory direct shear box test. A detailed testing procedure that gives a complete description of the test practice has been presented by Lutenegger (1987a). The test is also described in ISO/WD 22476-16. Lutenegger & Timian (1987) demonstrated that the test results obtained by different operators were not statistically different.

8.4.1 Multistage Testing

The normal test procedure used to perform the BST is called multistage testing in that shearing is performed repeatedly in the same soil mass, at successively higher normal stresses. This is similar to laboratory multistage testing. This procedure improves test precision, saves time, and encourages the dissipation of pore water pressures. This procedure is applicable to most soils if the consolidated strength of the soil close to the shear plates exceeds that of the adjacent undisturbed soil; i.e., the previously sheared soil consolidates and builds up a cake of soil that rides with the shear plates during successive shearing. In this way, successive shear surfaces move outward to engage fresh soil with each higher increment of normal stress. This procedure is repeated until 5 or 6 pairs of data points are obtained (normal stress and shear stress), and the shear strength envelope is defined. The result shows

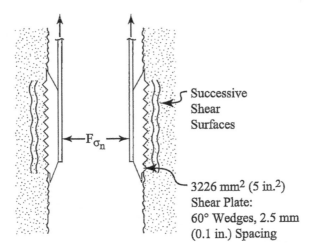

Figure 8.5 Geometry of standard BST shear plates and progressive shear planes in multistage testing.

an exceptionally linear failure envelope in sands, silts, and soft to medium consistency clays. Figure 8.5 shows a schematic of the standard BST shear plates and the formation of successive shear planes.

The use of *multistage testing* provides for a very rapid testing since the shear head does not have to be removed and cleaned after each test point. Using a typical consolidation time of 10 min for each normal stress increment, a complete test yielding a full failure envelope can be performed in about an hour. Figure 8.6 shows soil adhered to the shear plates after removing the shear head following multistage testing.

As previously noted, multistage testing is intended for use in most soils wherein the failure plane is driven progressively outward into the fresh unsheared soil with each increment of normal stress. In some soils, this doesn't occur simply because the soil is too strong to allow standard shear plates to bite into the soil and grip the material sufficiently to shear a soil-soil shearing surface. Handy & Fox (1967) justified the use of multistage testing by showing a close agreement between tests in soils in which both multistage and single-stage tests had been performed. These soils primarily included insensitive softer silty clays, clays, and sand.

Figure 8.6 Soil buildup on shear plates.

Handy (1975) further rationalized the multistage testing by stating "one possibility is that during or after shearing, the major principal stress causes sufficient compression to 'seal' the shear plane and causes it to move outward to a lower stress region. From the typical test behavior, we infer that, after shear failure, the thin layer of soil grains participating in the failure is compacted and added to the cake adhering to the pressure plates. If this is correct, subsequent shearing should occur at the outer surface of the shear cake, in a zone of normally consolidated but otherwise unaltered soil".

8.4.2 Single-Stage "Fresh" Testing

Schmertmann (1975) pointed out that the multistage testing procedure may not be applicable to all soils. In sensitive or structured soils, the failure envelope may become tipped downward at progressively higher values of normal stress as a result of the accumulation of shear stresses on the same surface. In the single-stage testing, the test is conducted in exactly the same manner as the initial data point for a multistage test. However, at the end of the shearing phase, after a peak value of shear stress has been obtained, the normal stress is released by decreasing the gas regulator on the control console and venting the pneumatic pressure. The shear plates are retracted and the shear head is then removed from the borehole. Once the shear plates have been cleaned, the shear head is lowered back down into the borehole to a new adjacent test location (either just above or just below the initial point) to obtain another test point using a different normal stress.

Several studies have shown that in softer soils, the geometry of the shear plates does not influence the test results substantially (e.g., Lutenegger et al. 1978; Demartinecourt & Bauer 1983). In stiff highly overconsolidated clays and some very dense glacial clay tills, the standard shear plates may not penetrate the soil and the shear teeth will tend to slide over the soil surface and only scrape deeper with each successive pull until the teeth are filled. As a result, no soil cake builds up and the test measures some components of sliding friction of soil on steel or disturbed soil on undisturbed soil. The resulting failure envelope is the result of the successive deepening of the "bite" brought about by higher normal stresses, and filling of the teeth with soil. In hard soils and soft rocks, high-pressure shear plates, as shown in Figure 8.7, should be used.

Lutenegger & Timian (1987) presented the results of single-stage tests conducted in a marine clay and found that there was no statistical difference between the mean friction angle and cohesion obtained with this procedure when compared to adjacent multistage tests in the same material. The linearity of all the tests was also excellent, although the single-stage tests gave slightly lower regression coefficients.

The initial normal stress and subsequent increments of normal stress should be chosen prior to testing so that a sufficient number of data points will be obtained throughout the test to define the failure envelope. In other words, if too high of an initial normal stress increment is chosen, the soil may experience a "bearing capacity" failure from a rapid punching of the shear plates and the test will be meaningless. This can especially be a problem in soft normally consolidated clays. On the other hand, if too low of an initial normal stress increment is chosen, the teeth on the shear plates won't bite into the soil and the plates will just drag along walls of the soil, and no soil failure will occur. Table 8.1 presents the recommended values of the initial normal stress and normal stress increments for use in different soils.

Table 8.2 gives the suggested consolidation times for various soil conditions. Regardless of the soil and groundwater conditions, the author recommends that a minimum consolidation time of 5 min should be used in all cases to "seat" the shear plates even in those situations where pore water pressures are not expected.

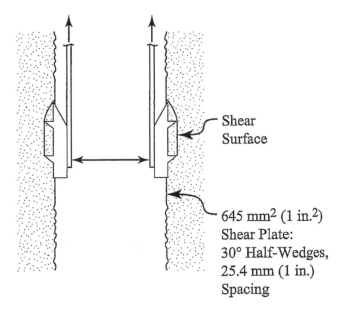

Shear
Surface

645 mm² (1 in.²)
Shear Plate:
30° Half-Wedges,
25.4 mm (1 in.)
Spacing

Figure 8.7 BST high-pressure shear plates for hard soils/soft rocks.

Table 8.1 Recommended values of initial normal stress and normal stress increments

Soil type	Initial normal stress, psi (kPa)	Normal stress increment, psi (kPa)
Very soft clays	2–4 (14–28)	2–4 (14–28)
Soft clays	2–4 (14–28)	3–5 (21–34)
Medium stiff clays Loose sand	3–5 (21–34)	5–10 (34–69)
Stiff/dense	5–10 (34–69)	10–15 (69–103)
Very stiff/hard heavily Overconsolidated clays Cemented sands Clay shales	15–20 (103–138)	15 (103)

Table 8.2 Recommended consolidation times for different soils

Soil type/condition	Consolidation time
Moist sands and silts	5 min
Medium stiff to stiff clays compacted fill	5 min
Soft clays of low plasticity	10 min
Soft to very soft high-plasticity clays below the water table	20 min

8.5 BOREHOLE PREPARATION

In order to conduct the BST, it is necessary to place the shear head into an open borehole with a diameter between 3.0 and 3.125 in. (76 and 79 mm). The equipment used to prepare the borehole should minimize the amount of disturbance to the wall of the borehole. The test should be performed as soon as possible after the test cavity is formed. Table 8.3 gives some guidelines for selecting methods for the borehole preparation in different materials based on previous success.

Table 8.3 Recommended method of borehole preparation for BST

Soil type	Pushed 3 in. (76 mm) Shelby tube	3 in. (76 mm) Hand auger	2 7/8 in. (73 mm) Pilot hole and trimming	2 7/8 in. (73 mm) Rotary drilling	2 7/8 in. (73 mm) Core drilling	3 in. (76 mm) Flight auger
	Borehole preparation					
	Clayey soils					
Soft	X	X	X	X	NA	NA
Medium	X	X	X	X	NA	NA
Stiff	X	X	X	X	NA	X
	Silty soils					
Above GWL	X	X	X	NA	NA	X
Below GWL	X	NA	X	X	NA	NA
	Sandy soils					
Above GWL	NA	X	NA	X	NA	X
Below GWL	NA	NA	NA	X	NA	NA
Soft or weathered rock	X	NA	NA	X	X	X

8.6 INTERPRETATION OF TEST RESULTS

Test results from the BST are usually plotted in the field as the test proceeds, i.e., after each pair of normal and shear stress data points is obtained. This is usually done while waiting during the consolidation phase of the next increment of normal stress and allows the operator to troubleshoot any mechanical malfunctions in the equipment or problems with the test setup as well as keep track of the test. Adjustments to the test procedures in terms of consolidation times, normal stress increments, etc. can then be made during the test.

Provided that the test proceeds in a normal fashion, the data should approximate a linear relationship. It is recommended that a least-squares linear regression analysis be used to reduce the data to obtain individual values of φ' and c' after first discarding any obvious erroneous test points. The use of a linear regression analysis eliminates the subjective data interpretation by simply fitting a line to the data by eye.

The BST may be used to estimate the postpeak or residual strength envelope by observing the behavior of the shear stress gage located on the base plate after it reaches a peak value. With additional movement of the shear plates, the measured shear stress will normally drop off at each normal stress increment until a steady value is obtained at large shear deformation. These postpeak values of shear stress may be combined to give a residual failure envelope.

8.7 RANGE OF SOIL APPLICABILITY

The BST has a wide range of applicability and can be used in any soil formations that will maintain a nominal 76-mm (3 in.)-diameter borehole. This means that the easiest materials to test are medium stiff fine-grained soils and moist sands and silts. Table 8.4 gives a summary of the reported soil materials in which the BST has been successfully used.

Table 8.4 Reported use of the BST in different soils

Soil type	References
Sand, loam, clay, loess	Fox et al. (1966)
	Handy & Fox (1967)
Loess	Lohnes & Handy (1968a)
	Lutenegger et al. (1978)
	Donchev & Lutenegger (1984)
	Lutenegger (1987b)
Glacial till	Lutenegger et al. (1978)
Sand (SP-SM), silt (ML), clay (CL)	Singh & Bhargava (1979)
Clay, silty clay, silty sand	Johns (1980)
Alluvial silt and clay	Little et al. (1982)
Clay shale	Ruenkrairergsa & Pimsarn (1982)
	Yang et al. (2006)
Sensitive marine clay	Bauer & Demartinecourt (1985)
	Lutenegger & Timian (1987)
Marine clay, lime rock, shell	Handy et al. (1985)
Glacial till, shale, compacted clay	Handy (1986)
Volcanic soil	Millan & Escobar (1987)
Lacustrine clay	Lutenegger & Miller (1994)
Very stiff unsaturated clay	Miller et al. (1998)
	Irigoyen & Coduto (2015)
Very stiff London and Gault Clay	Lutenegger & Powell (2008)
Residual soil	Lohnes & Handy (1968b)
	Lutenegger & Adams (1999)
	Chang & Zhu (2004)
	Ogunro, et al. (2008)
Expansive clay	Kong et al. (2017)
Coal	Handy et al. (1976)
	Haramy & Demarco (1983)

8.8 FACTORS AFFECTING TEST RESULTS

A number of factors can affect the results obtained with the BST, and as with other tests, care should be taken to perform the test using the recommended standard procedure to eliminate these variables. There have been several studies to investigate variables related to both variations in the equipment, such as shear plate dimensions and shear teeth geometry, and variations in test procedures, including consolidation time, *multistage* vs. *single-stage* testing, shearing rate, and pore water pressure influences. Table 8.5 gives a summary of several important factors that may influence the test results.

8.9 INTERFACE SHEAR TESTS

A practical use of the BST is in the determination of shearing characteristics between soil at the borehole wall and a smooth interface like steel. This can be accomplished by substituting the standard serrated plates with smooth mild steel plates having no shear teeth. This is done to promote a shear failure at the interface between the smooth plate and the soil rather than at a soil-soil interface. Measurement of this interface characteristic may be useful in some soils for designing temporary steel casing using an effective stress analysis.

Table 8.5 Factors affecting BST results

Factor	Comments
Consolidation time	Longer consolidation time promotes more drainage in saturated clays. No effect in soils above the water table. Allow more time if results are in question.
Shearing rate	Slower shearing rate is better to reduce shear-induced pore water pressures in saturated clays. In all other soils, the normal recommended shearing rates have a minor effect on results.
Borehole disturbance	Borehole should be prepared with as much care as possible. Severe disturbance at the borehole wall will be reduced by allowing consolidation. Disturbed soils will give lower strength envelope than undisturbed soils.
Shear plate geometry	Standard shear plates are recommended for most soils except hard clay and soft rock. High-pressure shear plates may be required in these materials to prevent plates from sliding and not shearing the soil.
Multistage testing	Multistage testing is preferred in most materials except for highly sensitive clays. Single-stage testing is used with high-pressure shear plates in hard soils and soft rock.

Handy et al. (1985) reported the results of interface friction tests obtained using smooth steel plates during a site investigation for a bridge in Florida. In all cases, the friction angle obtained using the smooth plates was lower than the friction angle obtained from adjacent tests using the standard shear plates with teeth. For three comparative tests in hard clay, they found that the ratio of soil-steel adhesion to soil cohesion, c'_s/c', varied from 0.06 to 0.17, averaging 0.11, whereas the comparable soil-steel sliding vs. soil internal friction ratio, φ'_s/φ', varied from 0.5 to 0.9, averaging 0.66.

The results from the smooth shear plate tests also demonstrate how the BST with regular serrated shear plates works to measure the shear strength of the soil. A smooth plate cannot engage the soil but only slides along the interface between the plate and the soil. Apparently, increases in normal stress do not push a failure surface outward into the soil; otherwise, the strength envelope would approach that obtained using serrated plates; i.e., repeated shear still takes place at the interface. This means that in order to actually fail the soil and obtain a soil-to-soil failure envelope, it is necessary to have shear plates with some forms of protrusion or teeth to engage the soil. Figure 8.8 shows the results of regular tests and adjacent smooth steel interface tests in silty sand.

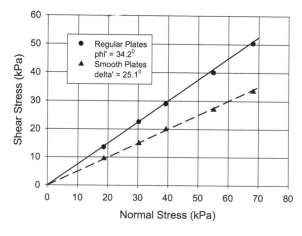

Figure 8.8 Results of regular and smooth steel interface tests in silty sand.

8.10 COMPARISON WITH LABORATORY TESTS

In the past, comparisons have been made between the BST results and either laboratory direct shear box test or triaxial compression test. This clearly can present some problems since even the results of different laboratory tests do not give the same result because of differences in stress conditions, stress paths, testing rates etc.

Early experience with the BST gave comparisons between BST results and direct shear box test or triaxial compression test (e.g., Handy & Fox 1967). Typically, the results obtained with the BST fell in between envelopes obtained from the laboratory tests. Other comparisons between BST and laboratory triaxial and shear box tests (e.g., Wineland 1975; Singh & Bhargava 1979; Lambrechts & Rixner 1981; Lutenegger & Hallberg 1981) have shown a close agreement, usually with φ' within 2°–3°, which is reasonable, considering differences in test procedures and soil variability.

8.11 EQUIPMENT MODIFICATIONS

Several modifications have been made to the BST equipment in order to expand the use of the test for particular applications. Modifications have included provisions for measuring pore water pressure at the shear plates (Demartinecourt & Bauer 1983; Lutenegger & Tierney 1986), measuring the shear deformation (Demartinecourt & Bauer 1983; Lutenegger 1983), measuring the shear head expansion (Johns 1980; Demartinecourt & Bauer 1983), and measuring the soil creep (Lohnes et al. 1972). Figure 8.9 shows the modification to the shear head for measuring pore water pressure, and Figure 8.10 shows a simple arrangement for measuring the shear deformation.

Demartinecourt & Bauer (1983) replaced the hand crank of the base plate with a small variable speed electric motor to provide a constant rate of shear deformation throughout the test. Ashlock & Lu (2012) described a new generation of BST equipment that consists of fully automated equipment.

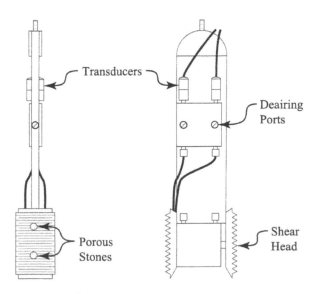

Figure 8.9 Modification to the shear head for measuring pore water pressure.

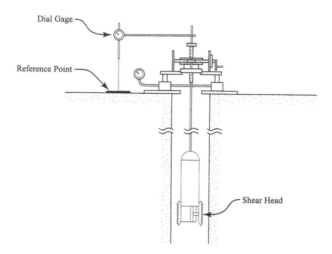

Figure 8.10 Arrangement for measuring shear deformation.

8.12 APPLICATIONS OF BST FOR DESIGN

The BST can be used in geotechnical design situations that require the use of the effective stress Mohr-Coulomb failure envelope. Limit-equilibrium problems involving input of φ' and c' represent the best application of the BST results. Problems in which soil deformation is not the controlling criteria in design may be approached using BST results provided that the test data are obtained properly and are representative of the soil involved, etc. Typical design problems include natural slope stability problems; cut slope stability; designed constructed slopes of embankment; foundation loading in which settlement is not an important factor, such as light pole or power pole structures; transmission tower foundations; and deep foundation side resistance. Chang & Zhu (2004) made a comparison of BST tests conducted in fresh boreholes and in boreholes with added water to determine the influence of dry and wet drilling on side resistance of bored piles. Results from unsaturated residual soils in Singapore showed that the shear strength decreased substantially when water was added. Table 8.6 summarizes a number of reported uses of the BST in geotechnical practice.

8.13 ADVANTAGES AND LIMITATIONS

As with all tests, the BST has a number of potential advantages; however, like all tests, it also has potential limitations and is not applicable in all soils and situations.

8.13.1 Advantages

The potential advantages of the BST include the following:

1. It is the most rapid and only method available for *in situ* evaluation of φ' and c', and a complete test usually requires on the order of 60 min. A large number of tests may be performed in a relatively short period of time;
2. Data from the test are reduced and plotted while the test is being conducted, enabling an immediate value judgment to be made and retesting if necessary;
3. The test may be used in a wide range of soils;

Table 8.6 Reported design applications of BST

Application	References
Slope stability	Handy & Williams (1967)
	Lohnes & Handy (1968a)
	Tice & Sams (1974)
	Little et al. (1982)
	Ruenkrairergsa & Pimsarn (1982)
	Handy (1986)
	Yang et al. (2006)
	Farouz et al (2007)
	Sakamoto et al. (2018)
Deep foundations	Wineland (1975)
	Lambrechts & Rixner (1981)
	Handy et al. (1985)
	Bechai et al. (1986)
	Lutenegger & Adams (1999)
	Chang & Zhu (2004)
	Xiao & Suleiman (2015)
Anchors	Lutenegger et al. (1988)
	Lutenegger & Miller (1994)
	Miller et al. (1998)
	Asoudeh & Oh (2014)
Streambank & Riverbank Stability	Thorne (1981)
	Casagli et al. (1999)
	Darby et al. (2000)
	Borg et al. (2014)

4. The test is not particularly sensitive to borehole disturbance;
5. The test is easy to perform, and the equipment can be transported by hand to test remote locations;
6. The equipment requires no external power;
7. Testing takes place in a small zone of soil, which reduces soil variability;
8. The test results require a minimal interpretation.

8.13.2 Limitations

The potential limitations of the BST are as follows:

1. A borehole is required;
2. If gravel content exceeds about 10%, it may be impossible to secure an adequate hole for testing or the test may give erratic results;
3. Cohesion exceeding about 70 kPa will keep the plate teeth from seating. In this case, φ' will generally be too high and c' too low. Special shear plates may be required;
4. Drainage conditions are inferred from the data and by retesting with different consolidation times. The alternative is to use a pore pressure-measuring device, which can be expensive;
5. Special testing procedures may be needed in sensitive soils.

8.14 SUMMARY OF BST

The BST is the only *in situ* test that is available for rapidly obtaining a direct measurement of the Mohr-Coulomb failure envelope of soils and soft rocks. The equipment is

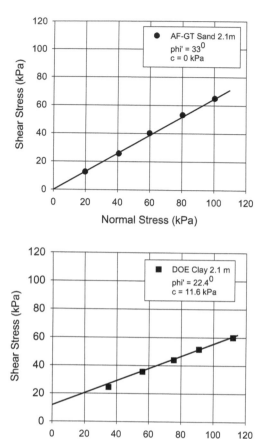

Figure 8.11 Typical BST results in sand and clay.

reliable, the test is quick and easy to conduct, and the results are usually very repeatable. Applications of the test results have demonstrated that the data are useful for a number of common geotechnical design situations. Figure 8.11 presents the results of BSTs obtained by the author in sand and clay.

REFERENCES

Ashlock, J. and Lu, N., 2012. Interpretation of Borehole Shear Strength Tests of Unsaturated Loess by Suction Stress Characteristic Curves. *GeoCongress 2012, ASCE*, pp. 2562–2571.

Asoudeh, A. and Oh, E., 2014. Strength Parameter Selection in Stability Analysis of Residual Soil Nailed Walls. *International Journal of Geomaterials*, Vol. 7, No. 1, pp. 950–954.

Bauer, G.E. and Demartinecourt, J.P., 1985. The Application of the Modified Borehole Shear Device to a Sensitive Clay. *Geotechnical Engineering*, Vol. 16, pp. 167–189.

Bechai, M., Law, K.T., Cragg, C.B.H., and Konard, J.M., 1986. In Situ Testing of Marine Clay for Towerline Foundations. *Proceedings of the 39th Canadian Geotechnical Conference*, pp. 115–119.

Borg, J., Dewoolkar, M.M. and Bierman, 2014. Assessment of Streambank Stability. *GeoCongress 2014, ASCE*.

Casagli, N., Rinaldi, M., Gargini, A. and Curini, A., Pore Water Pressure and Streambank Stability: Results from a Monitoring Site on the Sieve River, Italy. *Earth Surface Processes and Landforms*, Vol. 24, pp. 1095–1114.

Chang, M.F. and Zhu, H., 2004. Construction Effect on Load Transfer along Bored Piles. *Journal of Geotechnical and Geoenvironmental Engineering, ASCE*, Vol. 130, No. 4, pp. 426–437.

Darby, S.E., Gessler, D. and Thorne, C.R., 2000. Computer Program for Stability Analysis of Steep Cohesive Riverbanks. *Earth Surface Processes*, Vol. 25, pp. 175–190.

Demartinecourt, J.P. and Bauer, G.E., 1983. The Modified Borehole Shear Device. *Geotechnical Testing Journal, ASTM*, Vol. 6, No. 1, pp. 24–29.

Donchev, P. and Lutenegger, A.J., 1984. In Situ Shear Strength of North-Bulgaria Loess by Borehole Shear Test. *Proceedings of the 6th Budapest Conference on Soil Mechanics and Foundation Engineering*, pp. 45–50.

Farouz, E., Karnik, B., and Stanley, R., 2007. Case Study: Optimization and Monitoring of Slope Design in Highly Weathered Shale. *Proceedings of the 7th International Symposium on Field Measurements in Geomechanics*, pp. 1–12.

Fox, N.S., Lohnes, R.A., and Handy, R.L., 1966. Depth Studies of Wisconsin Loess in Southwestern Iowa: IV, Shear Strength. *Proceedings of the Iowa Academy of Sciences*, Vol. 73, pp. 198–204.

Handy, R.L., 1975. Discussion: Measurement of In-situ Shear Strength. *Proceedings of the Conference on In-Situ Measurement of Soil Properties, ASCE*, Vol. 2, pp. 143–149.

Handy, R.L., 1986. Borehole Shear Test and Slope Stability. *Use of In Situ Tests in Geotechnical Engineering, ASCE*, pp. 161–175.

Handy, R.L. and Fox, N.S., 1967. A Soil Borehole Direct Shear Test Device. Highway Research News, No. 27, pp. 42–51.

Handy, R.L. and Williams, W.W., 1967. Chemical Stabilization of an Active Landslide. *Civil Engineering*, Vol. 37, No. 8, pp. 62–65.

Handy, R.L., Pitt, J.M., Engle, L.E., and Klochow, D.E., 1976. Rock Borehole Shear Test. *Proceedings of the 17th Symposium on Rock Mechanics*, pp. 4B6-1 to B6–11.

Handy, R.L., Schmertmann, J.H., and Lutenegger, A.J., 1985. Borehole Shear Tests in a Shallow Marine Environment. ASTM Special Technical Publication 883, pp. 140–153.

Haramy, K.Y. and Demarco, M.J., 1983. Use of the Borehole Shear Tester in Pillar Design. *Proceedings of the 20th U.S. Symposium on Rock Mechanics*, pp. 639–644.

Irigoyen, A. and Coduto, D., 2015. Shear Strength of Unsaturated Soils Using the Borehole Shear Test. *Proceedings of the 16th European Conference on Soil Mechanics and Geotechnical Engineering*, pp. 2933–2938.

Johns, S.B.P., 1980. In-Situ Measurement of Soil Properties, Final Rept. No. FHWA-CA-TL-80-13.

Kong, L., Li, J., Guo, A., and Zhou, Z., 2017. Influence of Loading Mode and Flooding on In Situ Strength of Expansive Soils by Borehole Shear Tests. *Proceedings of the 19th International Conference on Soil Mechanics and Geotechnical Engineering*, pp. 1197–1200.

Lambrechts, J.R. and Rixner, J.J., 1981. Comparison of Shear Strength Values Derived from Laboratory Triaxial, Borehole Shear and Cone Penetration Tests, ASTM Special Technical Publication 740, pp. 551–565.

Little, W.C., Thorne, C.R., and Murphy, J.B., 1982. Mass Bank Failure Analysis of Selected Basin Streams. *Transactions of the American Society of Agricultural Engineers*, Vol. 25, No. 5, pp. 1321–1328.

Lohnes, R.A. and Handy, R.L., 1968a. Slope Angles in Friable Loess. *Journal of Geology*, Vol. 76, pp. 247–258.

Lohnes, R.A. and Handy, R.L., 1968b. Shear Strength of Some Hawaiian Latosols. *Proceedings of the 6th Annual Engineering Geology and Soils Engineering Symposium*, pp. 64–79.

Lohnes, R.A., Millan, A., Demirel, T., and Handy, R.L., 1972. Tests for Soil Creep. Highway Research Record No. 405, pp. 24–33.

Lutenegger, A.J., 1983. Discussion of The Modified Borehole Shear Device. *Geotechnical Testing Journal, ASTM*, Vol. 6, No. 3, p. 161.

Lutenegger, A.J., 1987a. Suggested Method for Performing the Borehole Shear Test. *Geotechnical Testing Journal, ASTM*, Vol. 10, No. 1, pp. 19–25.

Lutenegger, A.J., 1987b. In Situ Shear Strength of Friable Loess, CATENA Supplement No. 9, pp. 27–34.

Lutenegger, A.J. and Adams, M.T., 1999. Tension Tests on Bored Piles in Residual Soil. *Deep Foundations in Residual Soils, ASCE*, pp. 43–53.

Lutenegger, A.J. and Hallberg, G.R., 1981. Borehole Shear Test in Geotechnical Investigations. ASTM Special Technical Publication 740, pp. 566–578.

Lutenegger, A.J. and Miller, G.A., 1994. Uplift Capacity of Small Diameter Drilled Shafts in Stiff Clay. *Journal of Geotechnical Engineering, ASCE*, Vol. 120, No. 8, pp. 1362–1380.

Lutenegger, A.J. and Powell, J.J.M., 2008. Borehole Shear Tests in Stiff London and Gault Clay. *Proceedings of the 3rd International Symposium on Site Characterization*, pp. 714–723.

Lutenegger, A.J. and Tierney, K.F., 1986. Pore Pressure Effects in Borehole Shear Testing. *Use of In Situ Tests in Geotechnical Engineering, ASCE*, pp. 752–764.

Lutenegger, A.J. and Timian, D.A., 1987. Reproducibility of Borehole Shear Test Results in Marine Clay. *Geotechnical Testing Journal, ASTM*, Vol. 113, No. 1, pp. 13–18.

Lutenegger, A.J., Remmes, B.D., and Handy, R.L., 1978. Borehole Shear Test for Stiff Soils. *Journal of the Geotechnical Engineering Division, ASCE*, Vol. 104, No. GT11, pp. 1403–1407.

Lutenegger, A.J., Smith, B.L., and Kabir, M., 1988. Use of In Situ Tests to Predict Uplift Performance of Multihelix Anchors. *Special Topics in Foundations, ASCE*, pp. 93–109.

Millan, A.A. and Escobar, L.E., 1987. Use of the B.S.T. in Volcanic Soils. *Proceedings of the 8th Pan American Conference on Soil Mechanics and Foundation Engineering*, pp. 101–113.

Miller, G.A., Azad, S., and Hassell, C.E., 1998. Iowa Borehole Shear Testing in Unsaturated Soils. *Proceedings of the 1st International Conference on Site Characterization*, Vol. 2, pp. 1321–1326.

Ogunro, V., Anderson, B., Starnes, J., and Burrage, R., 2008. Characterization and Geotechnical Properties of Piedmont Residual Soils. *GeoCongress 2008, ASCE*, pp. 44–51.

Ruenkrairergsa, T. and Pimsarn, T., 1982. Deep Hole Lime Stabilization for Unstable Clay Shale Embankment. *Proceedings of the 7th Southeast Asian Geotechnical Conference*, pp. 631–645.

Sakamoto, M.Y., Guesser, L.H., Contessi, R.J., Higashi, R.A., Muller, V.S. and Sbrolglia, R.M., 2018. Use of the Borehole Shear Test Method for Geotechnical Mapping of Landslide Risk Areas. *Landslides and Engineered Slopes: Experience, Theory and Practice. Proceedings of the International Symposium on Landslides*, Naples, Italy, 9 pp.

Schmertmann, J.H., 1975. Measurement of In Situ Shear Strength. *Proceedings of the Conference on In Situ Measurement of Soil Properties, ASCE*, Vol. 2, pp. 57–138.

Singh, A. and Bhargava, S.N., 1979. Shear Test in a Bore Hole. *Proceedings of the International Symposium on In Situ Testing of Soil and Rocks and Performance of Structures*, Roorkee, pp. 417–421.

Thorne, C.R., 1981. Field Measurements of Rates of Bank Erosion and Bank Material Strength. *Erosion and Sediment Transport Measurement*, pp. 503–512.

Tice, J.A. and Sams, C.E., 1974. Experiences with Landslide Instrumentation in the Southeast. Transportation Research Record, No. 482, pp. 18–29.

Wineland, J.D., 1975. Borehole Shear Device. *Proceedings of the Conference on In Situ Measurement of Soil Properties, ASCE*, Vol. I, pp. 511–522.

Xiao, S. and Suleiman, M., 2015. Investigation of Thermo-Mechanical Load Transfer (t-z curves) Behavior of Soil-Energy Pile Interface Using Modified Borehole Shear Tests. *Proceedings IFCEEE 2015*, pp. 1658–1667.

Yang, H., White, D.J., and Schaefer, V., 2006. In-Situ Borehole Shear Test and Rock Borehole Shear Test for Slope Investigations. *Site and Geomaterial Characterization, ASCE*, pp. 293–298.

Chapter 9

Plate Load Test (PLT) and Screw Plate Load Test (SPLT)

9.1 INTRODUCTION

Plate Load Tests (PLTs) are one of the simplest of all *in situ* tests to perform and to interpret, and yet they are perhaps one of the most underutilized tests by the profession. Historically, a rigid circular plate has been used near the surface to apply a vertical load to the soil, as shown in Figure 9.1. Prior to the development of "modern" techniques for investigating soils and determining soil properties, PLTs were used to make a direct assessment of the load-bearing characteristics of soils for shallow foundations. Reports of load tests are available beginning at the turn of the century and continuing up to the present day (e.g., ENR 1922, 1932).

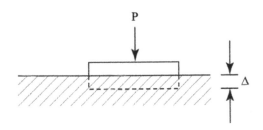

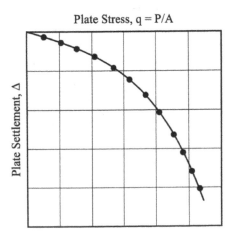

Figure 9.1 Concept of plate load test (PLT).

PLTs are considered both a specific property measurement test and a prototype test in which the load vs. deformation performance of a rigid plate is determined for use in the design of shallow foundations or end bearing of deep foundations. Screw plate load tests (SPLTs) have generally been used for specific property measurement because of their smaller size.

One advantage of PLTs is the ability to evaluate the mass soil behavior since a large volume of soil is involved in the plate response. Although PLTs may be used in virtually any soil type, the test is particularly useful for evaluating the behavior of structurally dependent soils and soils that may not allow undisturbed sampling or where scale effects may significantly influence the interpretation of strength and stiffness.

9.2 PLATE LOAD TEST

The PLT is highly versatile; the plates may be either round or square; they may vary in dimension to suit the soil and the project; they may be constructed of steel or concrete. Tests may be conducted at the ground surface or below grade. Typical sizes of plates for most routine testing range from 0.15 m (6 in.) to 0.91 m (36 in.). The equipment and procedure for conducting the PLT is described by ASTM Method D1194 *Standard Test Method for Bearing Capacity of Soil for Static Load and Spread Footings*. Table 9.1 gives a summary of some reported uses of PLTs in different soil materials.

9.2.1 Equipment

The equipment used to conduct a PLT is simple, and consists of a rigid loading plate, a reaction system, a loading jack (usually hydraulic), a load cell, and a reference beam and settlement measuring system. Steel plates may be used provided that they are sufficiently stiff to represent a rigid loading. ASTM D1194 requires that steel plates must be at least 25.4 mm thick. Rigid loading may be accomplished by using a series of stacked plates or by using plates with stiffening ribs, as shown in Figure 9.2.

Reaction can be provided by a number of different systems, as shown in Figure 9.3. In most cases, the reaction is provided by dead load (Figure 9.3a); anchor piles or grouted anchors (Figure 9.3b); and helical anchors (Figure 9.3c). In the case of large plate tests, in which a prototype-scale or full-scale concrete footing is used, the reaction may be provided by a central internal anchor, as shown in Figure 9.3d.

The applied load is measured using an independent calibrated load cell. The pressure being applied to the hydraulic jack may also be monitored using a pressure gauge; however, this should be considered a backup measurement of load and not the primary measurement of load. Settlement of the plate is usually measured by analog or digital dial gauges or by LVDTs. A reference beam or some other form of independent reference is needed to measure the plate deformation during the test. The use of electronic load and settlement equipment also allows for an automated data acquisition; however, the test can be kept very simple.

9.2.2 Test Procedures

The procedure used to conduct the PLTs is relatively simple; prepare the soil surface so that the plate rests uniformly on the soil; arrange a loading system to provide a sufficient reaction; and provide a means for measuring plate deformation. Normally, the test is conducted

Table 9.1 Summary of some reported uses of plate load tests in different materials

Soil type	References
Residual soils	Chin (1983)
	Filho & Celso (1983)
	Ferreir & Teixeira (1989)
	Consoli et al. (1998)
Gravels	Wrench (1984)
	Wrench & Nowatzki (1986)
Sand	Ismael (1987)
	Ismael (1996)
	Kesharwani et al. (2015)
Dense glacial till	Klohn (1965)
	Soderman et al. (1968)
	Radhakrishna & Klym (1974)
	McKinlay et al. (1974)
	Fisher (1983)
Soft clay	Housel (1929)
	Bergado & Chang (1986)
Stiff clay	Burland et al. (1966)
	Ertel (1967)
	Lo et al. (1969)
	Marsland 1971)
	Marsland (1977)
	Bauer et al. (1973)
	Marsland & Randolph (1977)
	Tand et al. (1986)
	Lefebvre et al. (1987)
	Bauer & Tanaka (1988)
	Marsland & Powell (1991)
Peat	Landva (1986)
	Nichols et al. (1989)
Waste fill	Eliassen (1942)
	Landva & Clark (1990)
	Watts & Charles (1990)
	Van Impe (1998)
Bedrock	Ward et al. (1968)
	Seychuk (1970)
	Rozsypal (1983)
	Marsland & Butcher (1983)
	Lo & Cooke (1989)

in a load-controlled manner with loads being applied incrementally and the deformation monitored over time under each load increment. ASTM D1194 suggests that equal load increments of no more than 95 kPa (1.0 tsf) should be used or alternatively, loads of not more than 10% of the estimated ultimate bearing capacity may be used. The load is to be maintained for a minimum of 15 min.

ASTM recommends that the test is typically continued until "a peak load is reached or until the ratio of load increment to settlement increment reaches a minimum, steady magnitude". Unless a distinct, well-defined failure occurs, the test should be conducted to a point where a minimum settlement of 10% of the plate diameter or width has occurred.

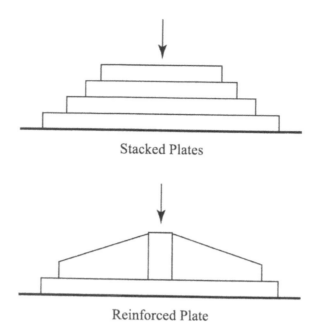

Figure 9.2 Typical geometry of steel plates for PLT.

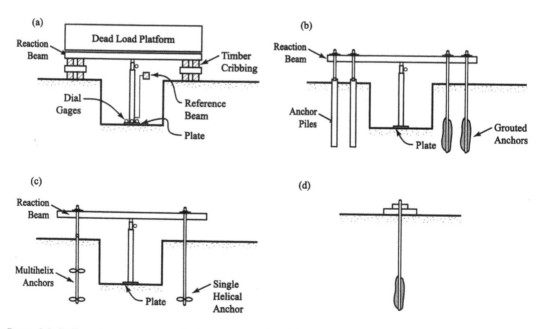

Figure 9.3 Different arrangements for load reaction for PLT.

An alternative deformation-controlled loading procedure is also allowed by ASTM as follows: apply the load increments corresponding to approximately 0.5% of the plate diameter; after applying each settlement increment, measure the load at some fixed time intervals, e.g., 30 s, 1, 2, 4, 8, 15 min, until the variation of the load stops or until the rate of the variation of the load on a load vs. log-time scale becomes linear.

9.2.2.1 Tests on the Ground Surface

Tests performed on the ground surface are the easiest to arrange since all of the equipment is readily accessible. Most surface tests are used to evaluate the deformation characteristics of fills and for the support of structural slabs. In some cases, the dead weight reaction may be supplied by heavy equipment, such as a fully loaded dump truck, provided that the vehicle does not interfere with the plate and influence results.

9.2.2.2 Tests in an Excavation/Test Pit

Occasionally, PLTs may be performed in excavations, as shown in Figure 9.4b. ASTM D 1194 states that "the distance between test locations shall not be less than five times the diameter of the largest plate used in the tests". In general, the test equipment and procedure is the same as used for surface tests. The primary purpose of performing tests in an excavation is to be able to evaluate the soil behavior in specific strata or at specific elevations that may correspond to the anticipated foundation locations.

9.2.2.3 Tests in Lined Borings

A special type of PLT may be performed in lined borings, and used to determine soil properties at various depths, as shown in Figure 9.4c. Marsland (1975) described the deep large-diameter plate tests performed using special test equipment, shown in Figure 9.5. A large-diameter steel casing would typically be used as the liner. Using this equipment, tests may be performed at any depth as the drilling proceeds to evaluate specific soil behavior in different strata.

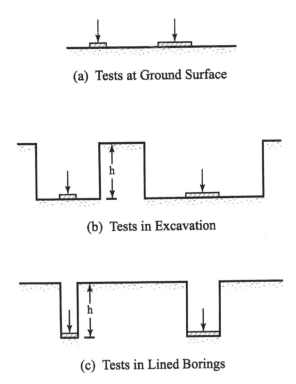

(a) Tests at Ground Surface

(b) Tests in Excavation

(c) Tests in Lined Borings

Figure 9.4 PLTs performed (a) on ground surface; (b) in excavations, or (c) in lined borings.

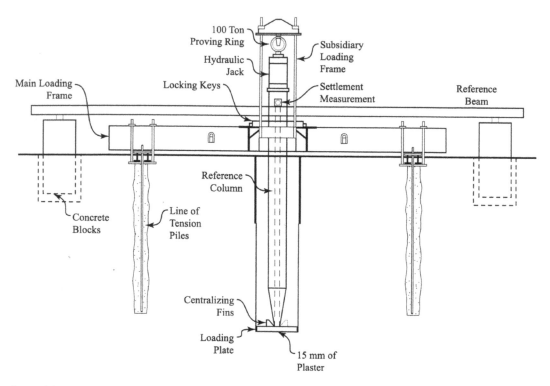

Figure 9.5 Equipment for PLTs in deep boreholes. (After Marsland 1971.)

9.2.2.4 Horizontal Plate Load Tests

In some cases, it may be desirable to conduct PLTs in the horizontal direction, either in an open trench, along the walls of a bored hole, or in tunnels to obtain a measure of soil deformation characteristics in the horizontal direction, as shown in Figure 9.6. In this case, the test is performed by jacking against opposite sides of the excavation walls. The use of lateral PLTs to evaluate lateral soil stiffness has been reported by a number of investigators

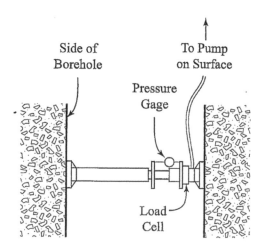

Figure 9.6 Typical arrangement for performing horizontal PLTs.

(e.g., Slack & Walker 1970; Sherif & Strazer 1973; Radhakrishna & Klym 1974; Reddy et al. 1979; Wrench & Nowatzki 1986).

9.3 SCREW PLATE TESTS

One potential problem with traditional plate tests is that in some soils, the depth of investigation may be limited, simply because an excavation may be difficult. A variation of the PLT, the SPLT, has been used to overcome this problem and evaluate engineering properties in a wide range of soils. The SPLT allows loading to be performed without an excavation or a borehole, although sometimes the test is performed below an open borehole. In this test, the plate is constructed of a single helical flight with a shallow pitch that is screwed into the ground. According to Strout & Senneset (1998), the SPLT originated in Norway in 1953 where it is known as the "field compressometer test".

Normally, the SPLT is controlled at the ground surface, and the load and displacement are measured just as in the conventional PLT. Schmertmann (1970b) presented a suggested method for the SPLT; however, it appears that this test was never adopted by ASTM as a standard. Table 9.2 gives a summary of different soils in which SPLTs have been reported.

SPLTs may be particularly suited to soils where undisturbed sampling is difficult. Tests in clays may require long testing times to obtain a drained behavior because of the slow rate of consolidation. Rapid loading tests may be performed in clays for evaluating undrained shear strength.

9.3.1 Equipment

Figure 9.7 shows a typical arrangement of the SPLT. The equipment usually consists of some types of surface reaction frame and a helical screw plate of one revolution of helix attached to a load rod. Usually, either a hydraulic jack or a mechanical screw jack is used to apply the load. The load is measured with a load cell or a proving ring and deformation is measured using an independent reference beam. The test setup is similar in many respects to the conventional PLT.

Table 9.2 Summary of some reported uses of screw plate tests

Soil type	References
Very soft sensitive clay	Janbu & Senneset 1973 Schwab & Broms (1977) Selvadurai et al. (1980) Selvadurai & Nicholas (1981)
Soft clay	Bergado et al. (1986) Bergado & Huan (1987) Bergado et al. (1990)
Stiff clay	Kay & Mitchell (1980) Kay & Avalle (1982) Kay & Parry (1982) Powell & Quarterman (1986) Bauer & Tanaka (1988) Noor et al. (2019)
Sand	Kummeneje & Eide (1961) Webb (1969) Schmertmann (1970a) Janbu & Senneset (1973) Dahlberg (1974)

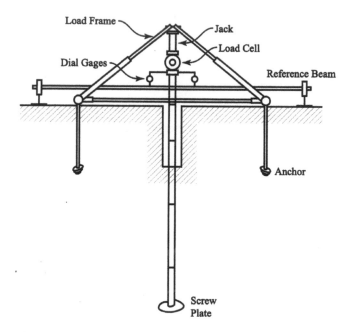

Figure 9.7 Schematic of typical screw plate load test.

Different diameter plates with different thickness and pitch have been used, generally depending on the anticipated stiffness of the material being tested. The thickness of the plate must be sufficient to provide for a rigid loading. Plate diameters range from 60 mm (2.36 in.) to 350 mm (13.75 in.), although screw plates on the order of 150 mm (6 in.) appear to be the most common. Table 9.3 summarizes some reported sizes of different screw plates used in different soils.

Kay & Avalle (1982) and Bergado et al. (1986) described test equipment that included a screw mechanism with the same pitch as the helical plate, as shown in Figure 9.8. Advance of the plate using this "pitch-matched" arrangement helps reduce disturbance of the soil.

The diameter of the plate must be large enough so that the test results accurately determine the soil behavior but small enough to be applicable over a wide range of soil conditions and still be installed with reasonable ease. Experience suggests that a plate with a projected horizontal area on the order of 0.093 m² (1 ft²) (diameter = 172 mm (0.56 ft)) will meet this criteria and provide a sufficient rigidity.

9.3.2 Test Procedures

SPLTs are usually conducted as incremental loading tests, where a constant load is applied and the deformation is recorded at various time intervals. The test can include one or more unload-reload cycles in order to more accurately estimate elastic modulus. The test may also be performed as a continuous loading test. Schmertmann (1970b) recommended that to avoid disturbance, the advance of the screw plate should be controlled such that the downward movement of the plate for each complete revolution should be equal to the pitch. It was also recommended that a vertical testing interval of at least three plate diameters be used.

Table 9.3 Different-sized screw plates used

Soil	Plate diameter (mm)	Plate thickness (mm)	Pitch	References
Soft clay	160 & 300	N/A	N/A	Schwab & Broms (1977)
	150 & 180	N/A		Berg & Olsson (1978)
	350		N/A	Selvadurai et al. (1980)
	295	N/A		Selvadurai & Nichols (1981)
	254	6	0.12D	Bergado et al. (1986) Bergado & Huan (1987) Bergado et al. (1990)
	150 & 300	N/A	N/A	Mital & Bauer (1989)
Stiff clay	88	N/A	N/A	Kay & Mitchell (1980)
	100	N/A	0.2D	Kay & Parry (1982) Kay & Avalle (1982)
	100	N/A	0.2D	Powell & Quarterman (1986)
	100	N/A	0.2D	Epps (1986)
	100		0.2D	Marsland & Powell (1991)
	200	N/A	N/A	Noor et al. (2019)
Sand	252	N/A	0.02D	Kummeneje & Eide (1961)
	152, 229 & 381	N/A	N/A	Webb (1969)
	172	4.3	0.2D	Schmertmann (1970a)
	160	N/A	0.28D	Janbu & Senesset (1973)
Clay & sand	75	1	0.24D	Lee et al. (2009) Kim et al. (2014)

9.4 PRESENTATION OF TEST RESULTS

The results of PLTs and SPLTs are typically presented as settlement vs. applied plate stress. Figure 9.9 shows the results of four PLTs performed on a compacted sand using plates of different diameter. It can be seen from these results that the response of the plates is different and that both the stiffness and the ultimate bearing capacity appear to be dependent on the size of the plate. The absolute settlement at a given level of applied stress is different for each plate.

It may also be useful to present results in a normalized form where the plate settlement is normalized by the plate diameter, also shown in Figure 9.10. This approach allows a comparison of tests made using plates of different diameter.

9.5 INTERPRETATION OF RESULTS

As previously indicated, the results of PLTs may be used as a prototype-scale shallow foundation or they may be used to provide estimates of specific soil properties, such as stiffness and strength. The use of PLTs as prototype foundation tests will be discussed in Section 9.6. In this section, the evaluation of specific properties will be described.

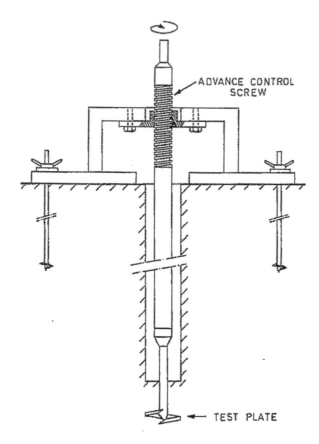

Figure 9.8 Screw mechanism for advancing screw plate. (From Kay & Avalle 1982.)

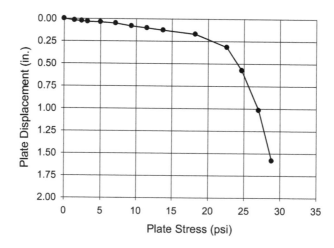

Figure 9.9 Typical PLT in compacted sand.

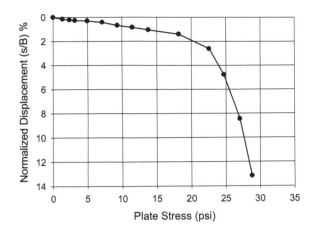

Figure 9.10 Normalized results from PLT.

9.5.1 Subgrade Reaction Modulus

The modulus of subgrade reaction, k_s, may be determined directly from the results of PLTs. This value may be useful for the design of mat foundations and pavements, and has been extensively used by structural engineers. The modulus of subgrade reaction is defined as the slope of the stress vs. settlement curve as:

$$k_s = q/\delta \qquad (9.1)$$

where

    q = applied plate stress
    δ = plate settlement

Units of k_s are in force/length^3.

Subgrade Reaction Modulus is not really a constant but depends on the definition. Several methods have been used for calculating k_s based on the results of PLTs. It appears that there is no standard procedure. For example, the U.S. Army Corps of Engineers has defined the value of k_s at a fixed applied plate stress of 68.9 kPa (10 psi), while the Portland Cement Association has defined the value of k_s at a fixed plate deflection of 1.27 mm (0.05 in.). Neither of these approaches truly gives the elastic response of the plate for all soil conditions, and therefore, neither may be desirable.

In some cases, the value of the subgrade reaction modulus has been defined as a secant slope at plate stresses corresponding to ½ to 1/3 the ultimate stress, but this obviously requires that the ultimate stress be evaluated and therefore may be sensitive to the definition of ultimate bearing stress. Alternatively, the initial straight-line portion of the load curve may be used. However, this requires a number of small load increments at the beginning of the test. In any case, the definition of k_s appears at this time to be more or less arbitrary. Using the results from the 0.30-m (1 ft)-diameter plate shown in Figure 9.9, the calculated values of k_s are 142 lbs/in.^{3} for PCA and 111 lbs/in.^{3} for USACE.

There are two problems with the application of Equation 9.1: (1) the load-displacement curve of the most PLTs is nonlinear, and therefore, the value of k_s will be dependent on the settlement or stress level where it is defined; (2) test results show that the value of k_s, as defined by Equation 9.1, is dependent on the absolute size of the plate and that k_s decreases as the plate width increases.

To account for the decrease in subgrade reaction modulus with increasing plate size, Terzaghi (1955) suggested that the subgrade reaction modulus of full-scale foundations with width B, k_B, could be obtained from the results of a PLT conducted on a plate with a width of 0.3 m (1ft). This means that a value of k_s for any size foundation could be estimated on the basis of test results obtained with a 0.3-m (1 ft)-wide plate. For foundations on sandy soils, it was suggested that:

$$k_B = k_{0.3}\left[(B+0.3)/2B\right]^2 \tag{9.2}$$

where

$k_{0.3}$ = modulus of subgrade reaction determined using a 0.3-m (1 ft) plate
B = footing width (m)

For foundations on clays, Terzaghi (1955) suggested:

$$k_B = k_{0.3}(0.3/B) \tag{9.3}$$

A comparison of some actual test data with the recommendation of Terzaghi (1955) for a series of PLTs conducted by the author on compacted sand using the UASCE definition for k_s is shown in Figure 9.11. Figure 9.12 shows a comparison of the normalized values of k_s for these tests compared to Terzaghi's suggestion (Equation 9.2). Except for the smallest plate, the fit appears reasonable.

According to elastic theory and the Boussinesq equation, the settlement of a plate may be related to the average applied stress as:

$$\delta = \left[qB\left(1-\upsilon^2\right)\right]\!/E_s \tag{9.4}$$

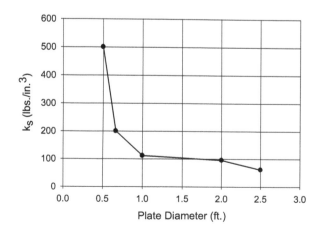

Figure 9.11 Values of k_s from PLTs conducted on compacted sand using different-sized plates.

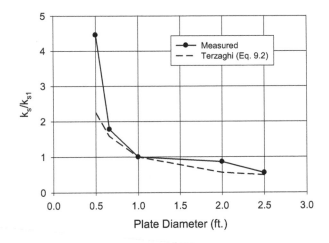

Figure 9.12 Normalized subgrade reaction modulus measured and predicted by Terzaghi (1955).

Since the coefficient of the subgrade reaction is defined from Equation 9.1 as:

$$k_s = q/\delta \qquad (9.5)$$

the subgrade reaction may be written in terms of the elastic modulus as:

$$k_s = E/\left[\delta B\left(1-\upsilon^2\right)\right] \qquad (9.6)$$

Equation 9.6 shows that the coefficient of the subgrade reaction depends on both the elastic properties of the soil and the size of the plate.

The approximation of Terzaghi (1955) assumes a constant value of elastic modulus, which is a simplification to the condition where the elastic modulus varies with depth. This nonhomogeneity has been addressed by a number of studies, including those of Carrier & Christian (1973), Rowe (1982), and Horvath (1983).

9.5.2 Elastic Modulus

9.5.2.1 Plate Load Test

The settlement of a rigid circular or square plate resting on the surface of an isotropic elastic material may be reasonably obtained from:

$$\delta = \left[qB\left(1-\upsilon^2\right)\right]/E_S \qquad (9.4)$$

where
  E_S = modulus of elasticity
  δ = settlement
  q = applied plate stress = load/area
  B = diameter of plate
  υ = Poisson's ratio

or

$$E_S = \left[(qB)(1 - v^2) \right] / \delta \tag{9.7}$$

For tests performed rapidly in saturated clays, the value of $v = 0.5$ and Equation 9.7 gives:

$$E_S = 0.75(qB)/\delta \tag{9.8}$$

Typically, the results of PLTs and footing load tests show that at bearing stresses up to about one-third of the ultimate capacity, the load-displacement behavior is approximately linear. Therefore, the elastic modulus should be obtained within this region of applied stress. Additionally, it may be advantageous to perform at least one unload-reload cycle in this region in order to more accurately obtain the reload modulus.

In sands, the value of Poisson's ratio may typically vary from about 0.1 to 0.33. For a value of $v = 0.3$ and for surface tests, Equation 9.8 gives:

$$E_S = 0.91(qB)/\delta \tag{9.9}$$

Equations 9.7–9.9 are applicable to PLTs performed on the surface. If the plate is embedded below the surface, then the surface equations must be modified by a depth reduction factor, μ_0. For a rigid plate, Equation 9.9 becomes:

$$E = \mu_0 \left(qB(1 - v^2) \right) / E_S \tag{9.10}$$

where
  μ_0 = depth reduction factor

Different values of μ_0 have been suggested by Burland (1970), Pells & Turner (1979), Donald et al. (1980), and Pells (1983). For surface tests, $\mu_0 = 1.0$. For PLTs performed at depths greater than about four times the plate diameter, the value of μ_0 may be taken as 0.87 (Burland 1970). The values of μ_0 as shown in Figure 9.13 are for PLTs performed in excavations that are the same size as the plate. If the excavation is larger than the plate, the reduction is less. Pells (1983) has suggested one solution for such a reduction as shown in Figure 9.14.

9.5.2.2 Screw Plate Test

Selvadurai et al. (1980) summarized a variety of numerical results derived from various theoretical models of the SPLT. In clays, the test is usually performed rapidly enough that full drainage does not occur. In this case, the *in situ* undrained elastic modulus, E_u, may be obtained from:

$$E_u = Kqr/\delta \tag{9.11}$$

where
  q = applied plate stress
  r = radius of the screw plate
  δ = plate settlement
  K = undrained modulus factor

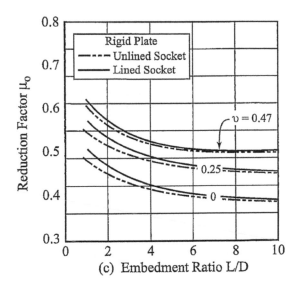

Figure 9.13 Modulus reduction factor for embedded plate having the same diameter as the shaft. (After Pells 1983.)

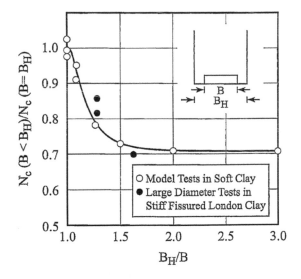

Figure 9.14 Modulus reduction factor for the embedded plate having larger diameter as the shaft. (After Pells 1983.)

Theoretical values of K range from 0.525 to 0.750 (Bergado & Huan 1987); however, a more realistic range, which is more applicable to the conditions of the SPLT, is K = 0.60 to 0.75.

Selvadurai & Nicholas (1979) recommended that the undrained elastic modulus from the SPLTs be defined as a secant modulus and taken at the point where a secant slope passes through a point on the bearing stress vs. settlement curve corresponding to one-half of the ultimate bearing stress. Using a value of K = 0.66 gives:

$$E_u = 0.66qr/\delta = 0.33qB/\delta \tag{9.12}$$

For the drained loading of clays, a drained value of Poisson's ratio = 0.20 may be more reasonable, which gives an expression for the drained elastic modulus as:

$$E_u = 0.42qB/S_{100} \tag{9.13}$$

The value of S_{100} is the settlement taken at the end of the test corresponding to 100% consolidation.

9.5.3 Shear Modulus

Marsland & Powell (1990, 1991) suggested that the shear modulus be calculated from the PLTs as:

$$G = 0.85(\Delta q/\Delta \delta)(\pi/8)(B)(1 - \upsilon) \tag{9.14}$$

where
  G = secant shear modulus
  $\Delta\delta$ = change in displacement for a given change in stress, Δq
  B = plate diameter
  υ = Poisson's ratio

Since the results of PLTs are nonlinear, the modulus decreases with increasing applied stress. Therefore, it is necessary to define the stress range over which the moduli are determined.

9.5.4 Undrained Shear Strength of Clays

The interpretation of PLT and SPLT results for estimating the undrained shear strength of clays requires the interpretation of the load vs. settlement curve to give the ultimate bearing stress, q_{ult}. In the absence of an obvious plunging failure condition, several methods have been suggested to evaluate the ultimate bearing stress from PLTs.

One method of interpreting the ultimate stress is to take the bearing pressure at some fixed relative plate displacement as the ultimate bearing pressure. For example, Skempton (1951) showed that for remolded clay, the ultimate bearing capacity of shallow foundations in clays is achieved at a displacement on the order of 3%–5% of the plate diameter. Powell & Quaterman (1986) used a bearing stress producing a settlement of 15% of the plate diameter for SPLTs. Most footing tests that are taken to large displacements indicate that the bearing capacity may be conveniently defined as the bearing stress producing a settlement of 10% of the footing width. The author suggests that this be applied to PLTs as well. Results from a 0.3-m (1 ft)-diameter PLT conducted in a stiff clay are shown in Figure 9.15. The 10% criterion indicates an ultimate load of about 19,750 lbs (q_{ult} = 25,950 lbs/ft²).

An alternative approach is to use a model that reasonably describes the pressure-settlement results and predicts an ultimate bearing pressure. This is a form of curve fitting, most commonly done using a transformed hyperbolic model. If the plate settlement/plate

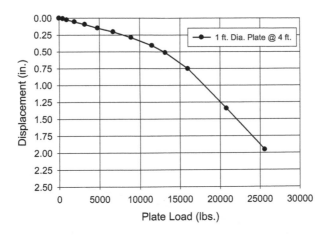

Figure 9.15 Results of PLTs in stiff clay.

stress is plotted as a function of settlement, the inverse slope of the linear portion of the curve gives the ultimate bearing stress. This model has previously been used to describe the load-displacement behavior of PLTs and footing tests (e.g., Chin 1983; Wrench & Nowatzki 1986; Wiseman & Zeitlan 1994; Thomas 1994). Figure 9.16 shows the PLT data from Figure 9.15 using this transformed model, indicating an ultimate load of 41,800 lbs. (q_{ult} = 52,900 lbs/ft²).

Kay & Parry (1982) used the hyperbolic model to extrapolating the load-displacement curve to obtain the ultimate plate capacity without actually plotting the data. By measuring the plate displacement at two points on the stress-displacement curve, they estimated the ultimate plate capacity as:

$$q_{ult} = 2.054\ q_2 - 1.54\ q_{1.5} \tag{9.15}$$

q_{ult} = ultimate stress
$q_{1.5}$ = plate stress at a displacement of 1.5% of the plate diameter
q_2 = plate stress at a displacement of 2% of the plate diameter

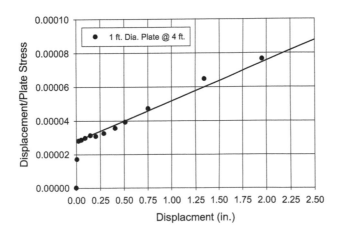

Figure 9.16 Transformed axes for hyperbolic estimation of ultimate plate bearing stress.

Using the interpreted ultimate bearing stress, the undrained shear strength may be estimated from a traditional bearing capacity equation, e.g., Skempton (1951). For PLTs performed on the surface:

$$s_u = (q_{ult})/N_C \qquad (9.16)$$

where

s_u = undrained shear strength
q_{ult} = interpreted ultimate bearing stress
N_C = shallow bearing capacity factor

The bearing capacity factor, N_C, for a surface footing is approximately 6.0 for a square or round plate.

For embedded plate tests or SPLTs beyond a relative embedment of about five times the plate width (diameter), the undrained shear strength is obtained from:

$$s_u = (q_{ult} - \sigma_v)/N_C \qquad (9.17)$$

where

s_u = undrained shear strength
q_{ult} = interpreted ultimate bearing stress
σ_v = total overburden stress at the depth of the test
N_C = deep bearing capacity factor

Equation 9.16 is also sometimes given as Equation 9.17 for SPLTs.

The bearing capacity factor for deep plate tests has traditionally been taken as $N_C = 9$, from the work of Skempton (1951); however, more rigorous analyses suggest that N_C varies from about 5.69 to 11.35 (e.g., Selvadurai et al. 1980; Bergado & Huan 1987). Several studies have shown that the undrained shear strength obtained from the SPLT or PLT is compared well with results from laboratory or other *in situ* tests.

9.5.5 Coefficient of Consolidation

Janbu & Senneset (1973) suggested that the results of SPLTs could be used to estimate the *in situ* coefficient of consolidation by using the time-settlement measurements. It was suggested that since the drainage is predominantly radial, the results provide an estimate of the horizontal coefficient of consolidation, c_h. In this procedure, the settlement is plotted vs. the square root of time for each load increment, as shown in Figure 9.17; a straight line is fitted to the initial portion of the curve and is projected back to the axis to give the corrected theoretical zero point. From this point, a second straight-line offset from the first line and having a slope 1.3:1 is constructed. This line intersects the curve at a point that approximately represents 90% primary consolidation, with the time corresponding to t_{90}. The value of c_h is obtained from:

$$c_h = (T_{90}R^2)/t_{90} = 0.335(R^2/t_{90}) \qquad (9.18)$$

where

c_h = coefficient of radial consolidation
T_{90} = theoretical time factor for 90% consolidation

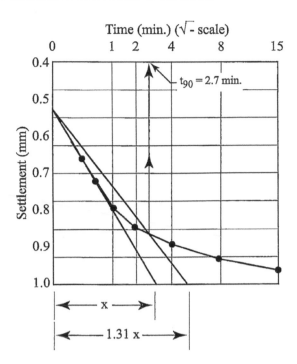

Figure 9.17 Graphical estimation of coefficient of consolidation from screw plate.

R = plate radius
t_{90} = interpreted time for 90% consolidation

Kay & Avalle (1982) observed that in many cases, the test data show a more S-shaped curve. They felt that the radial drainage model was not appropriate and suggested that interpretations of the coefficient of consolidation be made using three-dimensional isotropic consolidation. The suggested construction procedure presented by Kay & Avalle (1982) using the settlement vs. square root of time plot is as follows:

1. Ignore the initial data points and extrapolate the reverse curve portion of the plot to zero on the time axis. This represents the point of zero percent drained settlement, t_0.
2. Draw a straight line from t_0 tangent to the settlement vs. square root of time plot.
3. Construct a line from t_0 with a slope of 1.28:1 flatter to the first line. This line intersects the curve at a point representing 70% consolidation, t_{70}, and gives t_{70}.

The coefficient of consolidation is given as:

$$c_v = \left(1.24R^2\right)/t_{70} \qquad (9.19)$$

This procedure is illustrated in Figure 9.17. Since the test results may be evaluated for each loading increment, the results give an interpretation of the coefficient of consolidation over a range of applied stresses. This means that the results may be plotted as a coefficient of consolidation vs. stress and may be used to estimate values for any stress level (Figure 9.18).

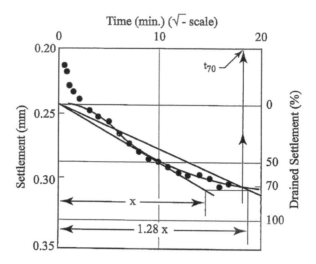

Figure 9.18 Graphical estimation of coefficient of consolidation from screw plate.

9.6 PLATE LOAD AS A PROTOTYPE FOOTING

In the past, the results of PLTs have occasionally been used to estimate the response of footings by the use of the subgrade reaction modulus, as previously described. The use of the PLT might seem attractive for predicting the behavior of shallow foundations, particularly on granular soils, since the plate more or less acts as a prototype foundation and load is applied in the same direction as anticipated by the foundation. In general, the settlements measured in the PLT are extrapolated to larger footings, and depending on any assumptions made, there may be errors associated with the extrapolation. The zone of influence under a small loaded plate is shallow as compared to a full-sized footing, and therefore, the results from a small plate may not include any influence of soil nonuniformity.

A direct method of applying PLTs to foundation design may be obtained by applying the concepts of subgrade reaction presented by Terzaghi (1955) and was suggested by Terzaghi & Peck (1967), who proposed a relationship between the settlement of a footing of width B and the settlement of a 0.3-m (1ft)-wide plate under the same applied stress as:

$$\delta_B/\delta_{0.3} = \left(2B/(B+0.3)\right)^2 \tag{9.20}$$

where
  B = footing width (meters)
  δ_B = settlement of a footing of width B
  $\delta_{0.3}$ = settlement of a plate with width = 0.30 m (1 ft)

Equation 9.20 indicates that for very large footings, on the order of B > 8 m, the settlement ratio tends to a maximum value of about 4.

Extrapolation of the behavior from a small plate to a large footing using Equation 9.20 uses the settlement ratio of different size footings or plates obtained at the same stress level. However, it should be expected that the same stress level will represent different conditions depending on the plate size. Similarly, the absolute settlement observed by different size footings has a different meaning depending on the plate size, especially for granular soils. This suggests that the use of normalized behavior may be a better approach to estimating the full-sized production footing performance from the small-sized plate.

Figure 9.19 (upper) shows the results of two of different-sized PLTs conducted on a uniform compacted sand. Figure 9.19 (lower) shows the same tests with settlement normalized by the plate diameter and applied stress normalized by the failure stress (taken as the stress producing a settlement of 10% of the plate diameter). The test results from different plate sizes show a more general behavior when represented in this manner. This means that the load-settlement behavior of any size plate could be obtained from a test conducted on any other size. A discussion of this approach has been presented by Lutenegger & Adams (2003).

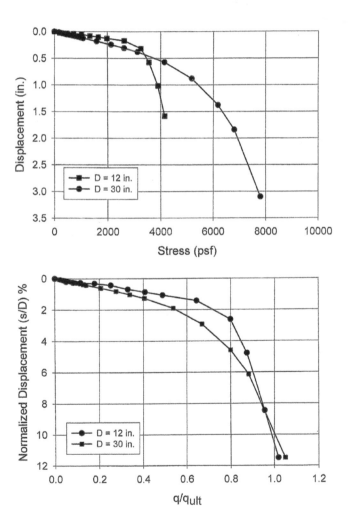

Figure 9.19 Stress-displacement behavior of two PLTs on compacted sand (upper), and normalized behavior (lower).

9.7 SUMMARY OF PLT AND SPLT

PLTs and SPLTs offer the potential for evaluating load-deformation behavior for use in both specific property evaluation and prototype shallow foundations. The test equipment and procedures are relatively simple and can be applied to a wide range of site and soil conditions. The interpretation of the test results is straightforward. Engineers may find the tests very useful in situations where deformation is the primary design issue.

REFERENCES

Bauer, G.E. and Tanaka, A., 1988. Penetration Testing of a Desiccated Clay Crust. *Proceedings of the 1st International Symposium on Penetration Testing*, Vol. 1, pp. 477–488.

Bauer, G.E., Scott, J.D., and Shields, D.H., 1973. The Deformation Properties of a Clay Crust. *Proceedings of the 8th International Conference on Soil Mechanics and Foundation Engineering*, Vol. 1, pp. 31–38.

Berg, G. and Olsson, O., 1978. Screw Plate Loading Tests on Gothenburg Clay – Deformation and Strength Properties Under Static and Repeated Loading. Swedish Geotechnical Institute Internal Report.

Bergado, D. and Chang, C., 1986. Correlation of Soil Parameters from Pressuremeter, Plate Load Test, and Screw Plate Tests Including Index Properties and Compressibility Characteristics of Soft Bangkok Clay. *Specialty Geomechanics Symposium*, Adelaide, pp. 10–16.

Bergado, D. and Huan, N., 1987. Undrained Deformability and Strength Characteristics of Soft Bangkok Clay by the Screw Plate Test. *Geotechnical Testing Journal, ASTM*, Vol. 10, No. 3, pp. 113–122.

Bergado, D., Khaleque, M., Neeyapan, R., and Chang, C., 1986. Correlations of In Situ Tests in Bangkok Subsoils. *Geotechnical Engineering*, Vol. 17, pp. 1–37.

Bergado, D.T., Chong, K.C., Daria, P.A.M., and Alfaro, M.C., 1990. Deformability and Consolidation Characteristics of Soft Bangkok Clay Using Screw Plate Tests. *Canadian Geotechnical Journal*, Vol. 27, pp. 531–545.

Burland, J.B., 1970. Contribution to Discussion. *Proceedings of the Conference on In Situ Investigations in Soils and Rocks*, pp. 61–62.

Burland, J., Butler, F., and Dunican, P., 1966. The Behaviour and Design of Large Diameter Bored Piles in Stiff Clay. *Proceedings of the Symposium on Large Bored Piles, ICE*, pp. 51–71.

Carrier, W.D. and Christian, J.T., 1973. Rigid Circular Plate Resting on a Non-Homogeneous Elastic Half-Space. *Geotechnique*, Vol. 23, No. 1, pp. 67–84.

Chin, F.K., 1983. Bilateral Plate Bearing Tests. *Proceedings of the International Symposium on In Situ Testing of Soil and Rock*, Paris, Vol. 2, pp. 39–41.

Consoli, N.C., Schnaid, F., and Milititsky, J., 1998. Interpretation of Plate Load Tests on Residual Soil Site. *Journal of Geotechnical and Geoenvironmental Engineering, ASCE*, Vol. 124, No. 9, pp. 857–867.

Dahlberg, R., 1974. Penetration Pressure and Screw Plate Tests in a Preloaded Natural Sand Deposit. *Proceedings of the 1st European Symposium on Penetration Testing*, Vol. 2.2, pp. 69–87.

Donald, I.B., Sloan, S.W., and Chiu, H.K., 1980. Theoretical Analysis of Rock Socketed Piles. *Proceedings of the Symposium on Structural Foundations on Rock*, pp. 303–316.

Eliassen, R., 1942. Load-Bearing Characteristics of Landfills. *Engineering News Record*, Vol. 108, pp. 103–105.

Engineering News Record, 1922. Foundation Tests for Nebraska State Capitol. *ENR*, Vol. 89, No. 15, pp. 606–609.

Engineering News Record, 1932. Load Tests on Hardpan. *ENR*, Vol. 108, p. 330.

Epps, R., 1986. A Comparison of Methods of Testing at a Site in London Clay. Geological Survey, Engineering Special Publication No. 2, pp. 207–212.

Ertel, W., 1967. Determination of the Initial Strength of Still Tertiary Frankfort Clay. *Proceedings of the Geotechnical Conference Oslo on Shear Strength Properties of Natural Soils and Rocks.* Vol. 1, pp. 109–111.

Filho, P.R. and Celso, R., 1983. Residual Soil Elastic Properties from In Situ Tests. *Proceedings of the International Symposium on In Situ Testing of Soil and Rock*, Paris, Vol. 2, pp. 27–34.

Fisher, W.R., 1983. Modulus of Elasticity of a Very Dense Glacial Till Determined by Plate Load Tests. *Canadian Geotechnical Journal*, Vol. 20, pp. 186–191.

Ferreira, S.R.M. and Teixeira, D.C.L., 1989. Collapsible Soil – A Practical Case in Construction (Pernambuco Brazil). *Proceedings of the 12th International Conference on Soil Mechanics and Foundation Engineering*, Vol. 1, pp. 603–606.

Horvath, J.S., 1983. Modulus of Subgrade Reaction: New Perspective. *Journal of Geotechnical Engineering, ASCE*, Vol. 109, No. 12, pp. 1591–1596.

Housel, W., 1929. A Practical Method for the Selection of Foundations Based on Fundamental Research in Soil Mechanics. Engineering Research Bulletin No. 13, University of Michigan, Ann Arbor.

Ismael, N., 1987. Coefficient of Subgrade Reaction for Footings on Desert Sands. Transportation Research Record 1137, pp. 82–91.

Ismael, N., 1996. Loading Tests on Circular and Ring Plates on Very Dense Cemented Sands. *Journal of Geotechnical Engineering, ASCE*, Vol. 122, No. 4, pp. 281–287.

Janbu, N. and Senneset, K., 1973. Field Compressometer – Principles and Applications. *Proceedings of the 8th International Conference on Soil Mechanics and Foundation Engineering*, Vol. 1.1, pp. 191–198.

Kay, J.N. and Avalle, D., 1982. Application of Screw-Plate to Stiff Clays. *Journal of the Geotechnical Engineering Division, ASCE*, Vol. 108, No. GT1, pp. 145–154.

Kay, J.N. and Mitchell, P., 1980. A Down Hole Plate Load Tests for In Situ Properties of Stiff Clays. *Proceedings of the Australia-New Zealand Conference on Geomechanics*, Vol. 1, pp. 255–259.

Kay, J.N. and Parry, R.H.G., 1982. Screw Plate Tests in a Stiff Clay. Ground Engineering, Vol. 15, No. 6, pp.22–27.

Kesharwani, R., Sahu, A., and Khan, N., 2015. Load Settlement Behaviour of Sandy Soil Blended with Coarse Aggregate. *Journal of Asian Scientific Research*, Vol. 5, No. 11, pp. 499–512.

Kim, T., Kang, G., and Hwang, W., 2014. Developing a Small Size Screw Plate Load Test. *Marine Georesources and Geotechnology*, Vol. 32, pp. 222–238.

Klohn, E.J., 1965. The Elastic Properties of a Dense Glacial Till Deposit. *Canadian Geotechnical Journal*, Vol. 11, No. 2, pp. 116–140.

Kummeneje, O. and Eide, E., 1961. Investigation of Loose Sand Deposits by Blasting. *Proceedings of the 5th International Conference on Soil Mechanics and Foundation Engineering*, Vol. 1, p. 491.

Landva, A.O., 1986. In Situ Testing of Peat. *Use of In Situ Tests in Geotechnical Engineering, ASCE*, pp. 191–205.

Landva, A.O. and Clark, J.I., 1990. Geotechnics of Waste Fill. ASTM Special Technical Publication 1070, pp. 86–103.

Lee, Y., Hwang, W., Choi, Y., and Kim, T., 2009. Development and Calibration of Screw Plate Load Tests. *Proceedings of the 19th International Offshore and Polar Engineering Conference*, pp. 236–241.

Lefebvre, G., Pare, J.-J., and Dascal, O., 1987. Undrained Shear Strength in a Surficial Weathered Crust. *Canadian Geotechnical Journal*, Vol. 24, No. 1, pp. 23–34.

Lo, K.Y. and Cooke, B., 1989. Foundation Design for the Skydome Stadium, Toronto. *Canadian Geotechnical Journal*, Vol. 26, No. 1, pp. 22–33.

Lo, K.Y., Adams, J., and Seychuk, J., 1969. The Shear Behavior of a Stiff Fissured Clay. *Proceedings of the 7th International Conference on Soil Mechanics and Foundation Engineering*, Vol. 1, pp. 249–255.

Lutenegger, A and Adams, M.T., 2003. Characteristic Load-Displacement Curves of Shallow Foundations. *Proceedings of the International Conference on Shallow Foundations*, Paris, France, Vol. 2, pp. 381–393.

Marsland, A., 1971. The Use of In Situ Tests in a study of the Effects of Fissures on the Properties of Stiff Clays. *Proceedings of the 1st Australia-New Zealand Conference on Geomechanics*, Vol. 1, pp. 180–189.

Marsland, A., 1975. In Situ and Laboratory Tests on Glacial Clays at Redcar. *Proceedings of the Symposium on Behavior of Glacial Materials, Midland Soil Mechanics and Foundations Society*, pp. 164–180.

Marsland, A., 1977. The Evaluation of the Engineering Design Parameters for Glacial Clays. *Quarterly Journal of Engineering Geology*. Vol. 10, No. 1, pp. 1–26.

Marsland, A. and Randolph, M.F., 1977. Comparison of the Results from Pressuremeter Tests and Large In Situ Plate Tests in London Clay. *Geotechnique*, Vol. 27, No. 2, pp. 217–243.

Marsland, A. and Butcher, A., 1983. In Situ Tests on Highly Weathered Chalk Near Luton, England. *Proceedings of the International Symposium on In Situ Testing*, Vol. 2, pp. 84–88.

Marsland, A. and Powell, J., 1990. Pressuremeter Tests on Stiff Clays and Soft Rocks: Factors Affecting Measurements and Their Interpretation. Field Testing in Engineering Geology. Geological Society Engineering Geology Special Publication No. 6, pp. 91–110.

Marsland, A. and Powell, J., 1991. Field and Laboratory Investigation of the Clay Tills at the Test Bed Site at the Building Research Establishment, Gartson, Hertfordshire. Quaternary Engineering Geology. Geological Society Engineering Geology Special Publication No. 7, pp. 229–238.

McKinlay, D., Tomlinson, M., and Anderson, W., 1974. Observations on the Undrained Strength of a Glacial Till. *Geotechnique*, Vol. 24, No. 4, pp. 503–516.

Mital, S. and Bauer, G., 1989. Screw Plate Test for Drained and Undrained Soil Parameters. *Foundation Engineering: Current Principles and Practices, ASCE*, Vol. 1, pp. 67–79.

Nichols, N., Benoit, J., and Prior, F., 1989. In Situ Testing of Peaty Organic Soils: A Case History. Transportation Research Record 1235, pp. 10–23.

Noor, S., Haider, S., and Islam, S., 2019. Screw Plate Load Test in the Estimation of Allowable Bearing Capacity in Cohesive Soil Deposit. *Lecture Notes in Civil Engineering*, Vol. 9, Springer Nature, Singapore, pp. 1247–1256.

Pells, P.J.N., 1983. Plate Loading Tests on Soil and Rock. *In Situ Testing for Geotechnical Investigations*, pp. 73–86.

Pells, P.J.N. and Turner, R.M., 1979. Elastic Solutions for the Design and Analysis of Rock Socketed Piles. *Canadian Geotechnical Journal*, Vol. 16, pp. 481–487.

Powell, J.J.M. and Quarterman, R.S.T., 1986. Evaluating the Screw Plate Test in Stiff Clay Soils in the U.K. *Proceedings of the Special Geomechanics Symposium*, Adelaide, pp. 128–133.

Radhakrishna, H.S. and Klym, T.W., 1974. Geotechnical properties of a Very Dense Glacial Till. *Canadian Geotechnical Journal*, Vol. 11, pp. 396–408.

Reddy, A.S., Pao, K.N., and Srunivasan, R.J., 1979. Horizontal Plate Load Test in Anisotropic and Nonhomogeneous Clays. *Proceedings of the International Symposium on In Situ Testing of Soils and Rocks and Performance of Structures*, Roorkee, pp. 178–183.

Rowe, R.K., 1982. The Determination of Rock Mass Modulus Variation with Depth for Weathered or Jointed Rock. *Canadian Geotechnical Journal*, Vol. 19, No. 1, pp. 29–43.

Rozsypal, A., 1983. Instrumented Large Scale Plate Loading Test on Limestones. *Proceedings of the International Symposium on In Situ Testing*, Vol. 2, pp. 133–137.

Schmertmann, J.H., 1970a. Static Cone to Compute Static Settlement over Sand. *Journal of the Soil Mechanics and Foundation Division, ASCE*, Vol. 96, No. SM3, pp. 1011–1043.

Schmertmann, J.H., 1970b. Suggested Method for Screw-Plate Load Test. ASTM Special Technical Publication 479, pp. 81–85.

Schwab, E. and Broms, B., 1977. Pressure-Settlement-Time Relationship by Screw Plate Tests In Situ. *Proceedings of the 9th International Conference on Soil Mechanics and Foundation Engineering*, Vol. 1, pp. 281–288.

Selvadurai, A. and Nichols, T., 1979. Theoretical Assessment of the Screw Plate Test. *Proceedings of the 3rd International Conference on Numerical Methods in Geomechanics*, Vol. 3, pp. 1245–1252.

Selvadurai, A. and Nichols, T., 1981. Evaluation of Soft Clay Properties by the Screw Plate Test. *Proceedings of the 10th International Conference on Soil Mechanics and Foundation Engineering*, Vol. 2, pp. 567–572.

Selvadurai, A., Bauer, G., and Nichols, J., 1980. Screw Plate Testing of a Soft Clay. *Canadian Geotechnical Journal*, Vol. 17, No. 4, pp. 465–472.

Seychuk, J., 1970. Load Tests on Bedrock. *Canadian Geotechnical Journal*, Vol. 7, No. 4, pp. 464–470.

Sherif, M. and Strazer, R., 1973. Soil Parameters for Design of Mt. Baker Ridge Tunnel in Seattle. *Journal of the Soil Mechanics and Foundations Division, ASCE*, Vol. 99, No. SM1, pp. 111–122.

Skempton, A., 1951. The Bearing Capacity of Clays. *Proceedings of the Building Research Congress*, London. pp. 180–189.

Slack, D.C. and Walker, J.N., 1970. Deflections of Shallow Pier Foundations. *Journal of the Soil Mechanics and Foundation Division, ASCE*, Vol. 96, No. SM4, pp. 1143–1157.

Soderman, L., Kim, Y., and Milligan, V., 1968. Field and Laboratory Studies of Modulus of Elasticity of a Clay Till. Highway Research Board Publication No. 243.

Strout, J. and Senneset, K., 1998. The Field Compressometer Test in Norway. *Proceedings of the 1st International Conference on Site Characterization*, Vol. 1, p. 863–868.

Tand, K.E., Funegard, E.G., and Briaud, J.-L., 1986. Bearing Capacity of Footings on Clay – CPT Method. *Use of In Situ Tests in Geotechnical Engineering, ASCE*, pp. 1017–1033.

Terzaghi, K., 1955. Evaluation of Coefficient of Subgrade Reaction. *Geotechnique*, Vol. 5, No. 4, pp. 297–326.

Terzaghi, K. and Peck, R.B., 1967. *Soil Mechanics in Engineering Practice*. John Wiley and Sons, New York.

Thomas, D., 1994. Spread Footing Prediction Event at the National Geotechnical Experimentation Site on the Texas A&M University Riverside Campus. *Predicted and Measured Behavior of Five Footings on Sand, ASCE*, pp. 149–152.

Van Impe, W.F., 1998. Dynamic Compaction to Improve Geotechnical Properties of MSW in Landfilling. *Proceedings of the 7th International Conference on Piling and Deep Foundations*, pp. 3.2.1–3.2.21.

Ward, W., Burland, J., and Gallois, R., 1968. Geotechnical Assessment of a Site at Munford, Norfolk for a Large Proton Accelerator. *Geotechnique*, Vol. 18, pp. 399–431.

Watts, K.S. and Charles, J.A., 1990. Settlement of Recently Placed Domestic Refuse Landfill. *Proceedings of the Institute of Civil Engineers*, Part 1, Vol. 88, pp. 971–993.

Webb, D.L., 1969. Settlement of Structures on Deep Alluvial Sandy Sediments in Durban, South Africa. *Proceedings of the Conference on In Situ Investigations in Soils and Rocks, BGS*, pp. 181–187.

Wiseman, G. and Zeitlen, J.G., 1994. Predicting the Settlement of the Texas A&M Spread Footings on Sand. *Predicted and Measured Behavior of Five Spread Footings on Sand, ASCE*, pp. 129–132.

Wrench, B.P., 1984. Plate Tests for the Measurement of Modulus and Bearing Capacity of Gravels. *The Civil Engineer in South Africa*, Vol. 26, No. 9, pp. 429–437.

Wrench, B.P. and Nowatzki, E.A., 1986. A Relationship between Deformation Modulus and SPT N for Gravels. *Use of In Situ Tests in Geotechnical Engineering, ASCE*, pp. 1163–1177.

Chapter 10

Other *In Situ* Tests

10.1 INTRODUCTION

There are a number of other *in situ* tests that may be useful for evaluating specific soil properties, but are perhaps less common than those described in previous chapters. In this chapter, a brief description is given for several of these tests. Interested readers will wish to consult specific references given for details on the different tests. The purpose here is to only acquaint engineers with other potentially valuable tests that may be used for different projects. Tests described in this chapter include Large-Scale Shear Box Tests, Hydraulic Fracture Tests (HFTs), and Push-in Earth Pressure Cell Tests.

10.2 LARGE-SCALE IN-PLACE SHEAR BOX TESTS

10.2.1 Background

In highly structured soils, weathered rock, or gravelly soils that are difficult to sample without significant disturbance to the structure, the in-place Shear Box Tests may be valuable for determining *in situ* shear strength parameters. These tests resemble smaller-scale laboratory direct shear box tests, but are of a much larger size in order to reduce scaling effects that can significantly influence the strength response of such geomaterials. This type of test has been used for about the past 60 years and had been extensively reported in the literature (e.g., Schultze 1957; Serafim & Lopes 1961; Wallace & Olsen 1966; Jain & Gupta 1974; Gifford et al. 1986; Chu et al. 1988).

The use of Large-Scale *In Situ* Shear Box Tests has been reported in a wide range of materials. Hutchinson & Rolfsen (1963) reported on the use of both a 50 × 50 cm (19.7 × 19.7 in.) and a 20×20 cm (7.9 × 7.9 in.) field shear box to determine the strength of quick clay in Norway. Bishop (1966) described the use of a 60 by 60 cm (23.6 × 23.6 in.) shear box to measure the strength of weathered London Clay. A portable in-place shear box was used by Brand et al. (1983) for determining *in situ* shear strength of residual soils in Hong Kong. Lefebvre et al. (1987) used Large-Scale In-Place Shear Box Tests to measure the undrained strength of a surficial clay crust. Large-scale Shear Box tests have also been reported for use in bedrock (e.g., Zeigler 1972; Franklin et al. 1974; Nicholson 1983). Table 10.1 gives a summary of some of the reported uses of large-Scale Shear Box Tests *in situ*.

Tests may also be performed to determine the interface strength between two materials (e.g., Schultze 1957) or between concrete and soil or rock by casting a concrete block directly on the surface and then conducting the test to force shearing at the interface.

Table 10.1 Summary of some reported uses of *In Situ* Large-Scale Shear Box Tests

Soil	Box dimensions	References
Soft clay	50 × 50 cm (19.7 ×19.7 in.) 20 × 20 cm (7.9×7.9 in.)	Hutchinson & Rolfsen (1963)
Stiff clay	61 by 61 cm (24×24 in.)	Bishop (1966)
	61×61 cm (24×24 in.)	Marsland & Butler (1967)
	61×61 cm (24×24 in.)	Marsland (1971)
	61 by 61 cm (24×24 in.)	Bishop & Little (1967)
	61 by 61 cm (24×24 in.)	Lo et al. (1969)
	61 by 61 cm (24×24 in.)	Radhakrishna & Klym (1974)
	61 by 61 cm (24×24 in.)	Lefebvre et al. (1987)
Residual soil	30×30 cm (12×12 in.)	Brand et al. (1983)
	30×30 cm (12×12 in.)	Cross (2010)
	50×50 cm (19.5×19.5 in.)	Li et al. (2014)
Gravel	1.5×1.5 m (59×59 in.)	Nichiporovitch & Rasskazov (1967)
	90×90 cm (35×35 in.)	Matsuoka et al. (2001)
	1.2 x 1.2 m (47 x 47 in.)	
Bedrock	70×70 cm (27.5×27.5 in.)	Serafim & Lopes (1961)
	60×60 cm (24×24 in.)	Sharma & Joshi (1983)
	70×70 cm (27.5×27.5 in.)	Baba (1983)
Rockfill	122.5×122.5 cm (48×48 in.)	Liu (2009)
Interface	1 × 1 m (39.4×39.4 in.)	Schultze (1957)
	71×71 mm (28×28 in.)	Dvorak (1957)
	70×70 cm (27.5×27.5 in.)	Serafim & Lopes (1961)
	60×60 cm (24×24 in.)	Sharma & Joshi (1983)
Soil roots	N/A	Wu et al. (1988)

10.2.2 Test Equipment

As is typically done in a laboratory direct Shear Box Test, it is necessary to have a system for applying a constant normal force on a test specimen and a system for applying a shear force perpendicular to the normal force to initiate a shear failure along a predesignated plane. Typically, the shear displacement and the contraction or dilation are also determined by measuring the displacement in the direction of the application of the normal and shear stresses.

Several different schemes for testing have been reported, all of which are similar in concept, but vary somewhat in the arrangement of the load application. In all cases, it is first necessary to isolate a block of material for testing. A typical arrangement of the test is shown in Figure 10.1.

The normal force is applied by a hydraulic jack that can be adjusted to maintain a constant force. The reaction can be provided by the dead load as indicated in Figure 10.1 or by anchor rods/piles. A rigid test box usually of steel channel is installed over the test block, which has been carefully trimmed to just fit inside the box. Any gap between the block and inside of the box can be filled with plaster of Paris, capping compound, or Portland cement mortar (neat cement).

In a tunnel, the normal force can be generated by jacking against the roof, as shown in Figure 10.2. The shear force is also applied by a hydraulic system. The reaction in this case is provided by the walls of the excavation. Load cells are used on both hydraulic systems to accurately measure the normal and the shear force. A roller-bearing system is inserted at the top of the test block to allow a free movement during shearing. The size of the test block

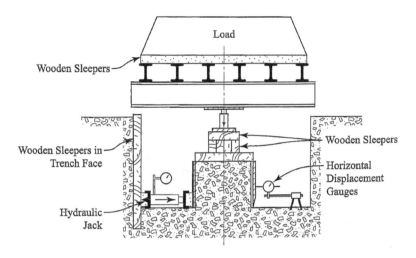

Figure 10.1 Typical arrangement for performing *In Situ* Large-Scale Shear Box Tests.

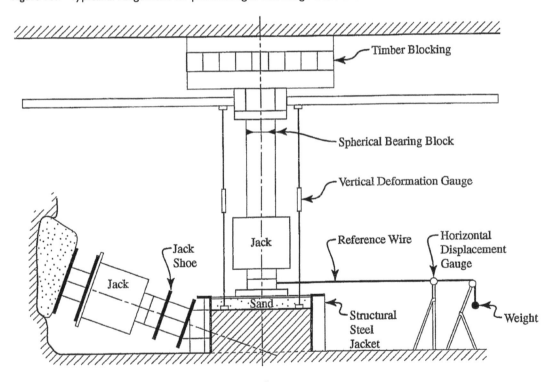

Figure 10.2 Typical arrangement for *In Situ* Shear Box Tests in a tunnel.

can vary between 0.3 and 1.5 m (1 and 5 ft), depending on the project. Displacements are measured with dial gages, LVDTs, or other appropriate devices. Since the test destroys the test surface, a series of tests on different blocks, at different normal stresses, is needed to define the failure envelope.

The author performed a series of in-place Shear Box Tests on the shale rock foundation at the base of Snell Lock on the St. Lawrence Seaway. The lock had been covered and heated for routine winter maintenance and dewatering of the lock chamber left the base rock exposed.

Normal force was provided by drilling and installing expandable rock bolts into the floor rock with a reaction beam and hydraulic system placed over the test block. Tests blocks were isolated by using a concrete saw to carve the material away from an individual block. A section was then cut out to allow the shear force reaction system to be placed horizontally again using the floor rock as the reaction. The setup is shown in Figure 10.3 and was successful in defining the strength envelope for use in a lock wall base sliding stability analysis.

10.2.3 Test Procedures

A procedure for conducting *in situ* Shear Box Tests in rock was recommended by Dodds (1980), and a standard test method has been suggested by ASTM as test method D4554 "Standard Test Method for *In Situ* Determination of Direct Shear Strength of Rock Discontinuities", which is essentially the same as suggested by the International Society of Rock Mechanics. Excellent summaries of in-place rock testing have been presented by the U.S. Army Engineer Waterways Experiment Station by Zeigler (1972) and Nicholson (1983). The test procedures are given in ASTM D4554 and can generally be used for most materials.

10.2.4 Results and Interpretation

Tests are normally interpreted in a manner similar to laboratory direct Shear Box Tests with the normal and shear stress plotted to define the failure envelope. In many cases, both a peak shear strength envelope and a residual envelope may be defined if the shear displacement goes far enough or if reversal of the shear direction after peak is performed.

10.3 HYDRAULIC FRACTURE TESTS (HFTs)

10.3.1 Background

HFTs were suggested for determining lateral stresses in clays by Bjerrum and Andersen (1972). The propagation of fractures in soils requires an increase in fluid pressure within the fracture or a decrease in the external stresses in the neighboring soil. The analysis of hydraulic fracture in clay has considered the compressibility of the soil to account for its plastic nature (Bjerrum et al. 1972). The test is limited to fine-grained soils of relatively

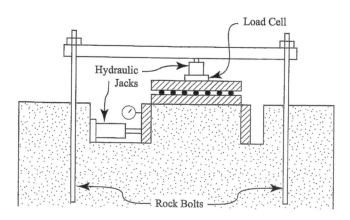

Figure 10.3 Large-Scale Shear Box Tests using rock bolts as reaction.

low permeability because soils with high permeability, e.g., sands and gravels, do not allow for a pressure head to be developed between the pressure source and the surrounding soil (Jaworski et al. 1981). The HFT is a simple inexpensive means for estimating limiting values of K_o in many cohesive soils.

10.3.2 Test Equipment

HFTs are generally conducted in two ways using different pieces of test equipment. In one case, the test is performed after first installing a push-in piezometer or slotted screen tip. In the other case, tests are performed in a drilled borehole by inserting a packer with a central pressure pipe.

10.3.2.1 Tests with Push-in Piezometer

For use in most soft to moderately overconsolidated clays, the most common test procedure for performing HFTs involves using a push-in piezometer tip. This is the procedure described by Bjerrum & Anderson (1972), and Bozozuk (1974). For the most part, at least up to about 1990, most HFTs conducted in this manner used a simple probe with a short filter element on or near the tip attached to a string of drill rods. In some cases, a Geonor M-206 piezometer tip, attached to E rods, has been used. This piezometer tip consists of a sintered bronze filter element with a nominal length of 30 cm (1 ft) mounted directly behind a conical tip, as shown in Figure 10.4.

Several investigations have shown that test results can be affected by both the length/diameter ratio of the filter element and the position of element behind the cone tip. In order to insure vertical fractures in the soil and the measurement of horizontal stresses, the filter

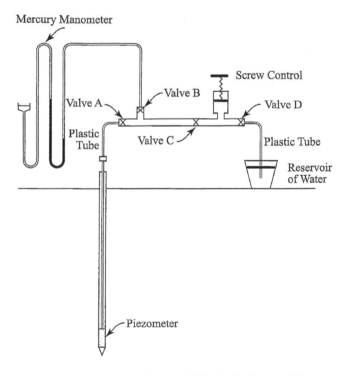

Figure 10.4 Push-in piezometer tip used to perform early Hydraulic Fracture Tests.

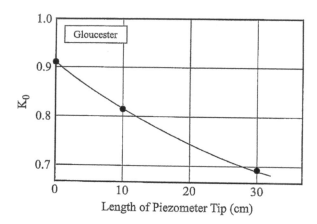

Figure 10.5 Influence of piezometer length on interpreted values of K_o in clay.

element should be located a minimum of about 5–6 diameters behind the tip and the L/D ratio of the element should be no less than about 6. The influence of piezometer tip length (really L/D) on the value of K_o is illustrated in Figure 10.5 using data from Lefebvre et al. (1981). Again, for a very short tip, the soil is fractured on horizontal planes giving essentially a measurement of the total vertical overburden stress tending toward a K_o value of 1.0. As the length of the filter element increases, and L/D increases, the results tend toward a constant value of K_o.

A control console at the ground surface is needed to force fluid into the piezometer tip at a controlled rate. Normally, a small hand-operated or electric screw pump is used. Tubing is used to connect the control console to the piezometer tip. A complete test arrangement is shown in Figure 10.6. The tubing and the entire system must be de-aired before testing. In some cases, twin tubing is used so that air can be flushed from the line. In other cases, a single line is used but is carefully de-aired prior to starting the test and the system is left connected for the duration of the test. Using a push-in piezometer probe, the test can be used as a profiling tool, i.e., by conducting tests at various depths at a single test location. Rad et al. (1988) have shown that HFTs may be performed in clays using a push-in Bengt A. Torstensson (BAT) probe.

10.3.2.2 Tests in an Open Borehole

Tests may also be conducted in stiff clays such as clay till or overconsolidated clay using an open borehole as the hole is advanced or after the entire length of the hole has been drilled. As a borehole is advanced, a single packer with a central fluid pipe is lowered into the borehole and held off the bottom of the hole as the packer is inflated. The packer seals off the lower part of the borehole. A typical test arrangement is shown in Figure 10.7. An alternative technique is to use a double packer system to isolate a test zone, as shown in Figure 10.8. In the case of either a single or a double packer, the central fluid pipe is often used to lower the device into the borehole.

10.3.3 Test Procedures

The procedure developed by Bjerrum & Andersen (1972), or modifications of the original procedure, is most often followed when conducting the HFT. The procedure was developed from *in situ* permeability tests using hydraulic piezometers that were pushed or driven into the ground. Tests are essentially conducted in the following sequence. The piezometer is installed to the test depth, and the excess pore water pressures generated during installation are allowed to dissipate to hydrostatic conditions. Once excess pore water pressure has

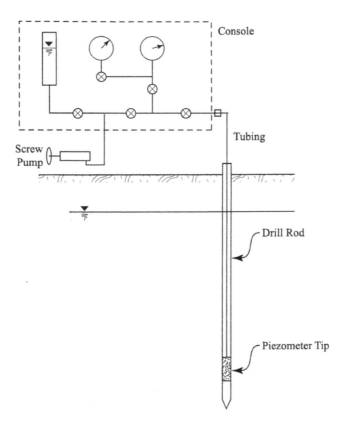

Figure 10.6 Test equipment for performing Hydraulic Fracture Tests.

dissipated, a screw pump is used to generate a constant head high enough to cause the soil to fracture. Pressure readings are taken during pumping to detect when the fracture has formed. Once fracture has occurred, the flow is stopped, and the dissipation of pore water pressure is monitored over time to determine when the fracture closes.

Using this procedure, it was postulated by Bjerrum & Andersen (1972) that the minor principal total stress at that depth could be determined by monitoring the flow of fluid into the soil, which was assumed to be the horizontal principal stress across a vertical fracture. Modifications to the apparatus and procedure were made by Bozozuk (1974) by replacing the mercury manometer with a pore water pressure transducer and taking readings automatically with a chart recorder.

10.3.4 Results and Interpretation

To perform the test, the pressure is incrementally increased and the volume of fluid pumped is recorded, as shown in Figure 10.9. Fracture is noted by a dramatic drop in the pressure or by leveling off in the pressure as the fracture allows the fluid to enter. Pumping is usually stopped at this point, and the change in pressure is recorded. The dissipation of pore water pressure after fracture is analyzed by plotting the flow of fluid into the soil as a function of the decreasing pressure.

It is assumed that the fracture closes when a significant decrease in the flow into the soil is observed. The pressure at this point is usually taken as the total stress acting across the fracture or close-up stress, U_C. Bjerrum & Andersen (1972) concluded that U_C was equal

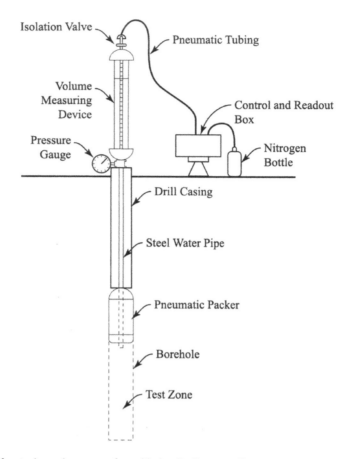

Figure 10.7 Use of a single packer to perform Hydraulic Fracture Tests.

to the minor principal total stress in the ground at the test depth, i.e., the horizontal total stress if a vertical fracture was formed. Bozozuk (1974) found that drawing two tangents through points prior to and following the break in the close-up curve gave good estimates of the minor principal stress, as illustrated in Figure 10.10.

After obtaining the close-up pressure following the initial fracture, it is often common practice to perform another fracture phase of the test. As shown in Figure 10.9, the pressure needed to reopen the fracture is also often obtained, typically indicated by the end of the linear pressure vs. volume curve. A second close-up curve is also sometimes obtained following the reopen part of the test. The complete sequence is shown in Figure 10.11.

Limitations to the HFT for determining lateral stresses in clays had been presented by Massarch et al. (1975), and Massarch & Broms (1976), as follows:

1. The uncertainty of the failure mode: varves, silt seams, fissures, and other nonuniformities may influence the failure mode;
2. The tensile strength of the clay is neglected in the interpretation of results;
3. The HFT is limited to normally consolidated clays;
4. The shape of the piezometer (i.e., cylindrical) produces significant disturbance and may create arching during reconsolidation after installation. This may affect the measured valve in the test, giving preference to fracture at the piezometer tip rather than producing a vertical fracture.

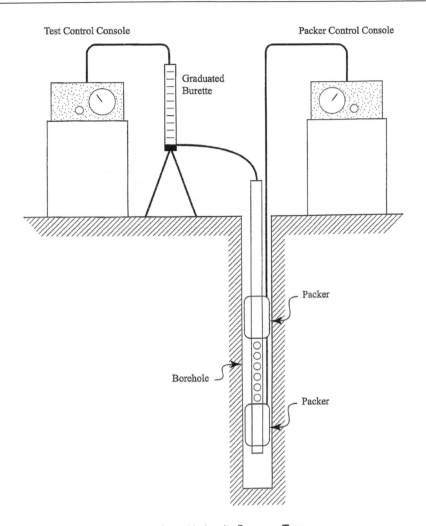

Figure 10.8 Use of a double packer to perform Hydraulic Fracture Tests.

Based on the work performed since these limitations were considered, the author would offer the following comments:

1. The use of longer filter elements on the piezometer and L/D ratio greater than about 5 appears to produce a failure mode generating a vertical fracture. Discontinuities such as fissures are typically present in the most fine-grained deposits to some degree but are more predominant in surficial clay crusts, and therefore, some difficulty may still be encountered in these deposits.
2. Since the normal test interpretation is to use either the close-up or reopen pressure, neglecting the soil tensile strength appears to be justified.
3. The HFT is not limited to normally consolidated clays ($K_o < 1.0$). Using longer filter tips positioned away from the piezometer tip has produced the reported K_o values of at least 3.0. Additionally, the author has experienced no difficulties installing push-in piezometers in very stiff overconsolidated clays and clay tills.
4. After complete reconsolidation, the value of horizontal stress measured by the HFT (when pushed) may be larger than that of *in situ* (before pushing) horizontal stress in

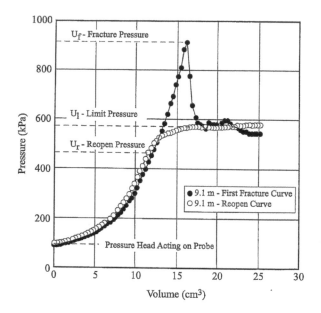

Figure 10.9 Typical results from Hydraulic Fracture Tests.

all but very soft normally consolidated clays. This is to be expected much in the way that piles driven into clays and allowed to reconsolidate will change the resulting stress field near the pile face. Therefore, the horizontal stress measured using HFTs with push-in piezometers should be referred to as the "effective hydraulic fracture pressure σ'_{hc}" and not σ'_{ho}.

HFTs are reasonably simple and inexpensive to perform, and the interpretation is relatively straightforward. Results obtained indicated that K_o values in the range of 0.5–3 may be determined. Table 10.2 gives a summary of some reported uses of HFTs.

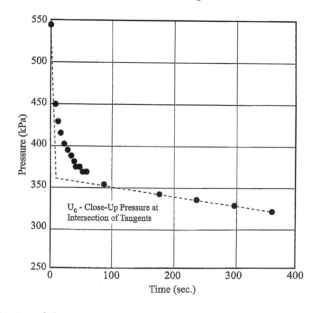

Figure 10.10 Determination of close-up pressure by tangent intersection.

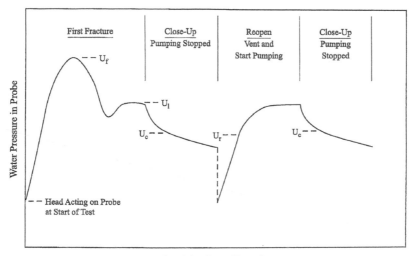

Figure 10.11 Sequence of Hydraulic Fracture Testing.

Table 10.2 Some reported results from Hydraulic Fracture Tests

Soil	Equipment	Range of K_o	References
Quick clay & soft clay	Push-in piezometer	0.4–0.6	Bjerrum & Andersen (1972)
Sensitive clay	Push-in piezometer	0.2–0.7	Bozozuk (1974)
Clay	Push-in piezometer	0.9–1.1	Palmer & Lo (1974)
Sensitive clay	Push-in piezometer	0.7–1.1	Tavenas et al. (1975)
Sensitive clay	Push-in piezometer	0.5–1.9	Massarch et al. (1975)
Sensitive clay	Push-in piezometer	0.6–4.0	Lefebvre et al. (1981)
Sensitive clay	Double packer on self-boring permeameter	0.5–1.6	Lafleur et al. (1987)
Very stiff compacted clay backfill	Packer in open borehole	1.7–2.0	Diviney (1990)
Sensitive clay	Push-in piezometer	0.7–3.3	Hamouche et al. (1995)

10.4 PUSH-IN EARTH PRESSURE CELLS

10.4.1 Background

Push-in earth pressure cells, sometimes referred to as "spade cells", are often considered as a tool for instrumentation, i.e., to monitor changes in lateral earth pressure, rather than as an *in situ* test. The author, however, considers spade cells as a bonafide *in situ* test that can be useful in helping to evaluate that current state of lateral stress in soils, whether at rest or as a result of some construction or change in stress. The first reported use of spade cells was in soft clays by Massarch (1975), Massarch et al. (1975), and Tavenas et al. (1975).

While push-in spade cells were initially used for soft clays, a number of applications described in the literature have been for stiff clays (e.g., Tedd & Charles 1981, 1983; Tedd et al. 1984; Lutenegger 1990; Ryley & Carder 1995). Spade cells have been used on several projects involving cut and cover tunneling or behind retailing structures to monitor the changes in lateral stress associated with construction (e.g., Tedd et al. 1985; Carder & Symons 1989; Symons & Carder 1992) and to measure lateral stresses in slopes in Sweden (Rankka 1990).

10.4.2 Test Equipment

The equipment used to conduct push-in spade cell tests is relatively simple and typically consists of a thin flat rectangular blade filled with hydraulic oil and a pressure transducer mounted at the top of the plate. The leading edge of the blade is usually either pointed or simply rectangular with an angled wedge. An adapter is attached to the back end of the blade to attach drill rods or pipe for installation. The blade and transducer are de-aired and sealed so that the system response to the applied stress will be very rapid and more direct. In some cases, a small amount of internal pressure is built into the system to insure good response. This means that in atmospheric pressure, the transducer will have some initial positive pressure or zero offset. It's important to know this offset prior to testing so that all subsequent readings can be corrected.

In addition to the spade and transducer, the only other equipment needed to conduct the test is a readout or control console to read the transducer. In some cases, twin tube pneumatic transducers are used and the control console is a simple gas control panel to apply pressure in order to read the transducer. In most cases, a small bottle of nitrogen gas is used to supply the control console. Vibrating wire transducers have also been used, in which case a vibrating wire readout is needed. Electrical resistance pressure transducers have also been used and require a simple bridge readout device or a constant power voltage supply and precision voltmeter. A schematic of a typical spade cell is shown in Figure 10.12, and a photo of a spade cell used at a number of sites by the author is shown in Figure 10.13.

Since the use of spade cells is not currently controlled by any regulating agency such as ASTM, there is no standard that defines the geometry of the spade cell or the exact test procedures. The test procedures are relatively straightforward; however, the geometry of the spade can vary, depending on the manufacturer and the user. Table 10.3 gives a summary of the different geometries of spade cells that have been reported in the literature. There is a noticeable difference in the thickness of spades used, ranging from 2 to 12 mm (0.1 to 0.5 in.).

The basic concept of the spade cell is to introduce a thin pressure cell into the ground with a minimal disruption and then monitor the change in stress with time until an equilibrium value is obtained. The spade should be as thin as possible, but still be able to be installed without bending or other damage. This suggests some practical lower bound on

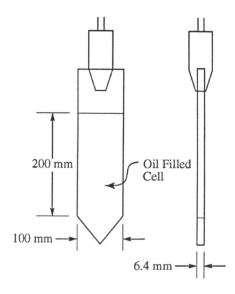

Figure 10.12 Geometry of typical push-in earth pressure cell.

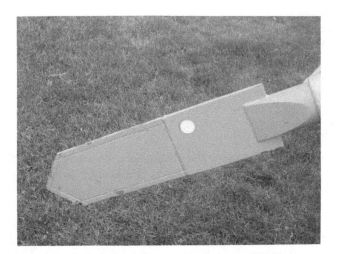

Figure 10.13 Photo of push-in earth pressure cell.

Table 10.3 Summary of some reported push-in earth pressure cell geometry

References	W (mm)	L (mm)	T (mm)
Massarch (1975)	100	200	4
Massarch et al. (1975)	100	200	4
Tavenas et al. (1975)	300	450	12
Massarch & Broms (1976)	N/A	N/A	3
Ladd et al. (1979)	114.5	565	9
Penman & Charles (1981)	100	200	2
Tedd & Charles (1981)	100	200	5
Fukuoka & Imamura (1983)	120	220	5
Ohya et al. (1983)	90	210	7
Chan & Morgenstern (1986)	200	200	6
Sully & Campanella (1990)	100	200	6.4
Lutenegger (1990, 2012)	100	200	6.4
Lutenegger (2013)	102	250	12.5

the thickness of the blade, probably in the range of 4–5 mm (0.16–0.20 in.) in order to have an application over a wide range of soil stiffness. The width and thickness should be small enough to be manageable and to fit into normal size boreholes, yet large enough to provide a reliable soil response over a sufficiently large area. The length should not be too long since a flat plate tends to drift from vertical when pushed into the ground. It appears that a width of 100 mm (3.9 in.) and a length on the order of 200–300 mm (7.9–11.8 in.) provide the necessary stiffness and area for use in most soils.

Because of the differences in plate geometry, it is likely that not all spades will give the same response in all soils. It appears that the most important factor in the soil response may be the aspect ratio or width/thickness ratio of the blade. That is, as the width becomes very large in relation to the thickness, the blade begins to look like a wide plate or sheet. As the width approaches the thickness, i.e., W/T = 1, this is the geometry of a circular probe.

Push-in earth pressure cells are total stress cells; i.e., when pushed vertically into the ground, they provide a response of the total horizontal stress. In some cases, spades have been equipped with a porous element and a pore water pressure transducer; however, the reported responses have not always been that good, and in most cases, the test is restricted to only giving total stress response. This means that an accurate estimate of the *in situ* pore water pressure must be obtained at each test location, preferably with piezometers, so that the final equilibrium effective stress may be determined. Considerable work has been conducted using push-in spade cells by the Geotechnical Section of the Building Research Establishment in the U.K. Much of this work has been summarized by Tedd et al. (1989).

10.4.3 Test Procedures

Unlike the DMT, which is pushed into the ground in a continuous manner without drilling a borehole, spade cells are typically installed at the bottom of a drilled hole. The reason for this is mostly practical as experience has shown that if the spade is pushed more than 1–1.5 m (3–5 ft) ahead of hole, there is a high probability of damage to the blade from bending. This means that in order to use the blade as a profiling tool, the testing will need to be planned out over several days or weeks, with the spade removed and borehole advanced only after the test has been completed by achieving an equilibrium stress. The final response varies with soil stiffness and is fastest for very stiff clays, on the order of several hours to one day, and slowest for very soft clays, on the order of several weeks.

In normal operation, the spade is pushed hydraulically into the ground at the base of the borehole and is never driven. Immediately after insertion, the test clock begins so that time zero for the test is the time that pushing stops and the load is removed. The pressure in the transducer is recorded at relatively close intervals at the beginning of the test and at progressively increasing intervals as the test proceeds. In the case of tests performed in soft clays, where it may take several weeks to obtain the final response, pressure readings every day or every other day are typical after obtaining the initial response throughout the first day.

After removing the spade from the test zone at the completion of the test, the blade is cleaned off and a zero reading obtained so that each test will have a before and after test zero reading. This will alert the operator of any potential problems with the cell. Cells are normally calibrated in a pressure chamber by the manufacturer.

10.4.4 Results and Interpretation

Since the test data obtained are total soil stress over time, the results of individual tests are normally presented graphically as shown in Figure 10.14. The test data shown in Figure 10.14 were obtained by the author at a clay site with OCR decreasing with depth. The curves all show a similar response and are generally "S" shaped when presented on a semi-log plot. These curves are similar to DMT "A" dissipation tests described in Chapter 6. The response for equilibrium is much faster in the overconsolidated zone than in the normally consolidated zone, which is to be expected. In this case, since the soil is a varved clay deposit with silt lenses and a relatively high horizontal hydraulic conductivity (10^{-6} cm/s), the final response in the soft clay occurs relatively fast, i.e., on the order of one week.

The equilibrium or final value of total horizontal stress is denoted as σ_{hf} and is used to define the horizontal stress ratio as:

$$K_C = (\sigma_{hf} - u_o)/\sigma'_{vo} \tag{10.1}$$

where

K_C = consolidated lateral earth pressure coefficient
u_o = *in situ* pore water pressure at the test depth
σ'_{vo} = *in situ* vertical effective stress at the test depth

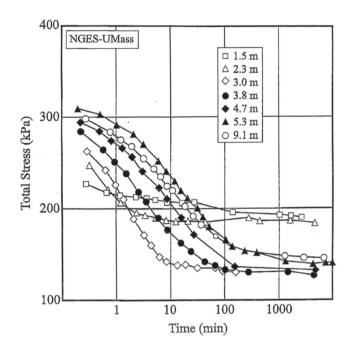

Figure 10.14 Typical push-in earth pressure cell test results.

The term K_C is used in Equation 10.1 and not K_o, the at-rest coefficient of earth pressure since the test measures the total lateral stress *after insertion of a flat blade*. The two terms should not be considered the same. Tedd & Charles (1981) suggested that a simple empirical correction factor could be applied to σ_{hf} to obtain σ_{ho}. Based on a series of tests where spades were used under known stress conditions, it was suggested that a stress equal to one-half of the undrained shear strength ($0.5\ s_u$) should be subtracted from the measured final total stress to account for the overstress created by inserting the blade. These data are shown in Figure 10.15. An updated summary of the available test data presented by Tedd et al. (1989) shows more scatter, which may be partly related to the selection of s_u and partly related to the reference test used for the true reading.

Ryley & Carder (1995) showed that for clays with undrained shear strength in the range of 70–150 kPa (1500–3000 psf), a more reasonable correction for the overstress would be $0.8\ s_u$. The use of a correction factor of $0.5\ s_u$ would be conservative for retaining walls as it would give higher stresses than may actually be present. On the other hand, this would be unconservative for the design of driven piles, suggesting lateral stresses that are higher than actual values.

A more direct approach to evaluating the results of spade cell tests may be to develop a functional relationship between K_C and K_o. Results taken from the literature for a number of different test sites and tests conducted by the author are shown in Figure 10.16 and demonstrate that K_C is related to the stress history through OCR. The scatter in the reported results is probably related to differences in blade geometry as previously discussed. Even for OCR=1, there is scatter in the test results. For a given clay, under simple unloading, there is a relationship between K_o and OCR (e.g., Mayne & Kulhawy 1982). Therefore, it is a relatively simple matter to establish a relationship between K_C and K_o since both appear to be related to OCR.

Spade cells should be considered as an adjunct testing program to provide additional test data for determining horizontal stresses in fine-grained soils. The tests are relatively inexpensive, easy to perform, and generally provide the reliable test data. On large projects

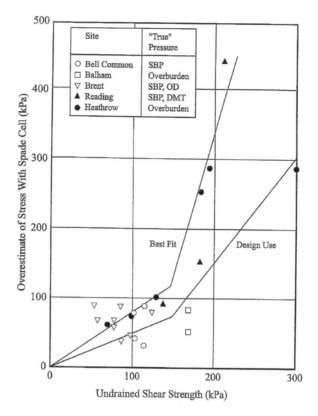

Figure 10.15 Overstress from push-in earth pressure cells as a function of undrained shear strength. (After Ryley & Carder 1995.)

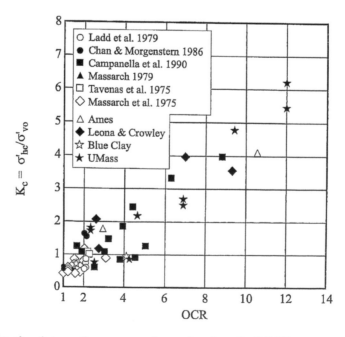

Figure 10.16 Results of push-in earth pressure cells as a function of soil OCR.

where the at-rest horizontal stress or K_o is of significant importance to the project, several approaches using both *in situ* and laboratory tests will likely be used. In conjunction with Dilatometer Tests, Pressuremeter Tests, or HFTs, the use of push-in spade cells is an attractive alternative.

REFERENCES

Baba, K., 1983. In Situ Tests of Dam Foundations. *Proceedings of the International Symposium on In Situ Testing of Soil and Rock*, Vol. 2, pp. 438–441.

Bishop, A.W., 1966. The Strength of Soils as Engineering Materials. *Geotechnique*, Vol. 16, No. 2, pp. 89–130.

Bishop, A. and Little, A., 1967. The Influence of the Size and Orientation of the Sample on the Apparent Strength of the London Clay at Malden, Essex. *Proceedings of the Geotechnical Conference Oslo on the Shear Strength Properties of Natural Soils and Rocks*, Vol. 1, pp. 89–96.

Bjerrum, L. and Andersen, K.H., 1972. In-Situ Measurement of Lateral Pressures in Clay. *Proceedings of the 5th European Conference on Soil Mechanics and Foundation Engineering*, Vol. 1, pp. 11–20.

Bjerrum, L., Nash, J.K.T.L., Kennard, R.M., and Gibson, R.E., 1972. Hydraulic Fracturing in Field Permeability Testing. *Geotechnique*, Vol. 22, No. 2, pp. 319–332.

Bozozuk, M., 1974. Minor Principal Stress Measurement in Marine Clay with Hydraulic Fracture Tests. *Proceedings of the Conference on Subsurface Exploration for Underground Excavation and Heavy Construction, ASCE*, pp. 333–349.

Brand, E.W., Phillipson, H.B., Borrie, G.W., and Clover, A.W., 1983. In Situ Shear Tests on Hong Kong Residual Soils. *Proceedings of the International Symposium on In Situ Testing of Soil and Rock*, Vol. 2, pp. 13–18.

Carder, D.R. and Symons, I.F., 1989. Long-Term Performance of an Embedded Cantilever Retaining Wall in Stiff Clay. *Geotechnique*, Vol. 39, No. 1, pp. 55–75.

Chan, A.C.Y. and Morgenstern, N.R., 1986. Measurement of Lateral Stresses in a Lacustrine Clay Deposit. *Proceedings of the 39th Canadian Geotechnical Conference*, pp. 285–290.

Chu, B.L., Hsu, T.W., and Lai, T.C., 1988. In Situ Direct Shear Tests for Lateritic Gravels in Taiwan. *Proceedings of the 2nd International Conference on Geomechanics in Tropical Soils*, Vol. 1, pp. 119–125.

Cross, M., 2010. The Use of a Field Open-Sided Direct Shear Box for the Determination of the Shear Strength of Shallow Residual and Colluvial Soils on Hillslopes in the South Pennines, Derbyshire. *North West Geography*, Vol. 10, pp. 8–18.

Diviney, J.G., 1990. Performance of Large Gravity Walls at Eisenhower and Snell Locks. *Design and Performance of Earth Retaining Structures, ASCE*, pp. 278–291.

Dodds, R.K., 1980. Suggested Method of Test for In Situ Shear Strength of Rock. ASTM Special Technical Publication 479, pp. 618–628.

Dvorak, A., 1957. Field Tests of Rocks on Dam Sites. *Proceedings of the 4th International Conference on Soil Mechanics and Foundation Engineering*, Vol. 2, pp. 221–224.

Franklin, J.A., Manailoglou, J., and Sherwod, D., 1974. Field Determination of Direct Shear Strength. *Proceedings of the 3rd Congress of the International Society of Rock Mechanics*, Vol. 2A, pp. 233–240.

Fukuoka, M. and Imamura, Y., 1983. Earth Pressure Measurement in Retaining Wall Backfill. *Proceedings of the International Symposium on In Situ Testing of Soil and Rock*, Vol. 2, pp. 49–53.

Gifford, A.B., Green, G.E., Buechel, G.J., and Feldman, A.I., 1986. In Situ Tests Aid Design of a Cylinder Pile Wall. *Use of In Situ Tests in Geotechnical Engineering, ASCE*, pp. 569–587.

Hamouche, K., Leroueil, S., Roy, M., and Lutenegger, A., 1995. In Situ Evaluation of K_o in Eastern Canada Clays. *Canadian Geotechnical Journal*, Vol. 32, pp. 677–688.

Hutchinson, J.N. and Rolfsen, E.N., 1963. Large Scale Field Shear Box Tests on Quick Clay. Norwegian Geotechnical Institute Publication No. 51, pp. 1–11.

Jain, S.P. and Gupta, R.C., 1974. In-Situ Shear Tests for Rock Fills. *Journal of the Geotechnical Engineering Division, ASCE*, Vol. 100, No. GT9, pp. 1031–1050.

Jaworski, G., Duncan, J., and Seed, H.B., 1981. Laboratory Study of Hydraulic Fracturing. *Journal of Geotechnical Engineering, ASCE*, Vol. 107, No. GT6, pp. 713–732.

Ladd, C.C., Germaine, J.T., Baligh, M.M., and Lacasse, S.M., 1979. Evaluation of Self-Boring Pressuremeter Tests in Boston Blue Clay. Report No. R79-4 to the Federal Highway Administration.

Lafleur, J., Giroux, F., and Huot, M., 1987. Field Permeability of the Weathered Champlain Crust. *Canadian Geotechnical Journal*, Vol. 24, No. 4, pp. 581–589.

Lefebvre, G., Philibert, A., Bozozuk, M., and Pare, J.-J., 1981. Fissuring from Hydraulic Fracture of Clays Soil. *Proceedings of the 10th International Conference on Soil Mechanics and Foundation Engineering*, Vol. 2, pp. 513–518.

Lefebvre, G., Pare, J.-J., and Dascal, O., 1987. Undrained Shear Strength in a Surficial Weathered Crust. *Canadian Geotechnical Journal*, Vol. 24, No. 1, pp. 23–34.

Li, X., Li, J. and Deng, H., 2014. In Situ Direct Shear Test Research of Rock and Soil of Typical Bank Slope in Three Gorges Reservoir Area. *Electronic Journal of Geotechnical Engineering*, Vol. 19, pp. 2523–2534.

Liu, S., 2009. Application of New In Situ Direct Shear Device to Shear Strength Measurements of Rockfill. *Water Science and Engineering*, Vol. 2, No. 3, pp. 48–57.

Lo, K., Adams, J., and Seychuk, J., 1969. The Shear Behavior of a Stiff Fissured Clay. *Proceedings of the 7th International Conference on Soil Mechanics and Foundation Engineering*, Vol. 1, pp. 249–255.

Lutenegger, A.J., 1990. Determination of In Situ Lateral Stresses in a Dense Glacial Till. Transportation Research Record No. 1278, pp. 194–203.

Lutenegger, A.J., 2012. A Push-In Earth Pressure Cell for Estimating Soil Properties. *Proceedings of the 4th International Symposium on Geotechnical and Geophysical Site Characterization*, pp. 561–564.

Lutenegger, A.J., 2013. Field Response of Push-In Earth Pressure Cells for Instrumentation and Site Characterization of Soils. *Geotechnical Engineering Journal of the Southeast Asian Geotechnical Society*, Vol. 43, No. 4, pp. 24–33.

Marsland, A., 1971. The Use of In Situ Tests in a study of the Effects of Fissures on the Properties of Stiff Clays. *Proceedings of the 1st Australia-New Zealand Conference on Geomechanics*, Vol. 1, pp. 180–189.

Marsland, A. and Butler, M., 1967. Strength Measurements on Stiff Barton Clay from Fawley (Hampshire). *Proceedings of the Geotechnical Conference Oslo on the Shear Strength Properties of Natural Soils and Rocks*, Vol. 1, pp. 139–145.

Massarch, K.R., 1975. New Method for Measurement of Lateral Earth Pressure in Cohesive Soils. *Canadian Geotechnical Journal*, Vol. 12, No. 1, pp. 142–146.

Massarch, K.R. and Broms, G., 1976. Lateral Earth Pressure at Rest in Soft Clay. *Journal of the Geotechnical Engineering Division, ASCE*, Vol. 102, No. GT10, pp. 1041–1047.

Massarch, K.R., Holtz, R.D., Holm, B.G., and Fredriksson, A., 1975. Measurements of Horizontal In Situ Stresses. *Proceedings of the Conference on In Situ Measurement of Soil Properties, ASCE*, Vol. 1, pp. 266–286.

Matsuoka, H., Liu, S., Sun, D., and Nishikata, U., 2001. Development of a New In-Situ Direct Shear Test. *Geotechnical Testing Journal, ASTM*, Vol. 24, No. 1, pp. 92–102.

Mayne, P.W. and Kulhawy, F.H., 1982. K_o – OCR Relationships in Soil. *Journal of the Geotechnical Engineering Division, ASCE*, Vol. 108, No. GT6, pp. 851–872.

Nichiporovitch, A. and Rasskazov, L., 1967. Shear Strength of Coarse Fragmental Materials. *Proceedings of the Geotechnical Conference Oslo on the Shear Strength Properties of Natural Soils and Rocks*, Vol. 1, pp. 225–229.

Nicholson, G.A., 1983. In Situ and Laboratory Shear Devices for Rock: A Comparison. U.S. Army Engineer Waterways Experiment Station, Technical Report GL-83-14.

Ohya, S., Imai, T., and Nagura, M., 1983. Recent Developments in Pressuremeter Testing, In-Situ Stress Measurement and S-Wave Velocity Measurement. *Recent Development in Laboratory and Field Tests and Analysis of Geotechnical Problems*, Bangkok, pp. 189–209.

Palmer, J. and Lo, K., 1974. In Situ Measurement of Lateral Pressures Using Hydraulic Fracturing Technique. Research Report SM-2-74, Faculty of Engineering Science, University of Western Ontario, 27 pp.

Penman, A.D.M. and Charles, J.A., 1981. Assessing the Risk of Hydraulic Fracture in Dam Cores. *Proceedings of the 10th International Conference on Soil Mechanics and Foundation Engineering*, Vol. 1, pp. 457–461.

Rad, N.S., Sullie, S., Lunne, T., and Torstensson, B.A., 1988. A New Offshore Soil Investigation Tool for Measuring the In Situ Coefficient of Permeability and Sampling Pore Water and Gas. *Proceedings of the 5th International Conference on the Behavior of Offshore Structures*, Vol. 1, pp. 409–417.

Radhakrishna, H. and Klym, T., 1974. Geotechnical Properties of a Very Dense Till. *Canadian Geotechnical Journal*, Vol. 11, pp. 396–408.

Rankka, K., 1990. Measuring and Predicting Lateral Earth Pressures in Slopes in Soft Clays in Sweden. Transportation Research Record No. 1278, pp. 172–182.

Ryley, M.D. and Carder, D.R., 1995. The Performance of Push-In Spade Cells Installed in Stiff Clay. *Geotechnique*, Vol. 45, No. 3, pp. 533–539.

Schultze, E., 1957. Large Scale Shear Tests. *Proceedings of the 4th International Conference on Soil Mechanics and Foundation Engineering*, Vol. 1, pp. 193–199.

Serafim, J.L. and Lopes, J.J.B., 1961. "In Situ" Shear Tests and Triaxial Tests of Foundation Rocks of Concrete Dams. *Proceedings of the 5th International Conference on Soil Mechanics and Foundation Engineering*, Vol. 1, pp. 533–539.

Sharma, V. and Joshi, A., 1983. Field Tests for the Feasibility of a Concrete Dam. *Proceedings of the International Symposium on In Situ Testing*, Vol. 2, pp. 143–14148.

Sully, J.P. and Campanella, R.G., 1990. Measurement of Lateral Stress in Cohesive Soils by Full-Displacement In Situ Test Methods. Transportation Research Record No. 1278, pp. 164–171.

Symons, I.F. and Carder, D.R., 1992. Field Measurements on Embedded Retaining Walls. *Geotechnique*, Vol. 42, No. 1, pp. 117–126.

Tavenas, F.A., Blanchete, G., Leroueil, S., Roy, M., and LaRochelle, P., 1975. Difficulties in the In Situ Determination of K_0 in Soft Sensitive Clays. *Proceedings of the Conference on In Situ Measurements of Soil Properties, ASCE*, Vol. 1, pp. 450–476.

Tedd, P. and Charles, J.A., 1981. In Situ Measurement of Horizontal Stress in Overconsolidated Clay Using Push-in Spade-Shaped Pressure Cells. *Geotechnique*, Vol. 31, No. 4, pp. 554–558.

Tedd, P. and Charles, J.A., 1983. Evaluation of Push-In Pressure Cell Results in Stiff Clay. *Proceedings of the International Symposium on Soil and Rock Investigation by In Situ Testing*, Vol. 2, pp. 579–584.

Tedd, P., Chard, B.M., Charles, J.A., and Symons, I.F., 1984. Behavior of Propped Embedded Retaining Wall in Stiff Clay at Bell Common Tunnel. *Geotechnique*, Vol. 34, No. 4, pp. 513–532.

Tedd, P., Charles, J.A., and Clarke, B.G., 1985. The Measurement of Horizontal Stress Changes in Stiff Clay Using Push-In Spade-Shaped Pressure Cells and a Self-Boring Pressuremeter. *Ground Engineering*, Vol. 18, No. 1, pp. 28–31.

Tedd, P., Powell, J.J.M., Charles, J.A., and Uglow, I.M., 1989. In Situ Measurements of Earth Pressures Using Push-In Spade-Shaped Pressure Cells; 10 Years Experience. *Proceedings of the Conference Geotechnical Instrumentation in Civil Engineering Projects*, pp. 701–716.

Wallace, G.B. and Olsen, O.J., 1966. Foundation Testing Techniques for Arch Dams and Underground Powerplants. ASTM Special Technical Publication 402, pp. 272–289.

Wu, T., Beal, P., and Lan, C., 1988. In Situ Shear Test of Soil-Root Systems. *Journal of Geotechnical Engineering, ASCE*, Vol. 114, No. 12, pp. 1376–1394.

Zeigler, T.W., 1972. In Situ Tests for the Determination of Rock Mass Shear Strength. U.S. Army Engineer Waterways Experiment Station, Technical Report S-72-12.

Index

anisotropy 182, 185
automatic hammer 17, 22

CBR 92, 230
coefficient of consolidation 150, 152, 230, 234
constrained modulus 132, 133, 149, 226, 238

deep foundations 60, 157, 284
dissipation tests 150, 151, 231–233

elastic modulus 34, 46, 131, 228, 239, 283, 319

friction angle 32, 33, 129, 131, 236, 237, 292, 304, 333

hydraulic conductivity 152–154

interface shear strength 300

lateral earth pressure 45, 147, 223–226, 237, 277, 338, 346
liquefaction 35, 38, 136, 241

overconsolidation ratio, OCR 43, 145, 218–222

preconsolidation stress 44, 141–145, 189, 222, 223, 280

rate 113, 177, 179
relative density 31, 32, 91, 125–127, 235
resilient modulus 93
rock testing 48–50, 280

seismic cone 133
seismic dilatometer 242
seismic SPT 53
sensitivity 140, 172
shallow foundations 59, 156, 282
shear modulus 37, 134, 135, 148, 229, 240, 280
shear wave velocity 36, 37, 47, 134, 148, 229
soil identification 119, 124, 211
SPT corrections 28–30
state parameter 128, 130, 236
stratigraphy 87, 118, 211
stress history 44, 189, 218
subgrade reaction 241, 317

torque 51, 52

undrained shear strength 40, 91, 137–141, 182, 213–218, 277–279, 322

353